Evidence-Based Climate Science

Evidence-Based Climate Science

Data Opposing CO₂ Emissions as the Primary Source of Global Warming

Don Easterbrook

ELSEVIER

AMSTERDAM • BOSTON • HEIDELBERG • LONDON • NEW YORK • OXFORD
PARIS • SAN DIEGO • SAN FRANCISCO • SYDNEY • TOKYO

Elsevier
The Boulevard, Langford Lane, Kidlington, Oxford OX5 1GB, UK
Radarweg 29, PO Box 211, 1000 AE Amsterdam, The Netherlands

First edition 2011

Library of Congress Cataloging-in-Publication Data
Evidence-based climate science : data opposing CO_2 emissions as the primary source of
global warming / Don Easterbrook [editor].
   p. cm.
   ISBN 978-0-12-385956-3
   1. Atmospheric carbon dioxide. 2. Greenhouse effect, Atmospheric. 3. Global
warming. 4. Climatic changes. I. Easterbrook, Don J., 1935-
   QC879.8.E94 2011
   551.6—dc22
                                          2011013163

British Library Cataloguing in Publication Data
A catalogue record for this book is available from the British Library

ISBN: 978-0-12-385956-3

For information on all Elsevier publications visit our web
site at elsevierdirect.com

Printed and bound in China

11 12 13 10 9 8 7 6 5 4 3 2 1

Contents

Part IV
Solar Activity

Part V
Modeling

The climatic changes that the Earth has experienced in the past several decades have led to an intense interest in their cause, with contentions by the IPCC (Intergovernmental Panel on Climate Change) that catastrophic global warming and sea level rise due to increased atmospheric CO_2 will occur by the end of the century or before. However, many scientists point to data strongly suggesting that climate changes are a result of natural cycles, which have been occurring for thousands of years. Unfortunately, many non-scientist activists and the news media have entered the debate and the arguments have taken on political aspects with little or no scientific basis.

So what is the physical evidence for the cause of global warming and cooling? Proponents of CO_2-caused warming contend that the coincidence of global warming since 1978 with rising CO_2 means that CO_2 is the cause of the warming. However, this is not proof of anything—just because two things happen coincidently doesn't prove one is the cause of the other. After 1945, CO_2 emissions soared for the next 30 years, but the climate cooled, rather than warmed, showing a total lack of correlation between CO_2 and climate. Then, in 1977, temperatures switched abruptly from cool to warm and the climate began to warm with no change in the rate of increase of CO_2.

Does CO_2 have any effect on climate? Physicists have shown that CO_2 is a greenhouse gas capable of warming the atmosphere. The question is *how much* warming is CO_2 capable of causing? CO_2 is a trace gas, making up only a tiny portion of the atmosphere—only 390 parts per million (or 0.039%), and accounts for only 3.6% of greenhouse gasses. The total increase in atmospheric CO_2 since 1945 has been only 0.009% above normal. How can such a small amount of CO_2 control global climate change? The truth is, it can't. Water vapor accounts for 95% of greenhouse gasses and is thus the dominant factor in greenhouse warming. The theory put forth by CO_2 advocates is that this miniscule increase in CO_2 raises atmospheric temperature a tiny amount and that small warming increases the water vapor content of the atmosphere and since water vapor accounts for 95% of atmospheric warming, the air warms. Computer modelers then insert an arbitrary water vapor factor in their models to get the catastrophic warming that they predict. The problem with this approach is that no evidence exists for increased water vapor content in the atmosphere to produce the warming their models call for. In fact, water vapor records indicate just the opposite—water vapor has decreased since 1948.

Because of the absence of any physical evidence that CO_2 causes global warming, the *only* argument for CO_2 as the cause of warming rests entirely in computer modeling. Thus, the question becomes, how good are the computer models in predicting climate? We can test this by comparing global warming predicted by the IPCC models against actual climate change over the past 10 years. In 2000, the IPCC models predicted a warming of 1 °F by 2010, 2 °F by 2040, and 10 °F by 2100. The 1 °F warming predicted by IPCC by 2010 did not happen—there has been *no* global warming since 1998. In fact, temperatures have cooled slightly since then.

If CO_2 is incapable of explaining global warming, what natural possibilities exist? A vast amount of physical evidence of climate change over the past centuries and millennia has been gathered by scientists. Significant climate changes have clearly been going on for many thousands of years, long before the recent rise in atmospheric CO_2. In order to understand modern climate changes, we need to look at the past history of climate changes. The past is the key to the future—to know where we are headed in the future, we need to know where we have been in the past. This volume is intended to document past climate changes and present physical evidence for possible causes. It includes data related to the causes of global climate change by experts in meteorology, geology, atmospheric physics, solar physics, geophysics, climatology, and computer modeling.

Time and nature will be the final judge of the cause of global warming. The next decade should tell us the answer. If CO_2 is the cause of global warming and the computer models are correct, then warming of 2 °F since 2000 should occur by 2040. If the climate continues to cool, then the computer models must be considered invalid, and we must look to other causes. As we enter the solar minimum predicted by solar physicists and cooling deepens in the next decade, as it did in 1800 and 1650, than a strong case can be made for solar variation as the main cause of climate change.

The reader is invited to toss aside all of the political rhetoric that has been introduced into the global warming debate, focus on the scientific evidence presented in the papers in this volume, and make your own conclusions. Dogma is an impediment to the free exercise of thought—it paralyses the intelligent. Conclusions based upon preconceived ideas are valueless—it is only the open mind that really thinks.

Don J. Easterbrook

Part I

Geologic Perspectives

Geologic Evidence of Recurring Climate Cycles and Their Implications for the Cause of Global Climate Changes—The Past is the Key to the Future

Don J. Easterbrook

Department of Geology, Western Washington University, Bellingham, WA 98225, USA

Chapter Outline

Evidence-Based Climate Science. DOI: 10.1016/B978-0-12-385956-3.10001-4

1. INTRODUCTION

Recent global warming (1978−1998) has pushed climate changes into the forefront of scientific inquiry with a great deal at stake for human populations. With no unequivocal, "smoking gun", cause-and-effect evidence that increasing CO_2 caused the 1978−1998 global warming, and despite the media blitz over the 2007 IPCC report, no tangible physical evidence exists that CO_2 is *causing* global warming. Computer climate models *assume* that CO_2 is the cause and computer model simulations are all based on that assumption.

The IPCC report has been hotly contested by many scientists (e.g., Idso and Singer, 2009; Spencer, 2010a,b; Horner, 2008). Abundant physical evidence from the geologic past provides a record of former periods of recurrent global warming and cooling that were far more intense than recent warming and cooling. These geologic records provide a clear evidence of global warming and cooling that could not have been caused by increased CO_2. Thus, we can use these records to project global climate in the future—the past is the key to the future.

2. IS GLOBAL WARMING REAL?

Little doubt remains that global temperatures have risen since the Little Ice Age several centuries ago. Although surface historic temperature records have been shown to be questionable due to poor siting practices and tampering of data by various governmental agencies, historical accounts, fluctuations of glaciers, and ice core records affirm that warming and cooling have indeed occurred. Temperatures have risen approximately a degree or so per century since the coldest part of the Little Ice Age ~500 years ago, but the rise has not been linear. Global temperatures have warmed and cooled many times in 25−35-year cycles, well before the atmospheric CO_2 began to rise significantly.

Two episodes of global warming and two episodes of global cooling occurred during the 20th century (Fig. 1). Overall, temperatures during the century rose about the same as the rate of warming per century since the Little Ice Age 500 years ago.

1880−1915 cool period: Atmospheric temperature measurements, glacier fluctuations, and oxygen isotope data from Greenland ice cores record a cool period from about 1880 to about 1915 (Fig. 2). Glaciers advanced, some nearly to the terminal positions reached during the Little Ice Age. Many cold temperature records in North America were set during this period. Temperatures reached a low point about 1890, rose slightly about 1910, and by about 1915 began to warm.

1915−1945 warm period: Global temperatures rose steadily in the 1920s, 1930s, and early 1940s (Fig. 3). By the mid-1940s, global temperatures were about 0.5 °C (0.9 °F) warmer than they had been at the turn of the century. More high temperature records for the century were recorded in the 1930s than in any other decade of the 20th century (Fig. 1).

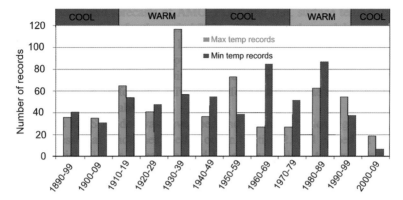

FIGURE 1   Warm and cool temperature records.

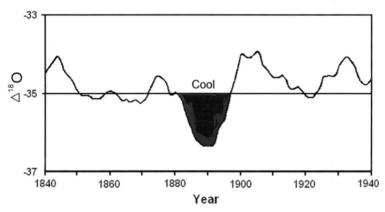

FIGURE 2   Cool period in the late 1800s and early 1900s from Greenland ice core oxygen isotope ratios.

Temperatures in the 1930s in the Arctic and Greenland were warmer than at present and rates of warming were higher, warming 4 °C (7 °F) in two decades. Greenland temperatures generally followed the global temperature pattern, warming in the 1920s, 1930s, and early 1940s, cooling until about 1977, and then rising again until the turn of the century. The average rate of warming from 1920 to 1930 was considerably higher than from 1980 to 2005 despite the fact that the 1920–1930 warming occurred before CO_2 could be a factor. Temperatures in Greenland during the Medieval Warm Period (900–1300 A.D.) were generally warmer than today.

1945–1977 cool period: Global temperatures began to cool in the mid-1940s at the point when CO_2 emissions began to soar. Global temperatures in the Northern Hemisphere dropped about 0.5 °C (0.9 °F) from the mid-1940s

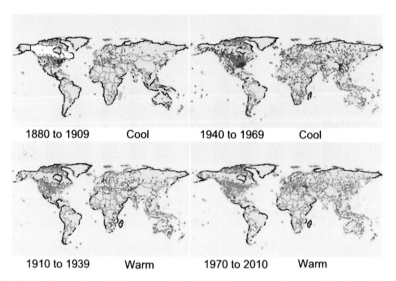

FIGURE 3 Temperatures during two periods of global warming and two periods of warming in this century. Blue dots are stations recording cooling, red dots are stations recording warming. (GISS raw data).

until 1977 and temperatures globally cooled about 0.2 °C (0.4 °F). Many of the world's glaciers advanced during this time, and recovered a good deal of the ice lost during the 1915–1945 warm period. However, cooling during this period was not as deep as in the preceding cool period (1880–1915).

1977–1998 warm period: The global cooling that prevailed from 1945 to 1977 (Fig. 3) ended abruptly in 1977 when the Pacific Ocean shifted from its cool mode to its warm mode (Fig. 4) and global temperatures began to rise, initiating two decades of global warming. 1977 has been called the year of the "Great Climate Shift". During this warm period, alpine glaciers retreated, Arctic sea ice diminished, and renewed melting of the Greenland Ice Sheet occurred.

The abruptness of the shift in Pacific sea surface temperatures and corresponding change from global cooling to global warming in 1977 is highly significant and strongly suggests a cause-and-effect relationship. The rise of atmospheric CO_2, which accelerated after 1945 shows no sudden change that could account for the "Great Climate Shift".

The global warming from 1977 to 1998 has received much attention in the news media and represents the period now popularly called "global warming". Previously, warming during the entire 20th century was referred to as the time of "global warming" but when it became apparent that increasing atmospheric CO_2 could not explain warming and cooling prior to 1977, advocates of CO_2 as the cause of the warming restricted what is now labeled as "global warming" to the post-1977 warming.

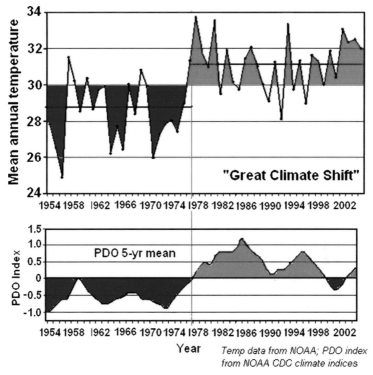

FIGURE 4   Mean annual temperatures for Anchorge, Fairbanks, and Nome (upper graph) and Pacific Decadal Oscillation (PDO) Index (lower graph) reflecting Pacific sea surface temperatures. The sudden switch from cool to warm PDO caused the "Great Climate Shift" in 1977 that initiated the latest global warming period.

2.1. Does Global Warming Prove that it was Caused by Increasing CO_2?

The news media has bombarded the public with countless pictures of melting glaciers as proof that CO_2 is causing global warming. No one disputes that glaciers have retreated from their maximums around 1910—1915, but does that prove it was caused by increased CO_2? Just because two things happened during the same time period does not prove that one is the cause of the other. During the 1880—1915 cool period, alpine glaciers advanced almost to their Little Ice Age (1300—1915) maximums, then retreated strongly during the 1915—1945 warm period with no significant change in atmospheric CO_2. Glaciers readvanced again from 1945 to 1977 in a cool period during which CO_2 emissions soared dramatically—just the opposite of what should have happened if CO_2 causes global warming. The lame excuse that sulfur emissions during the cool period caused the cooling is not credible because the cool

FIGURE 5   Greenland ice sheet (Photo by Austin Post).

period came to an abrupt halt in 1977 with no change in atmospheric sulfur or CO_2. The cause had to be something other than either CO_2 or sulfur.

3. MELTING GLACIERS

3.1. Ice Sheets

3.1.1. Greenland Ice Sheet

During the 1977–1998 warm period, the Greenland Ice Sheet (Fig. 5) experienced some melting and the news media has featured many stories that melting of the ice sheet was occurring at an unprecedented, extreme rate that threatened to raise sea level rapidly. However, past climatic records show that Greenland is following a perfectly predictable, normal pattern of warming and cooling. Melting of Greenland ice has waxed and waned repeatedly in the past, following both global temperatures and the warming/cooling patterns in the oceans (Chylek et al., 2004, 2006).

Temperatures in Greenland were cooler during the 1880–1915 global cool period (Figs. 6, 7). From 1915 to 1945, Greenland warmed faster than it did from 1977 to 1998 and was actually *warmer* in the 1930s than at present (Figs. 6, 7),

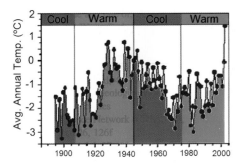

FIGURE 6 Temperature changes in Angmagssalik, Greenland over the past century (modified from Chylek et al., 2004, 2006).

so the 1977–1998 warming is not at all unusual, much less 'extreme'. Until 'The Great Climate Shift' in 1977 that initiated global warming from 1977 to 1998, Greenland had been cooling for the previous three decades (1945–1977) (Figs. 6, 7) and the ice sheet grew.

3.1.2. Antarctic Ice Sheet

The Antarctic continent is 1.4 times bigger than the USA and has 90% of the Earth's ice and 70–80% of its fresh water. The Antarctic ice sheet (Fig. 8) reaches thicknesses of 15,670 feet with a mean thickness of 7,300 feet. The lowest point is 8,188 feet below sea level. The average winter temperature at the South Pole is −60 °C (−76 °F) and the average summer temperature is

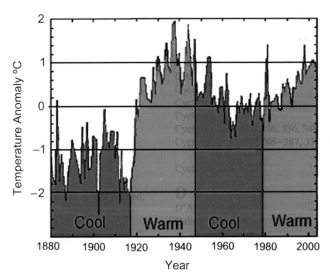

FIGURE 7 Temperature changes in the Arctic over the past century, 70–90° N latitude, −180 to 180 longitude (modified from Chylek et al., 2004, 2006).

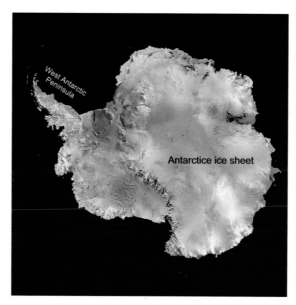

FIGURE 8   Antarctica (NASA).

−27.5 °C (−17.5 °F). The lowest temperature ever recorded on Earth was −89.2 °C (−128.6 °F) at Vostok on the ice sheet. The average daily temperature is −55.1 °C (−67.2 °F) at Vostock and −49.4 °C (−57 °F) at the south pole. In order to get any significant amount of Antarctic ice to melt, temperatures would have to rise above the melting point, i.e., more than 100 °F. Thus, claims of large scale melting of the Antarctic ice sheet are highly exaggerated.

The west Antarctic Peninsula extends northwestward from the main continent (Fig. 8) and contains a small proportion of the total ice in Antarctica. Warm ocean water around the west Antarctic Peninsula has caused some melting of ice and breaking off of shelf ice, but the volume is a small percentage of the main Antarctic ice sheet. The main Antarctic ice sheet is cooling (Fig. 9) and ice there is increasing, not melting. The paucity of weather stations in Antarctica makes interpretation of temperature distributions difficult. Steig et al. (2009) attempted to project temperatures from the West Antarctica Peninsula, where more data are available, to the main Antarctic ice sheet and contended that all of Antarctic was warming. This conclusion was hotly disputed by O'Donnell et al. (2010) who showed that warming over the period of 1957–2006 was concentrated in the West Antarctic Peninsula and the main Antarctic ice sheet varied from cooling to only slightly warmer. Both studies used statistical projection of far distant weather stations for their reconstructions. Temperature records from 1957 to 2007 at the South Pole and at Vostock (Fig. 10) on the main ice sheet show no warming in 50 years of record.

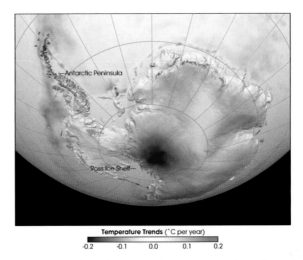

FIGURE 9   Temperature trends in Antarctica. Blue areas are getting colder, red areas warmer (NOAA, 2004).

3.1.3. Arctic and Antarctic Sea Ice

Is polar sea ice disappearing? Assertions that Arctic sea ice (Fig. 11) is vanishing at an accelerating rate and that the Arctic Ocean will soon be ice free appear almost daily in the news media. The news media frequently carries stories of breaking off of large pieces of shelf ice, which is claim of to be proof accelerating warming in the polar regions. But what does the data show?

Measurements of Arctic and Antarctic sea ice from satellite images began in 1979, just after the Great Climate Shift of 1977—1978 so the entire record covers only the most recent period of global warming (1978—1998). No satellite images are available from the 1945—1977 period of global cooling.

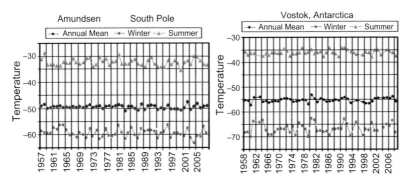

FIGURE 10   Temperature records from the South Pole and Vostock stations. Neither shows any indication of warming since 1957.

FIGURE 11 Arctic sea ice. There is no ice cap at the north pole so melting of thin, floating sea ice there has no effect whatsoever on sea level (NOAA photo).

Time-lapse images show substantial expansion and contraction of Arctic sea ice with the seasons (Fig. 12), so no single image is representative of annual changes. In general, Arctic sea ice diminished between 1979 and 2010, reaching a low in 2007, then rebounding in 2008, 2009, and 2010 (Figs. 12, 13). It underwent normal melting during the 1977−1998 warm cycle and was aided in large part by warm ocean water entering the Arctic thru Bering Strait as a result of the 1977−1998 warm cycle.

According to data from the University of Illinois, Antarctic sea ice area (Fig. 14) is nearly 30% above normal and the anomaly has reached 1 million km^2, an area of excess sea ice equal to Texas and California (or 250 Rhode

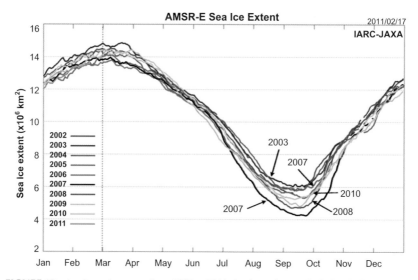

FIGURE 12 Arctic sea ice extent from 2002 to 2011. Arctic sea ice reached a low in 2007 but has rebounded in 2008, 2009, and 2010 (IARC-JAXA Information System).

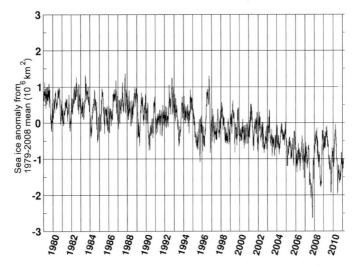

FIGURE 13 Arctic sea ice has declined since 1979 (Arctic Climate Research, University of Illinois; NOAA data).

Islands). The net global sea ice anomaly is also positive, 850,000 km^2 above the normal and arctic ice is the highest level since 2002. Figure 15 shows a comparison of Arctic and Antarctic sea ice from 1979 to 2009. Although Arctic sea ice has declined, Antarctic sea ice has increased (Fig. 16). The overall total of sea ice globally has increased since 1979.

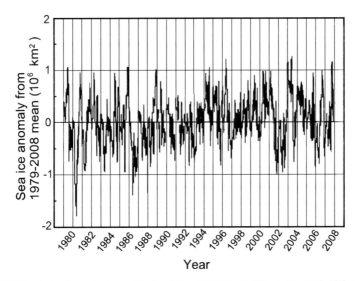

FIGURE 14 Antarctic sea ice has not declined since 1979 (Arctic Climate Research, University of Illinois; NOAA data).

Arctic and Antarctic Standardized Anomalies and Trends
Jan 1979 - Jul 2009

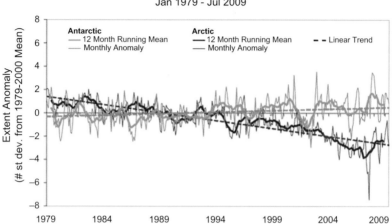

FIGURE 15   Comparison of sea ice extent in the Arctic and Antarctic from 1979 to 2009. Arctic sea ice extent declined, whereas Antarctic sea ice increased (National Snow and Ice Data Center, University of Colorado).

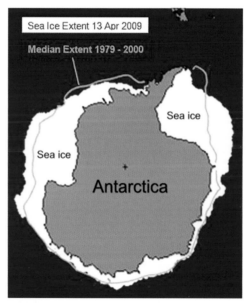

FIGURE 16   Extent of Antarctic sea ice in 2009 compared to the median extent from 1979 to 2000 (National Snow and Ice Data Center, University of Colorado).

3.2. Alpine Glaciers

Because their ice volume is not large and they are in equilibrium with local climate, Alpine glaciers record climatic changes by retreating during warm periods and receding during cool periods. The news media has made much of the glacier recession resulting from the 1977 to 1998 warm period as proof of warming allegedly caused by rising atmospheric CO_2, but ignore the glacial advances that occurred during the 1945–1977 cool period when CO_2 rose dramatically. During the past century, alpine glaciers expanded during the 1880 to ~1915 cool period, retreated during the ~1915 to ~1945 warm period, expanded again during the ~1945 to 1977 cool period, and retreated during the 1977–1998 warm period. Thus, three of the four glacial oscillations occurred before significant rise of CO_2 (or advanced during rising CO_2) and cannot have been caused by changes in CO_2.

Alpine glaciers advanced far downvalley during the Little Ice Age (~1300 to late 1800s) and have generally retreated upvalley during the warming following the cooler climates of the Little Ice Age. Thus, they are well upvalley from their Little Ice Age maximums.

4. ATMOSPHERIC CARBON DIOXIDE

Atmospheric CO_2 is a non-toxic, colorless, odorless gas that constitutes a tiny portion of the Earth's atmosphere, making up only ~0.038% of the atmosphere (Fig. 17). In every 100,000 molecules of air, 78,000 are nitrogen,

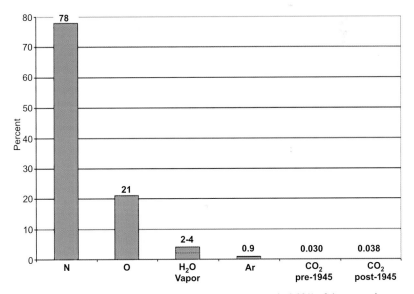

FIGURE 17 Composition of the atmosphere. CO_2 makes up only 0.03% of the atmosphere.

21,000 are oxygen, 2,000–4,000 are water vapor, and only 30 are carbon dioxide.

CO_2 is soluble in water and varies with the temperature of the water. Cold water can hold more CO_2 than warm water and seawater contains about 75 times as much CO_2 as fresh water. As water temperature increases, the solubility of CO_2 decreases and is given off into the atmosphere to establish a new equilibrium between the air and water. At 25 °C, water contains about 50 times as much CO_2 as air. The high solubility and chemical reactivity of CO_2 permit ready exchange of CO_2 between the atmosphere and oceans. CO_2 solubility depends on temperature, so changes in sea surface temperature affect CO_2 exchange with the atmosphere.

When global temperatures rise, as during interglacial periods, atmospheric CO_2 rises and when temperatures decline, as during Ice Ages, atmospheric CO_2 declines. Measurements of CO_2 from air trapped in polar ice cores over tens of thousands of years show that atmospheric CO_2 concentrations typically vary from about 260 to 285 ppm, averaging about 280 ppm. Higher CO_2 levels during the past interglacial periods do not indicate that CO_2 is the *cause* of the warmer interglacials because the CO_2 increase *lagged* Antarctic warming by 600 to 800 ± 200 years (Fischer et al., 1999; Caillon et al., 2003).

Water vapor accounts for about 95% of the greenhouse gas, with CO_2, methane, and a few other gases making up the remaining 5%. The greenhouse effect from CO_2 is only about 3.6%. Most of the greenhouse warming effect takes place within the first 20 ppm of CO_2. After that, the effect decreases exponentially (Fig. 18) so the rise in atmospheric CO_2 from 0.030% to 0.038%

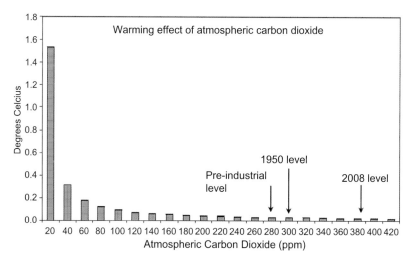

FIGURE 18   Warming effect of CO_2 (from D. Archibald).

from 1950 to 2008 could have caused warming of only about 0.01 °C. The total change in CO_2 of the atmosphere amounted to an addition of only 1 molecule of CO_2 per 10,000 molecules of air.

Atmospheric CO_2 rose slowly from the mid-1700s to 1945. Emissions began to soar abruptly in 1945 after World War II (Fig. 19). CO_2 has risen at a fairly constant rate since then, going from about 300 ppm in 1955 to about 385 ppm in 2007.

At the abrupt 1977 'Great Climate Shift' when the global climate shifted from cooling to warming, no significant change occurred in the rate of increase of CO_2 (Fig. 20), suggesting that CO_2 had nothing to do with the shifting of the climate.

CO_2 which makes up only 0.038% of the atmosphere and constitutes only 3.6% of the greenhouse effect has increased only 0.008% since emissions began to soar after 1945. How can such a tiny, tiny increment of CO_2 cause the 10 °F increase in temperature predicted by CO_2 advocates? The obvious answer is that it can't. Computer climate modelers build into their models a high water vapor component, which they claim is due to increase atmospheric water vapor caused by very small warming from CO_2, and since water vapor makes up 95% of the greenhouse effect, they claim the result will be warming. The problem is that there is no evidence whatsoever of any increase in atmospheric water vapor content. Until the modelers can prove that water vapor has increased, their models are not credible.

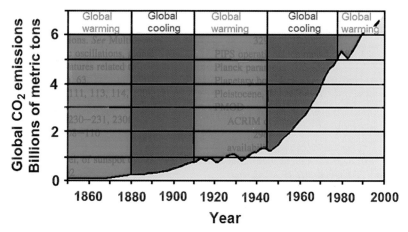

FIGURE 19 CO_2 emissions from 1850 to 2000. Note that CO_2 emissions were low during the global warming from 1850 to 1880 and rose slowly during the deep global cooling from 1880 to about 1915. Emissions were fairly constant during the strong global warming from 1915 to 1945. While emissions were soaring from 1945 to 1977, the global climate cooled, rather than warmed as it should have if CO_2 was the cause of global warming.

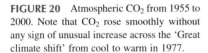

FIGURE 20 Atmospheric CO_2 from 1955 to 2000. Note that CO_2 rose smoothly without any sign of unusual increase across the 'Great climate shift' from cool to warm in 1977.

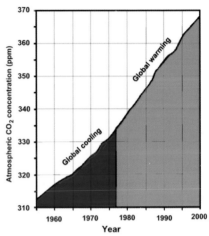

4.1. Global Warming and CO_2 During the Past Century

Atmospheric temperature measurements, glacier fluctuations, and oxygen isotope data from Greenland ice cores all record a cool period from about 1880 to about 1915, reaching a low about 1890. During this period, global temperatures were about 0.9 °C (1.6 °F) cooler than at present. From 1880 to 1890, temperatures dropped 0.35 °C (0.6 °F) in only 10 years. From 1890 to 1900, temperatures rose 0.25 °C (0.45 °F) in 10 years, after which temperatures dipped slightly (0.15 °C (0.3 °F)) until about 1915.

4.1.1. Global Warming from 1915 to 1945 could not be Caused by Atmospheric CO_2

From 1915 to 1945, global temperatures rose 0.4 °C (0.7 °F), half of the total temperature rise for the past century. As expected, glaciers during this period retreated and, in general, followed the warming climate pattern. All of this occurred before CO_2 emissions began to soar (after 1945) (Figs. 19), so at least half of the warming of the past century cannot have been caused by manmade CO_2.

4.1.2. Global Cooling Occurred from 1945 to 1977 While CO_2 Emissions Soared

Global temperatures began to cool in the 1940s at the point when CO_2 emissions began to soar (Fig. 19). For 30 years thereafter temperatures declined 0.2 °C (0.4 °F) globally and 0.5 °C (0.9 °F) in the Northern Hemisphere. During this 30-year period (1945−1977), glaciers ceased the recession of the preceding ~30 years and advanced. By 1977, many advancing glaciers had recovered much of the length lost in the previous ~30 years of warming. Many examples of glacial recession during the past century cited in the news media show contrasting terminal positions beginning with the maximum extent at the end of a ~30-year cool period (1915 or 1977) and ending with the minimum extent of the present 30-year warm period (1998). A much better gauge of the effect of climate on glaciers would be to compare glacier terminal positions between the ends of successive cool periods or the ends of successive warm periods.

Even though CO_2 emissions rose sharply from 1945 to 1977, global temperature dropped during that 30-year period. If CO_2 causes global warming, temperature should have risen, rather than declined, strongly suggesting that rising CO_2 does not cause significant global warming. Clearly the climate was driven by natural causes.

4.1.3. Global Warming from 1977 to 1998

In 1977, global temperatures, which had been declining since the late 1940s, abruptly reversed and began to rise. This sudden reversal of climate has been termed as "The Great Climate Shift" because it happened so abruptly (Miller et al., 1994). Global temperatures rose ~0.5 °C (0.9 °F), alpine glaciers have retreated, Arctic sea ice has diminished, melting of the Greenland Ice Sheet has accelerated, and other changes have occurred. The warmest year was 1998, after which global temperatures declined slightly until 2007 when sharp cooling began.

During this time, atmospheric CO_2 has continued to rise, the only period in the past century when global warming and atmospheric CO_2 have risen together. However, this doesn't prove a cause-and-effect relationship—just because two things happened together doesn't prove that one is the cause of the other.

4.2. Is Global Warming Caused by Rising CO_2?

No tangible, physical evidence exists that proves a cause-and-effect relationship between global climate changes and atmospheric CO_2. The fact that CO_2 is a greenhouse gas and that CO_2 has increased doesn't prove that CO_2 has caused global warming. Ninety five percent of greenhouse gas warming is due to water vapor and there is no evidence that atmospheric water vapor has increased. Only 3.6% of the greenhouse effect is due to CO_2.

As shown by isotope measurements from ice cores in Greenland and Antarctica and by measurements of atmospheric CO_2 during El Nino warming oceans emit more CO_2 into the atmosphere during climatic warming. The ice core records indicate that after the last Ice Age, temperatures rose for about 600—800 years *before* atmospheric CO_2 rose, showing that climatic warming caused CO_2 to rise, not vice versa.

5. LESSONS FROM PAST GLOBAL CLIMATE CHANGES

Those who advocate CO_2 as the cause of global warming have stated that never before in the Earth's history of has climate changed as rapidly as in the past century and that proves global warming is being caused by anthropogenic CO_2. Statements such as these are easily refutable by the geologic record. Figure 21 shows temperature changes recorded in the GISP2 ice core from the Greenland Ice Sheet. The global warming experienced during the past century pales into insignificance when compared to the magnitude of the profound climate reversals over the past 15,000 years.

The GISP2 Greenland ice core isotope data have proven to be a great source of climatic data from the geologic past. Paleo-temperatures for thousands of years have been determined by Minze Stuiver and Peter Grootes from nuclear accelerator measurements of thousands of oxygen isotope ratios ($^{16}O/^{18}O$), and these data have become a world standard (Grootes and Stuiver, 1997). Oxygen isotope ratios are a measure of paleo-temperatures at the time snow fell that was later converted to glacial ice. The age of such temperatures can be accurately measured from annual layers of accumulation of rock debris marking each summer's melting of ice and concentration of rock debris on the glacier. The top of the core is 1987.

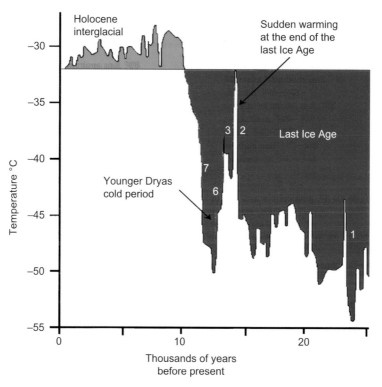

FIGURE 21 Greenland temperatures over the past 25,000 years recorded in the GISP2 ice core. Strong, abrupt warming is shown by nearly vertical rise of temperatures, strong cooling by nearly vertical drop of temperatures (modified from Cuffey and Clow, 1997).

The ratio of ^{18}O to ^{16}O depends on the temperature at the time snow crystals formed, which were later transformed into glacial ice. Ocean volume may also play a role in $\delta^{18}O$ values, but $\delta^{18}O$ serves as a good proxy for temperature. The oxygen isotopic composition of a sample is expressed as a departure of the $^{18}O/^{16}O$ ratio from an arbitrary standard

$$\delta^{18}O = \frac{\left(^{18}O/^{16}O\right)_{sample} - \left(^{18}O/^{16}O\right)_{standard} \times 10^3}{\left(^{18}O/^{16}O\right)_{standard}}$$

where $\delta^{18}O$ is the of ratio $^{18}O/^{16}O$ expressed in per mil (0/00) units.

The $\delta^{18}O$ data clearly show remarkable swings in climate over the past 100,000 years. In just the past 500 years, Greenland warming/cooling temperatures fluctuated back and forth about 40 times, with changes in every 25–30 years (27 years on the average). None of these changes could have been caused by changes in atmospheric CO_2 because they predate the large CO_2 emissions that began about 1945. Nor can the warming of 1915–1945 be related to CO_2, because it pre-dates the soaring emissions after 1945. Thirty years of global cooling (1945–1977) occurred during the big post-1945 increase in CO_2.

5.1. Magnitude and Rate of Abrupt Climate Changes

But what about the *magnitude* and *rates of climates change*? How do past temperature oscillations compare with recent global warming (1977–1998) or with warming periods over the past millennia. The answer to the question of magnitude and rates of climate change can be found in the $\delta^{18}O$ and ice core temperature data.

Temperature changes in the GISP2 core over the past 25,000 years are shown in Fig. 21 (from Cuffey and Clow, 1997). The temperature curve in Fig. 21 is a portion of their original curve. Color has been added to make it easier to read. The horizontal axis is time and the vertical axis is temperature based on the ice core $\delta^{18}O$ and borehole temperature data. Details are discussed in their paper. Places where the curve becomes nearly vertical signify times of very rapid temperature change. Keep in mind that these are temperatures in Greenland, not global temperatures. However, correlation of the ice core temperatures with worldwide glacial fluctuations and correlation of modern Greenland temperatures with global temperatures confirms that the ice core record does indeed follow global temperature trends and is an excellent proxy for global changes. For example, the portions of the curve from about 25,000 to 15,000 represent the last Ice Age (the Pleistocene) when huge ice sheets, thousands of feet thick, covered North America, northern Europe, and northern Russia and alpine glaciers readvanced far downvalley.

How do the magnitude and rates of change of modern global warming/cooling compare to warming/cooling events over the past 25,000 years? We can compare the warming and cooling in the past century to approximate 100-year periods in the past 25,000 years. The scale of the curve doesn't allow enough accuracy to pick out exactly 100-year episodes directly from the curve, but that can be done from the annual dust layers in ice core data. Thus, not all of the periods noted here are exactly 100 years. Some are slightly more, some are slightly less, but they are close enough to allow comparison of magnitude and rates with the past century.

Temperature changes recorded in the GISP2 ice core from the Greenland Ice Sheet (Fig. 1) (Cuffey and Clow, 1997) show that the global warming experienced during the past century pales into insignificance when compared to the magnitude of profound climate reversals over the past 25,000 years. In addition, small temperature changes of up to a degree or so, similar to those observed in the 20th century record, occur persistently throughout the ancient climate record.

Some of the more remarkable sudden climatic warming periods are listed below. Numbers correspond to the temperature curves in Fig. 21.

1. About 24,000 years ago, while the world was still in the grip of the last Ice Age and huge continental glaciers covered large areas, a sudden warming of about 10 °C (20 °F) occurred.
2. About 14,000 years ago, a sudden, intense, climatic warming (~13 °C; ~22 °F) caused dramatic melting of large Pleistocene ice sheets that covered Canada and the northern U.S., all of Scandinavia, and much of northern Europe and Russia.
3. Shortly thereafter, temperatures dropped abruptly about 10 °C (20 °F) and temperatures then remained cold for several thousand years but oscillated between about 4 °C (8 °F) warmer and cooler.
4. About 13,000 years ago, global temperatures plunged sharply (~12 °C; ~21 °F) and a 1,300-year cold period, the Younger Dryas, began.
5. 11,500 years ago, global temperatures rose sharply (~12 °C; ~21 °F), marking the end of the Younger Dryas cold period and the end of the Pleistocene vIce Age. The end of the Younger Dryas cold period warmed by 5 °C (9 °F) over 30–40 years and as much as 8 °C (14 °F) over 40 years.

Figure 22 shows comparisons of the largest magnitudes of warming/cooling events per century over the past 25,000 years. At least three warming events were 20–24 times the magnitude of warming over the past century and four were 6–9 times the magnitude of warming over the past century. The magnitude of the only modern warming which might possibly have been caused by CO_2 (1978–1998) is insignificant compared to the earlier periods of warming.

5.1.1. Holocene Climate Changes (10,000 Years Ago to Present)

Most of the past 10,000 have been warmer than the present. Figure 23 shows temperatures from the GISP2 Greenland ice core. With the exception of a brief cool period about 8,200 years ago, almost all of the entire period from 1,500 to 10,500 years ago was significantly warmer than the present.

Another graph of temperatures from the Greenland ice core for the past 10,000 years is shown in Fig. 24. It shows essentially the same temperatures as

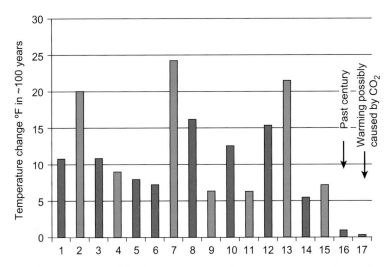

FIGURE 22 Magnitudes of the largest warming/cooling events over the past 25,000 years. Temperatures on the vertical axis are rise or fall of temperatures in about a century. Each column represents the rise or fall of temperature shown in Fig. 21. Event number 1 is about 24,000 years ago and event number 15 is about 11,000 years old. The sudden warming about 14,000 years ago caused massive melting of these ice sheets at an unprecedented rate (Steffensen et al., 2008).

Cuffey and Clow (1997) but with somewhat greater detail. What both of these temperature curves show is that virtually all of the past 10,000 years have been warmer than the present.

5.1.1.1. Early Holocene Climate Changes

8,200 years ago, the post-Ice Age interglacial warm period was interrupted by a sudden global cooling that lasted for a few centuries (Fig. 25). During this time, alpine glaciers advanced and built moraines. The warming that followed

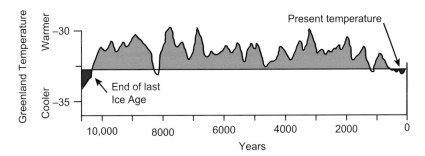

FIGURE 23 Temperatures over the past 10,500 years recorded in the GISP2 Greenland ice core. Red corresponds to temperatures warmer than the present (modified from Cuffey and Clow, 1997).

Greenland GISP2 Ice Core - Temperature Last 10,000 Years

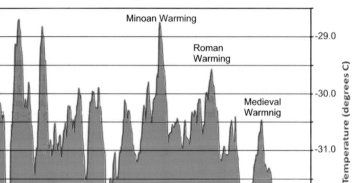

FIGURE 24   Temperatures over the past 10,000 years recorded in the GISP2 Greenland ice core (modified from Alley, 2000).

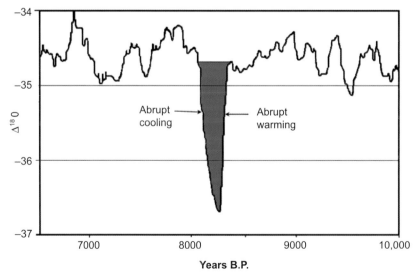

FIGURE 25   The 8,200-year B.P. sudden climate change, recorded in oxygen isotope ratios in the GISP2 ice core, lasted about 200 years.

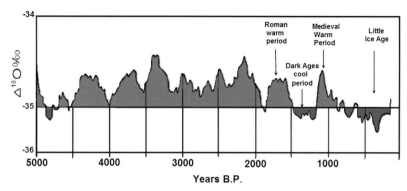

FIGURE 26 Oxygen isotope ratios for the past 5,000 years. Red areas are warm periods and blue areas are cool periods. Except for the Dark Ages Cool Period, almost all of the last 5,000 years was warmer than present until about 700 years ago.

the cool period was also abrupt. Neither the abrupt climatic cooling nor the warming that followed was preceded by atmospheric CO_2 changes.

The Greenland ice core isotope curve shows that almost all of the past 5,000 years, except for the Dark Ages Cool Period, were warmer than present until about 700 years ago (Fig. 26).

5.1.1.2. 750 B.C. to 200 B.C. Cool Period

Prior to the founding of the Roman Empire, Egyptians records show a cool climatic period from about 750 B.C. to 450 B.C. and the Romans wrote that the Tiber River froze and snow remained on the ground for long periods (Singer and Avery, 2007).

5.1.1.3. The Roman Warm Period (200 B.C. to 600 A.D.)

After 100 B.C., Romans wrote of grapes and olives growing farther north in Italy that had been previously possible and of little snow or ice (Singer and Avery, 2007).

5.1.1.4. The Dark Ages Cool Period (440 A.D. to 900 A.D.)

The Dark Ages were characterized by marked cooling. A particularly puzzling event apparently occurred in 540 A.D. when tree rings suggest greatly retarded growth, the sun appeared dimmed for more than a year, temperatures dropped in Ireland, Great Britain, Siberia, North and South America, fruit didn't ripen, and snow fell in the summer in southern Europe (Baillie in Singer, 2007). In 800 A.D., the Black Sea froze and in 829 A.D. the Nile River froze (Oliver, 1973).

5.1.1.5. Climate Changes Over the Past 500 Years

The oxygen isotope curve of the Greenland GISP ice core shows a remarkable oscillation of warm/cool periods since 1480 A.D. (Fig. 27). At least 40 periods

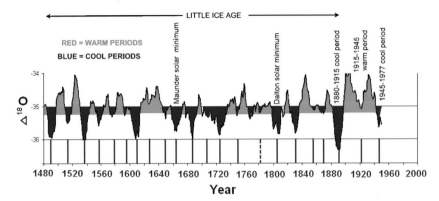

FIGURE 27   Greenland ice core isotope curve showing warming and cooling periods since 1480 A.D. The vertical lines at the bottom show peak cool periods which occurred with an average of every 27 years.

of warming and cooling have occurred since 1480 A.D., all well before CO_2 emissions could have been a factor. Historic warm and cool periods are well shown on the curve. The warm/cool cycles vary from about 25 to 30 years with an average of 27 years, almost identical to the typical Pacific Decadal Oscillation period (discussed later).

5.1.1.6. The Medieval Warm Period (900 A.D. to 1300 A.D.)

The Medieval Warm Period (MWP) was a time of warm climate from about 900 A.D. to 1300 A.D. when global temperatures were apparently somewhat warmer than at present. Its effects were evident in Europe where grain crops flourished, alpine tree lines rose, many new cities arose, and the population more than doubled. The Vikings took advantage of the climatic amelioration to colonize Greenland, and wine grapes were grown as far north as England where growing grapes is now not feasible and about 500 km north of present vineyards in France and Germany. Grapes are presently grown in Germany up to elevations of about 560 m, but from about 1100 A.D. to 1300 A.D., vineyards extended up to 780 m, implying temperatures warmer by about $1.0-1.4\,°C$ (Oliver, 1973). Wheat and oats were grown around Trondheim, Norway, suggesting climates about 1 °C warmer than present (Fagan, 2000).

Elsewhere in the world, prolonged droughts affected the southwestern United States and Alaska warmed. Sediments in central Japan record warmer temperatures. Sea surface temperatures in the Sargasso Sea were approximately 1 °C warmer than today, and the climate in equatorial east Africa was drier from 1000 A.D. to 1270 A.D. An ice core from the eastern Antarctic Peninsula shows warmer temperatures during this period.

Oxygen isotope studies in Greenland, Ireland, Germany, Switzerland, Tibet, China, New Zealand, and elsewhere, plus tree-ring data from many sites around the world all confirm the presence of a global Medieval Warm Period. Soon and

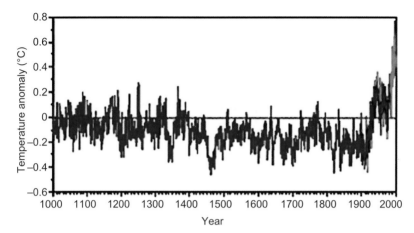

FIGURE 28   Mann et al. (1998) "hockey stick" graph of temperature change over the past 1,000 years based on tree rings.

Baliunas (2003) found that 92% of 112 studies showed physical evidence of the MWP, only two showed no evidence, and 21 of 22 studies in the Southern Hemisphere showed evidence of Medieval warming. Evidence of the MWP at specific sites is summarized in Fagan (2007) and Singer and Avery (2007). Evidence that the Medieval Warm Period was a global event is so widespread that one wonders why Mann et al. (1998) ignored it.

The Hockey Stick Trick   Over a period of many decades, several thousand papers were published establishing the Medieval Warm Period (MWP) from about 900 A.D. to 1300 A.D. and the Little Ice Age (LIA) from about 1300 A.D. to 1915 A.D. as global climate changes. Thus, it came as quite a surprise when Mann et al. (1998) (Fig. 28) concluded that neither the MWP nor the Little Ice Age actually happened on the basis of a tree-ring study and that became the official position of the 2001 Intergovernmental Panel on Climate Change (IPCC). The IPCC 3rd report (Climate Change 2001) then totally ignored the several thousand publications detailing the global climate changes during the MWP and the LIA and used the Mann et al. tree-ring study as the basis for the now famous assertion that "*Our civilization has never experienced any environmental shift remotely similar to this. Today's climate pattern has existed throughout the entire history of human civilization.*" (Gore, 2007). This claim was used as the main evidence that increasing atmospheric CO_2 was causing global warming so, as revealed in the 'Climategate' scandal, advocates of the CO_2 theory were very concerned about the strength of data that showed the Medieval Warm Period (MWP) was warmer than the 20th century and that global warming had occurred naturally, long before atmospheric CO_2 began to increase. The contrived elimination of

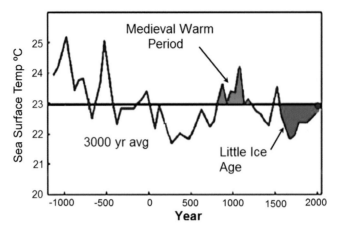

FIGURE 29   Surface temperatures of the Sargasso Sea reconstructed from isotope ratios in marine organisms (Keigwin, 1996).

the MWP and Little Ice Age by Mann et al. became known as "the hockey stick" of climate change where the handle of the hockey stick was supposed to represent constant climate until increasing CO_2 levels caused global warming, the sharp bend in the lower hockey stick.

The Mann et al. "hockey stick" temperature curve was at so at odds with thousands of published papers, including the Greenland GRIP ice core isotope data, sea surface temperatures in the Sargasso Sea sediments (Fig. 29) (Keigwin, 1996), paleo-temperature data other than tree rings (Fig. 30) (Loehle, 2007), and sea surface temperatures near Iceland (Fig. 31) (Sicre et al., 2008) one can only wonder how a single tree-ring study could purport to prevail over such a huge amount of data. At best, if the tree-ring study did not accord with so much other data, it should simply mean that the tree rings were

FIGURE 30   Reconstructed paleo-temperatures without tree ring data (Loehle, 2007).

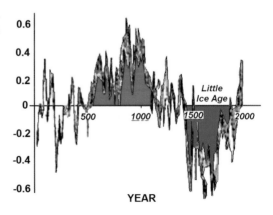

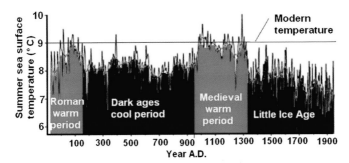

FIGURE 31   Summer sea surface temperatures near Iceland (Sicre et al., 2008).

not sensitive to climate change, not that all the other data were wrong. McIntyre and McKitrick (2003, 2005) evaluated the data in the Mann paper and concluded that the Mann curve was invalid *"due to collation errors, unjustifiable truncation or extrapolation of source data, obsolete data, geographical location errors, incorrect calculation of principal components and other quality control defects"*. Thus, the "hockey stick" concept of global climate change is now widely considered totally invalid and an embarrassment to the IPCC.

Why, then, did Mann's hockey stick persuade so many non-scientists and gain such widespread circulation? The answer is apparent in revelations from e-mails disclosed in the Climategate scandal (Mosher and Fuller, 2010; Montford, 2010). These e-mails describe how they tried to "hide the decline" in temperatures, using various "tricks" in order to perpetuate a dogmatic view of anthropogenic global warming.

5.1.1.7. The Little Ice Age (1300 A.D. to the 20th Century)

At the end of the Medieval Warm Period, ~1230 A.D., temperatures dropped ~4 °C (~7 °F) in ~20 years and the cold period that followed is known as the Little Ice Age. The colder climate that ensued for several centuries was devastating. Temperatures of the cold winters and cool, rainy summers were too low for growing of cereal crops, resulting in widespread famine and disease (Fagan, 2000; Grove, 2004). When temperatures declined during the 30-year cool period from the late 1940s to 1977, climatologists and meteorologists predicted a return to a new Little Ice Age.

Glaciers in Greenland began advancing and pack ice extended southward in the North Atlantic in the 13th century. Glaciers expanded worldwide. The population of Europe had become dependent on cereal grains as a food supply during the Medieval Warm Period and when the colder climate, early snows, violent storms, and recurrent flooding swept Europe, massive crop failures occurred. Three years of torrential rains that began in 1315 led to the Great Famine of 1315–1317. The Thames River in London froze over, the growing season was significantly shortened, crops failed repeatedly, and wine production dropped sharply.

Winters during the Little Ice Age were bitterly cold in many parts of the world. Advance of glaciers in the Swiss Alps in the mid-17th century gradually encroached on farms and buried entire villages. The Thames River and canals and rivers of the Netherlands frequently froze over during the winter. New York Harbor froze in the winter of 1780 and people could walk from Manhattan to Staten Island. Sea ice surrounding Iceland extended for miles in every direction, closing many harbors. The population of Iceland decreased by half and the Viking colonies in Greenland died out in the 1400s because they could no longer grow enough food there. In parts of China, warm weather crops that had been grown for centuries were abandoned. In North America, early European settlers experienced exceptionally severe winters.

In 1609, Galileo perfected the telescope, allowing observation of sunspots. From 1645 to1715, the number of sunspots observed activity was extremely low, with some years having no sunspots at all. This period of low sunspot activity, known as the Maunder Minimum, coincided with the thermal low of the Little Ice Age. The Spörer and Dalton sunspot minima also occurred during significant cold periods within the Little Ice Age. Low solar activity during the Little Ice Age is also shown by changes in the production rates of radiocarbon and ^{10}Be in the upper atmosphere.

Global temperatures have risen about 1 °F per century since the cooling of the Little Ice Age, but the warming has not been continuous. Numerous ~30-year warming periods have been interspersed with ~30-year cooling periods (Fig. 27). However, each warming period has been slightly warmer than the preceding one and cool period has not been quite as cool as the previous one.

During each warm cycle, glaciers retreated and during each cool cycle, glaciers advanced. However, because each warm cycle was slightly warmer than the previous one and each cool cycle not quite as cool as the previous one, glacier termini have progressively receded upvalley from their Little Ice Age maximums. These relationships are well shown on glaciers on Mt. Baker, Washington where large distinct Little Ice Age moraines mark the glacier termini well below present ice termini (Figs. 32, 33). The oldest Little Ice Age moraines have trees growing on them dating back to the 1500s. Successively higher moraines upvalley mark progressive advances and stillstands resulting from warm/cool cycles. The later moraines match the historic global climate changes.

A buried forest on the Coleman glacier moraine (Fig. 34), dated at 680 ± 80 and 740 ± 80 ^{14}C years B.P., grew during the Medieval Warm Period atop an older moraine (Easterbrook, 2010). The forest was later buried by a Little Ice Age moraine about 1300 A.D. Annual rings from trees growing on successively younger moraines upvalley show moraine—building episodes in the 1600s, ~1750, ~1790, ~1850, and ~1890, matching historic periods of global warming and cooling.

FIGURE 32   Ice retreat from Little Ice Age margins, Deming and Easton glaciers, Mt. Baker, Washington.

Ice margins of Mt. Baker glaciers are shown on air and ground photos dating back to 1943 (Figs. 35—40) (Easterbrook, 2010; Harper, 1993). Glaciers that had been retreating since at least the 1920s advanced during the 1947—1977 cool period to positions downvalley from their 1943 termini. They began to retreat once again at the start of the 1977—2007 warm cycle and present termini of the Easton and Boulder glaciers are about 1500 feet upvalley from their 1979 positions.

These glacier fluctuations closely follow the global cooling record and indicate that the ~30-year warming and cooling cycles seen in the glacial record mimic global climate changes. Thus, pre-historic glacial fluctuations also record global climate changes.

Glaciers on Mt. Baker show a regular pattern of advance and retreat which matches sea surface temperatures in the NE Pacific Ocean (the Pacific Decadal Oscillation) (Fig. 41). The glacier fluctuations are clearly driven by changes in the sea surface temperatures. An important aspect of this is that the sea surface temperature records extend to about 1900, but the glacial record goes back many centuries and can be used as a proxy for climate changes.

Climate changes are also recorded in the Greenland Ice Sheet isotope data. Regular warm/cool cycles of approximately the same duration as the glacial moraine record extend back 500 years (Fig. 27).

The importance of the various types of evidence of climate fluctuations is that they show long-standing evidence of cool/warm cycles over many

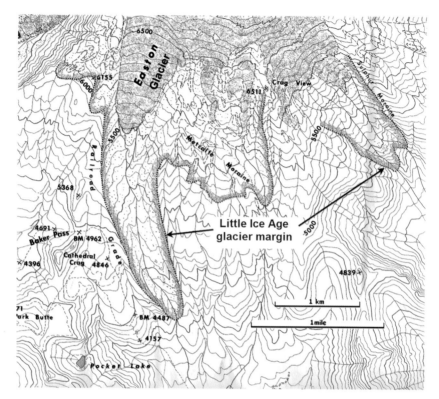

FIGURE 33   Little Ice Age glacier margins, Mt. Baker, WA.

centuries. Adding more recent, observed climatic fluctuations to the earlier records show that we are now right where we ought to be in this pattern, i.e., just past the end of the recent 20-year warm period. Extending this ongoing record into the future provides an opportunity to predict coming climate changes.

FIGURE 34   Little Ice Age moraine burying a forest that grew during the Medieval Warm Period, Coleman glacier, Mt. Baker, WA.

FIGURE 35 Easton glacier and Little Ice Age moraine, Mt. Baker, WA.

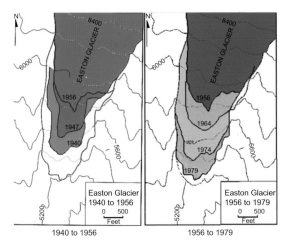

1940 to 1956   1956 to 1979

FIGURE 36 Retreat and advance of the Easton glacier (modified from Harper, 1993).

FIGURE 37 Retreat of the Easton glacier since 1979.

FIGURE 38 Boulder glacier, Mt. Baker, WA.

FIGURE 39 Retreat and advance of the Boulder glacier.

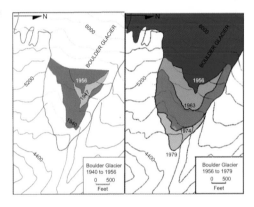

FIGURE 40 Retreat of the Boulder glacier since 1979, the beginning of the present 30-year warm cycle.

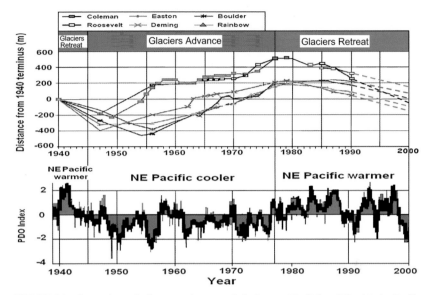

FIGURE 41   Comparison of advance and retreat of glaciers on Mt. Baker, WA with the Pacific Decadal Oscillation.

5.2. Significance of Previous Global Climate Changes

If CO_2 is indeed the cause of global warming, then global temperatures should mirror the rise in CO_2. For the past 1,000 years, atmospheric CO_2 levels have remained fairly constant at about 280 parts per million (ppm). Atmospheric CO_2 concentrations began to rise during the industrial resolution early in the 20th century. In 1945, atmospheric CO_2 began to rise sharply and by 1980 it had risen to about 340 ppm. During this time, however, global temperatures fell about 0.5 °C (0.9 °F) in the Northern Hemisphere and about 0.2 °C (0.4 °F) globally. In 1977, global atmospheric temperatures again reversed suddenly, rising about 0.5 °C (0.9 °F) above the 1945—1977 cool cycle in 25 years. If CO_2 is the cause of global warming, why did temperatures fall for 30 years while CO_2 was sharply accelerating? Logic dictates that this anomalous cooling cycle during accelerating CO_2 levels must mean either (1) rising CO_2 is *not* the *cause* of global warming or (2) some process other than rising CO_2 is capable of overriding its effect on global atmospheric warming and CO_2 is thus not significant. Temperature patterns since the Little Ice Age (~1600 A.D. to 1860 A.D.) show a very similar pattern—25—30 periods of alternating warm and cool temperatures. These temperature fluctuations took place well before any effect of anthropogenic atmospheric CO_2 and many were far greater. Most of the CO_2 from human activities was added to the air after 1945, so the early 20th century and earlier warming trends had to be natural and the recent trend in surface warming cannot be primarily attributable to human-made greenhouse gases. Thus, CO_2 cannot have been the cause of these climatic changes, so why should we suppose that the last one must be?

6. THE PACIFIC DECADAL OSCILLATION (PDO)

The Pacific Decadal Oscillation (PDO) refers to cyclical variations in sea surface temperatures in the Pacific Ocean. A summary of the PDO is given in D'Aleo (this volume). It was discovered in the mid-1990s by fisheries scientists studying the relationship between Alaska salmon runs, Pacific Ocean temperatures, and climate. Hare (1996), Zhang et al. (1997), and Mantua et al. (1997) found that cyclical variations in salmon and other fisheries correlated with warm/cool changes in Pacific Ocean temperatures that followed a regular pattern. Each warm PDO phase lasted about 25−30 years then switched to the cool phase and vice versa. The PDO differs from El Nino/La Nina warm/cool oscillations, which persist for only 6−18 months in an east-west belt near the equator.

Figure 42 shows the cold and warm modes of the PDO. During a typical PDO cold mode, cool sea surface temperatures extend from the equator northward along the coast of North America into the Gulf of Alaska. During a typical PDO warm mode, warm sea surface temperatures extend from the equator northward along the coast of North America into the Gulf of Alaska. As seen in the lower part of Fig. 42, the PDO was warm from about 1915 to about 1945, cool from about 1945 to 1977, warm from 1977 to 1998, and cool beginning in 1999 (interrupted by El Nino in 2005−2006).

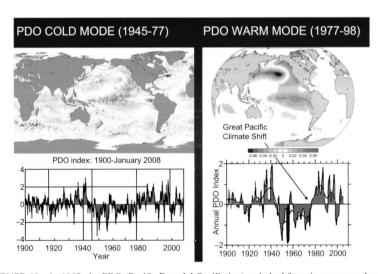

FIGURE 42   In 1945, the PDO (Pacific Decadal Oscillation) switched from its warm mode to its cool mode and global climate cooled from then until 1977, despite the sudden soaring of CO_2 emissions. In 1977, the PDO switched back from its cool mode to its warm mode, initiating what is regarded as 'global warming' (from 1977 to 1998).

Global temperatures are tied directly to sea surface temperatures. When sea surface temperatures are cool (cool phase PDO), as from 1945 to 1977, global climate cools. When sea surface temperatures are warm (warm phase PDO), as from 1977 to 1998, the global climate warms, regardless of any changes in atmospheric CO2 (Easterbrook, 2005, 2008a,b).

During the past century, global climates have consisted of two cool periods (1880−1915 and 1945−1977) and two warm periods (1915−1945 and 1977−1998). In 1997, the PDO switched abruptly from its cool mode, where it had been since about 1945, into its warm mode and global climate shifted from cool to warm. This rapid switch from cool to warm has become to known as "The Great Pacific Climatic Shift" (Fig. 4). Atmospheric CO_2 showed no unusual changes across this sudden climate shift (Fig. 20) and was clearly not responsible for it. Similarly, the global warming of 1915-1945 could not have been caused by increased atmospheric CO_2 because that time preceded the rapid rise of CO_2 after 1945 (Fig. 19) and when CO2 began to increase rapidly after 1945, 30 years of global cooling occurred (1945−1977).

The two warm and two cool PDO cycles during the past century have periods of about 25-30 years. Reconstruction of ancient PDO cycles by Verdon and Franks (2006) shows PDO warm and cool phases dating back to 1662 A.D.

6.1. The Cool Phase of the PDO is Now Entrenched and 'Global Warming' (1977−1998) is Over

'Global warming' (the term used for warming from 1977 to 1998) is over. No warming above 1998 temperatures has occurred (Fig. 43) and the winters of

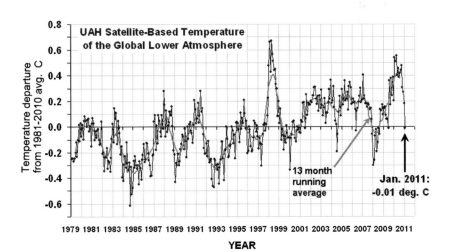

FIGURE 43   Satellite-based temperature of the global lower atmosphere from 1979 to 2011 (UAH, Spencer).

2006−2007, 2008−2009, 2010−2011 have set many records for snow and cold temperatures.

Figure 42 shows examples of the two temperature modes of the Pacific Ocean, its cool mode, which prevailed from about 1945 to 1977(left side) and its warm mode, which prevailed from 1977 to 1998. In each case, the global climate exactly followed the ocean temperature. The Pacific switches back and forth from warm to cool modes about every 30 years, a phenomenon known as the Pacific Decadal Oscillation (PDO).

Switching of the PDO back and forth from warm to cool modes has been documented by NASA satellite imagery. The satellite image from 1997 (Fig. 44) is typical of the warm PDO mode (1945−1977) with most of the eastern Pacific adjacent to North America showing shades of yellow to red, indicating warm water.

The satellite image from 1999 (Fig. 45) shows a strong contrast to the 1997 image, with deep cooling of the eastern Pacific and a shift from the PDO warm to the PDO cool mode. This effectively marked the end of 'global warming' (i.e., the 1977−1998 warm cycle). The images below (Figs. 45−48) show the switch of the PDO from its warm cycle to the present cool cycle.

Each time this has occurred in the past century, global temperatures have remained cool for about 30 years. Thus, the current sea surface temperatures not only explain why we have had no global warming for the past 10 years, but also assure that cool temperatures will continue for several more decades (Easterbrook, 2001, 2006a,b, 2007, 2008c).

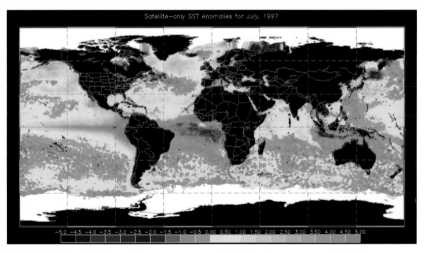

FIGURE 44   Satellite image of ocean temperature, 1997, showing strong warm PDO in the eastern Pacific. The deep red band at the equator is a strong El Nino that made 1997−1998 particularly warm.

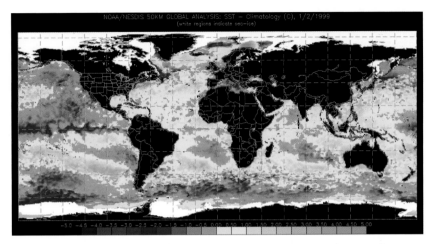

FIGURE 45   Satellite image of ocean temperature, 1999, showing the development of a strong cool PDO in the eastern Pacific that marked the end of 'global warming' and the beginning of the present cool cycle.

7. THE ATLANTIC MULTIDECADAL OSCILLATION (AMO)

The Atlantic Ocean also has multidecadal warm and cool modes with periods of about 30 years, much like the PDO. During warm phases, the Atlantic is warm in the tropical North Atlantic and far North Atlantic and relatively cool in the central area. During cool phases, the tropical area and far North Atlantic are cool and the central ocean is warm. For a more detailed discussion, see D'Aleo (this volume).

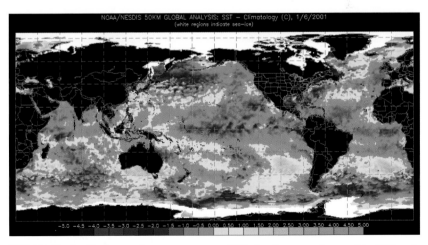

FIGURE 46   Satellite image of ocean temperature, 2001, showing entrenchment of the PDO cool cycle in the eastern Pacific off the coast of North America.

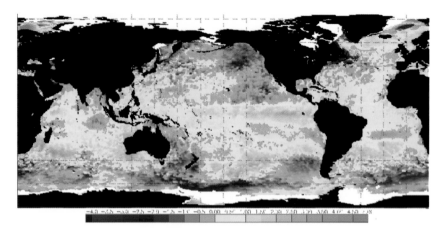

FIGURE 47   Satellite image of ocean temperature, 2006, showing continued entrenchment of the PDO cool cycle in the eastern Pacific off the coast of North America.

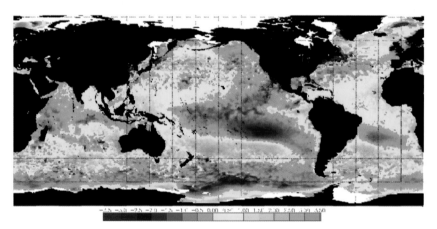

FIGURE 48   Satellite image of ocean temperature, February 22, 2011, showing the continued entrenchment of the PDO cool cycle in the eastern Pacific off the coast of North America.

Figures 49 and 50 show AMO and PDO cycles since 1900. Figure 51 shows the strong correlation between the PDO + AMO cycles and Arctic mean temperatures.

8. SOLAR VARIABILITY AND CLIMATE CHANGE

The global climate changes described above have coincided with changes in sunspot activity, solar irradiance, and rates of production of ^{14}C and ^{10}Be in the atmosphere by radiation, suggesting that the climate changes are caused by fluctuations in solar activity. A good example of the relationship between

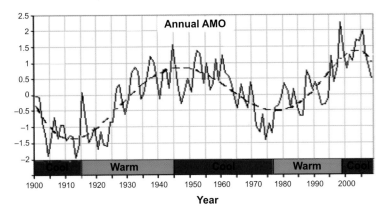

FIGURE 49 AMO cycles and global warming and cooling back to 1900 (modified from D'Aleo and Easterbrook, 2010).

solar activity and climate occurred during the Maunder Minimum. When Galileo perfected the telescope in 1609, scientists could see sunspots for the first time. They were of such interest that records were kept of the number of sunspots observed, and although perhaps not entirely accurate due to cloudy days, lost records, etc., the records show a remarkable pattern for more than a century (Fig. 52). From 1600 A.D. to 1700 A.D., very few sunspots were seen, despite the fact that many scientists with telescopes were looking for them, and reports of aurora borealis were minimal. This interval is known as

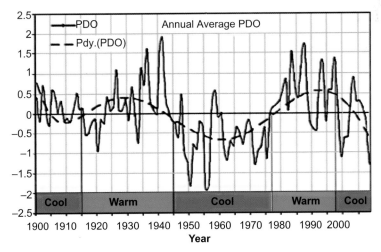

FIGURE 50 Annual average PDO back to 1900. Each warm/cool cycle lasts about 25–30 years and matches global climate changes (modified from D'Aleo and Easterbrook, 2010).

FIGURE 51 PDO + AMO oscillations since 1900 and arctic mean temperatures (modified from D'Aleo and Easterbrook, 2010).

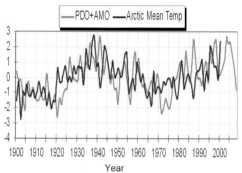

the Maunder Minimum (Maunder, 1894; Maunder, 1922; Eddy, 1976; Soon, 2005). Between 1650 A.D. and 1700 A.D., global climates turned bitterly cold (the Little Ice Age), demonstrating a clear correspondence between sunspots and cool climate. After 1700 A.D., the number of observed sunspots increased sharply from nearly zero to 50–100 (Fig. 52) and the global climate warmed.

The Maunder Minimum was preceded by the Sporer Minimum (~1410–1540 A.D.) and the Wolf Minimum (~1290–1320 A.D.) (Fig. 53). Each of these periods is characterized by low numbers of sunspots, significant changes in the rate of production of ^{14}C in the atmosphere, and cooler global climates. During the Maunder Minimum, almost no sunspots occurred. The Dalton sunspot minimum, which occurred between 1790 and 1830, was also a time of deep global cooling, as was the period from 1890 to 1915. A more modest global cooling from 1945 to 1977 was also a time of sunspot minima (Fig. 53).

The correlation between sunspots and global climate is remarkable. Unlike CO_2, which shows only a 4% correlation with climate changes over the past 500 years, the correlation of sunspots with climate change is close to 100% (Figs. 54, 55).

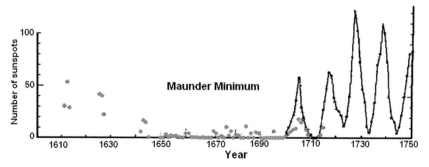

FIGURE 52 Sunspots during the Maunder Minimum (modified from Eddy, 1976).

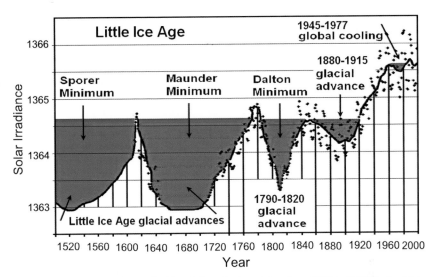

FIGURE 53   Correspondence of cold periods and solar minima from 1500 to 2000. Each of the five named solar minima was a time of sharply reduced global temperatures (blue areas).

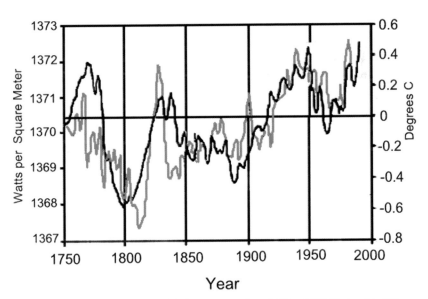

FIGURE 54   Solar irradiance and global temperature from 1750 to 1990. During this 250-year period, the two curves follow a remarkably similar pattern (modified from Hoyt and Schatten, 1997).

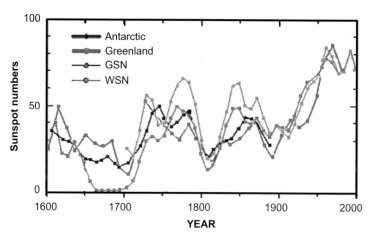

FIGURE 55   Correlation of sunspot numbers and temperatures in Greenland and Antarctica and cooling (modified from Usoskin et al., 2004).

8.1. Global Temperature Change, Sunspots, Solar Irradiance, ^{10}Be and ^{14}C Production

Good correlations can now be made between global temperature change, sunspots (Eddy, 1976, 1977; Stuiver and Quay, 1979), solar irradiance, and ^{10}Be (Beer et al., 1994, 1996, 2000; Beer et al., 2000) and ^{14}C production (Stuiver, 1961, 1994; Stuiver and Brasiunas, 1991, 1992; Stuiver et al., 1991, 1995) in the atmosphere. ^{10}Be is produced in the upper atmosphere by radiation bombardment of oxygen. Increased radiation results in increased ^{10}Be production. Plots of ^{10}Be production and sunspots indicate a good correlation between the two. Thus, ^{10}Be measurements can serve as a proxy for solar activity.

Figure 56 shows a remarkable correlation between temperature, as measured from oxygen isotope variation, and variation in the rate of production of radiocarbon by radiation in the upper atmosphere, suggesting that temperature variations are caused by changes in radiation.

For many years, solar scientists considered variation in solar irradiance to be too small to cause significant climate changes. However, Svensmark has proposed a new concept of how the sun may impact Earth's climate (Svensmark and Calder, 2007; Svensmark and Friis-Christensen, 1997; Svensmark et al., 2007). Svensmark recognized the importance of cloud generation as a result of ionization in the atmosphere caused by cosmic rays. Clouds highly reflect incoming sunlight and tend to cool the Earth. The amount of cosmic radiation is greatly affected by the sun's magnetic field, so during times of weak solar magnetic field, more cosmic radiation reaches the Earth. Thus, perhaps variation in the intensity of the solar magnetic field may play an important role in climate change.

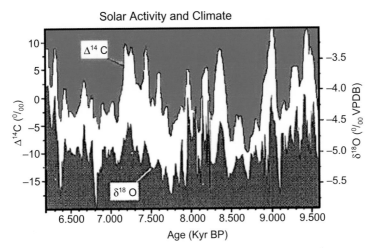

FIGURE 56 Correlation of temperature $(\delta^{18}O)$ and radiocarbon production $(\delta^{14}C)$ from a stalagmite in Oman (Matter et al., 2001).

9. WHERE ARE WE HEADED DURING THE COMING CENTURY?

9.1. IPCC Predictions

What does the century have in store for global climates? According to the Intergovernmental Panel on Climatic Change (IPCC), the Earth is in store for climatic catastrophe this century. Computer models predict global warming of as much as 5−6 °C (10−11 °F), predicated on the assumption that global warming is caused by increasing atmospheric CO_2 and that CO_2 will continue to rise.

The ramifications of such an increase in global warming are far reaching, even catastrophic in some areas. Such a rise of global surface temperatures would have devastating results. The Arctic Ocean would likely become free of its cover of sea ice, the Greenland Ice Sheet would diminish, and alpine glaciers would retreat rapidly, resulting in decreased water supply in areas that depend on snowmelt. Melting of Greenland and Antarctic ice would cause sea level to rise, flooding low coastal areas and submerging low coral islands in the oceans. Crops in critical agricultural areas would fail, resulting in widespread food shortages for people in agriculturally marginal areas. Wheat/grain belts, such as the mid-continent area of North America, would have to shift northward. Droughts would become increasingly severe in dry areas. Environmental impacts would be severe, resulting in extinction of some species and drastic population decreases in other.

IPCC computer models have predicted that global temperatures will rise 1 °F per decade for the next 10 decades and be 10 °F warmer by 2100 (Fig. 57).

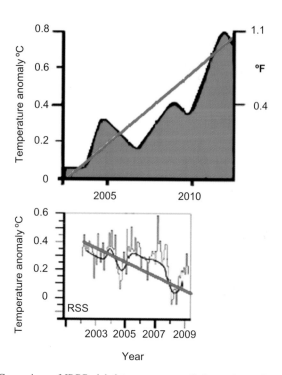

FIGURE 57 Comparison of IPCC global temperature prediction and actual temperature. Top: IPCC prediction in 2000 of 1 °F temperature increase per decade. Bottom: actual temperature for the decade.

According to their models, global temperature should have warmed 1 °F from 2000 to 2011, but global climates have actually cooled, not warmed, since 1998. Thus, the computer models have failed badly in predicting global climates and therefore must be considered unreliable.

9.2. Predictions Based on Past Climate Patterns

Predictions based on past warming and cooling cycles over the past 500 years accurately predicted the end of the 1977–1998 warm period and the establishment of cool Pacific sea surface temperatures. Past patterns strongly suggest that the next several decades will be cooler, not warmer.

Considering the positive correlations between solar activity and global climate change, what if the cause of global warming is solar, rather than atmospheric CO_2? Then all of the computer models are meaningless, and we can look to past natural climatic cycles as a basis for predicting future climate changes. The climatic fluctuations over the past few hundred years suggest ~30-year climatic cycles of global warming and cooling, on a general rising trend from the Little Ice Age cool period. The PDO over the past century (Fig. 58)

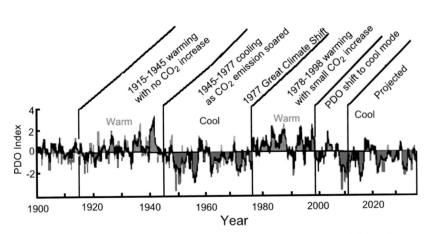

FIGURE 58   Pacific Decadal Oscillation pattern over the past century, extended into the next several decades.

has a consistent cyclic pattern that matches global climate changes. If the past climatic trend continues as it has for the past 500 years (Fig. 27) and if the PDO cyclic pattern continues as it has for the past century, global temperatures for the coming century might look like those shown in Fig. 59. When the Pacific Ocean switched from its warm mode to its cool in 1999 virtually assures global cooling for several decades, as it has with each mode switch in the past century.

The left side of Fig. 59 is the warming/cooling history of the past century. The right side of the graph shows the IPCC predicted temperature and several

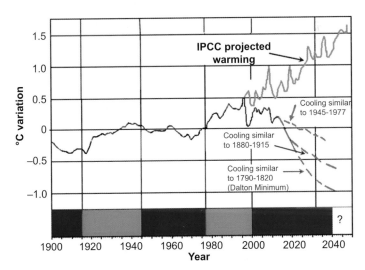

FIGURE 59   Projected climate for the century based on climatic patterns over the past 500 years and the switch of the PDO to its cool phase.

possible temperature patterns that we may well encounter if the cyclic climatic patterns of the past 400 years continue.

Three possible projections are shown: (1) moderate cooling (similar to the 1945−1977 cooling); (2) deeper cooling (similar to the 1880−1915 cooling); or (3) severe cooling (similar to the 1790−1830 cooling) during the Dalton Solar Minimum. A fourth possibility, very severe cooling similar to the Maunder Minimum, is also possible, but less likely. Time will tell which of these will be the case, but at the moment, the sun is behaving very similar to the Dalton Solar Minimum (Archibald, 2010), which was a very cold time. This is based on the similarity of sunspot cycle 23 to cycle 4 (which immediately preceded the Dalton Minimum).

We live in a most interesting time. As the global climate and solar variation reveal themselves in a way not seen in the past 200 years, we will surely attain a much better understanding of what causes global warming and cooling. Time will tell. If the climate continues its deepening cooling and the sun behaves in a manner not witnessed since 1800, we can be sure that climate changes are dominated by the sun and that atmospheric CO_2 has a very small role in climate changes. If the same climatic patterns, cyclic warming and cooling, that occurred over the past 500 years continue, we can expect several decades of global cooling.

REFERENCES

Alley, R.B., 2000. The Younger Dryas cold interval as viewed from central Greenland. Quaternary Science Reviews 19, 213−226.

Archibald, D., 2010. The Past and Future of Climate: Why the World is Cooling and Why Carbon Dioxide Won't Make a Detectable Difference. Rhaetian Management Pty Ltd, p. 142.

Beer, J., Mende, W., Stellmacher, R., 2000. The role of the sun in climate forcing. Quaternary Science Reviews 19, 403−415.

Beer, J., Mende, W., Stellmacher, R., White, O.R., 1996. Intercomparisons of proxies for past solar variability. In: Jones, P.D., Bradley, R.S., Jouzel, J. (Eds.), Climatic Variations and Forcing Mechanisms of the Last 2000 Years. Springer−Verlag, Berlin, pp. 501−517.

Beer, J., Joos, F., Lukasczyk, C., Mende, W., Rodriguez, J., Sikegenthaler, U., Stellmacher, R., 1994. ^{10}Be as an indicator of solar variability and climate. In: Nesme-Ribes, E. (Ed.), The Solar Engine and Its Influence on Terrestrial Atmosphere and Climate. Springer−Verlag, Berlin, pp. 221−233.

Caillon, et al., 2003. Timing of atmospheric CO_2 and Antarctic temperature changes across termination III: Science 299, 1728−1731.

Chylek, P., Box, J.E., Lesins, G., 2004. Global warming and the Greenland ice sheet. Climatic Change 63, 201−221.

Chylek, P., Dubey, M.K., Lesins, G., 2006. Greenland warming of 1920−1930 and 1995−2005. Geophysical Research Letters 33.

Cuffey, K.M., Clow, G.D., 1997. Temperature, accumulation, and ice sheet elevation in central Greenland through the last deglacial transition. Journal of Geophysical Research 102, 26, 383−26, 396.

D'Aleo, J., Easterbrook, D.J., 2010. Multidecadal tendencies in Enso and global temperatures related to multidecadal oscillations. Energy & Environment 21 (5), 436−460.

Easterbrook, D.J., 2001. The next 25 years: global warming or global cooling? Geologic and oceanographic evidence for cyclical climatic oscillations. Geological Society of America, Abstracts with Program 33, 253.

Easterbrook, D.J., 2005. Causes and effects of abrupt, global, climate changes and global warming. Geological Society of America, Abstracts with Program 37, 41.

Easterbrook, D.J., 2006a. Causes of abrupt global climate changes and global warming predictions for the coming century. Geological Society of America, Abstracts with Program 38, 77.

Easterbrook, D.J., 2006b. The cause of global warming and predictions for the coming century. Geological Society of America, Abstracts with Program 38, 235–236.

Easterbrook, D.J., 2007. Geologic evidence of recurring climate cycles and their implications for the cause of global warming and climate changes in the coming century. Geological Society of America, Abstracts with Programs 39 (6), 507.

Easterbrook, D.J., 2008a. Solar influence on recurring global, decadal, climate cycles recorded by glacial fluctuations, ice cores, sea surface temperatures, and historic measurements over the past millennium. Abstracts of American Geophysical Union Annual Meeting, San Francisco.

Easterbrook, D.J., 2008b. Implications of glacial fluctuations, PDO, NAO, and sun spot cycles for global climate in the coming decades. Geological Society of America, Abstracts with Programs 40, 428.

Easterbrook, D.J., 2008c. Correlation of climatic and solar variations over the past 500 years and predicting global climate changes from recurring climate cycles. Abstracts of 33rd International Geological Congress, Oslo, Norway.

Easterbrook, D.J., 2010. A Walk Through Geologic Time from Mt. Baker to Bellingham Bay. Chuckanut Editions, p. 330.

Eddy, J.A., 1976. The Maunder Minimum: Science 192, 1189–1202.

Eddy, J.A., 1977. Climate and the changing sun. Climatic Change 1, 173–190.

Fagan, B., 2000. The Little Ice Age. Basic Books, N.Y., p. 246.

Fagan, B., 2007. The Great Warming: Climate Change and the Rise and Fall of Civilizations. Bloomsbury Press, p. 283.

Fischer, et al., 1999. Ice core record of atmospheric CO_2 around the last three glacial terminations: Science 283, 1712–1714.

Gore, A., 2007, An Inconvenient Truth: Rodale, PA, p. 325.

Grootes, P.M., Stuiver, M., 1997. Oxygen 18/16 variability in Greenland snow and ice with 10^{-3}- to 10^5-year time resolution. Journal of Geophysical Research 102, 26455–26470.

Grove, J.M., 2004. Little Ice Ages: Ancient and Modern. Routledge, London, UK, p. 718.

Hare, S.R., 1996. Low frequency climate variability and salmon production. Ph.D. dissertation, School of Fisheries, University of Washington, Seattle, WA.

Harper, J.T., 1993. Glacier fluctuations on Mount Baker, Washington, U.S.A., 1940–1990, and climatic variations. Arctic and Alpine Research 4, 332–339.

Horner, C.C., 2008. Red Hot Lies: How Global Warming Alarmists Use Threats, Fraud, and Deception to Keep You Misinformed. Regnery Publishing, p. 407.

Hoyt, D.V., Schatten, K.H., 1997. The Role of the Sun in Climate Change. Oxford University. p. 279.

Idso, C., Singer, S.F., 2009. Climate Change Reconsidered: 2009 Report of the Nongovernmental Panel on Climate Change (NIPCC). The Heartland Institute, Chicago, IL, p. 855.

IPCC-AR4, 2007. Climate Change: The Physical Science Basis. Contribution of Working Group I to the Fourth Assessment Report of the Intergovernmental Panel on Climate Change. Cambridge University Press.

Keigwin, L.D., 1996. The Little Ice Age and medieval warm period in the Sargasso Sea. Science 274, 1504–1508.

Loehle, C., 2007. A 2000-year global temperature reconstruction based on non-tree-ring proxies. Energy and Environment 18, 1049−1058.

Mann, M.E., Bradley, R.S., Hughes, M.K., 1998. Global-scale temperature patterns and climate forcing over the past six centuries. Nature 392, 779−787.

Mantua, N.J., Hare, S.R., Zhang, Y., Wallace, J.M., Francis, R.C., 1997. A Pacific interdecadal climate oscillation with impacts on salmon production. Bulletin of the American Meteorological Society 78, 1069−1079.

Matter, A., Neff, U., Fleitmann, D., Burns, S., Mangini, A., 2001. 350,000 years of climate variability recorded in speleothems from Oman. GeoArabia (Manama) 6, 315−316.

Maunder, E.W., 1894. A prolonged sunspot minimum. Knowledge 17, 173−176.

Maunder, E.W., 1922. The prolonged sunspot minimum, 1645−1715. Journal of the British Astronomical Society 32, p. 140.

McIntyre, S., McKitrick, R., 2003. Corrections to Mann et al. (1998) proxy database and northern hemisphere average temperature series. Energy & Environment 14, 751−777.

McIntyre, S., McKitrick, R., 2005. Hockey sticks, principal components and spurious significance. Geophysical Research Letters 32.

Miller, A.J., Cayan, D.R., Barnett, T.P., Graham, N.E., Oberhuber, J.M., 1994. The 1976−77 climate shift of the Pacific Ocean. Oceanography 7, 21−26.

Montford, A.W., 2010. Hockey Stick Illusion: Climategate and the Corruption of Science. CreateSpace, p. 482.

Mosher, S., Fuller, T.W., 2010. Climategate: The Crutape Letters. Stacey International, p. 186.

O'Donnell, R., Lewis, N., McIntyre, S., Condon, J., 2011. Improved methods for PCA-based reconstructions: case study using the Steig et al. 2009 Antarctic temperature reconstruction. Journal of Climate 24, 2099−2115.

Oliver, J.E., 1973. Climate and Man's Environment. Wiley, N.Y., p. 365.

Sicre, M., Jacob, J., Ezat, U., Rousse, S., Kissel, C., Yiou, P., Eiriksson, J., Knudsen, K.L., Jansen, E., Turon, J., 2008. Decadal variability of sea surface temperatures off North Iceland over the last 2000 years. Earth and Planetary Science Letters 268, 137−142.

Singer, S.F., Avery, D., 2007. Unstoppable Global Warming Every 1,500 Years. Rowman & Littlefield Publishers, Inc. p. 278.

Soon, W., 2005. Variable solar irradiance as a plausible agent for multidecadal variations in the Arctic-wide surface air temperature record of the past 130 years. Geophysical Research Letters 32, L16712.

Soon, W., Baliunas, S., 2003. Proxy climatic and environmental changes of the past 1000 years. Climate Research 23, 89−110.

Spencer, R.W., 2010a. The Great Global Warming Blunder: How Mother Nature Fooled the World's Top Climate Scientists. Encounter Books, p. 180.

Spencer, R.W., 2010b. Climate Confusion: How Global Warming Hysteria Leads to Bad Science, Pandering Politicians and Misguided Policies that Hurt the Poor. Encounter Books, p. 215.

Steffensen, J.P., Andersen, K.K., Bigler, M., Clausen, H.B., Dahl-Jensen, D., Goto-Azuma, K., Hansson, M.J., Sigfus, J., Jouzel, J., Masson-Delmotte, V., Popp, T., Rasmussen, S.O., Roethlisberger, R., Ruth, U., Stauffer, B., Siggaard-Andersen, M., Sveinbjornsdottir, A.E., Svensson, A., White, J.W.C., 2008. High-resolution Greenland ice core data show abrupt climate change happens in few years. Science 321, 680−684.

Steig, E.J., Schneider, D.P., Rutherford, S.D., Mann, M.E., Josefino, C.C., Shindell, D.T., 2009. Warming of the Antarctic ice-sheet surface since the 1957 International Geophysical Year. Nature 457, 459−462.

Stuiver, M., 1961. Variations in radio carbon concentration and sunspot activity. Geophysical Research 66, 273–276.

Stuiver, M., 1994. Atmospheric ^{14}C as a proxy of solar and climatic change. In: Nesme-Ribes (Ed.), The Solar Engine and Its Influence on Terrestrial Atmosphere and Climate. Springer–Verlag, Berlin, pp. 203–220.

Stuiver, M., Brasiunas, T.F., 1991. Isotopic and solar records. In: Bradley, R.S. (Ed.), Global Changes of the Past. Boulder University, Corporation for Atmospheric Research, pp. 225–244.

Stuiver, M., Brasiunas, T.F., 1992. Evidence of solar variations. In: Bradley, R.S., Jones, P.D. (Eds.), Climate Since A.D. 1500. Routledge, London, pp. 593–605.

Stuiver, M., Quay, P.D., 1979. Changes in atmospheric carbon-14 attributed to a variable sun. Science 207, 11–27.

Stuiver, M., Braziunas, T.F., Becker, B., Kromer, B., 1991. Climatic, solar, oceanic, and geomagnetic influences on late-glacial and Holocene atmospheric ^{14}C/^{12}C change. Quaternary Research 35, 1–24.

Stuiver, M., Grootes, P.M., Brasiunas, T.F., 1995. The GISP2 δ^{18}O record of the past 16,500 years and the role of the sun, ocean, and volcanoes. Quaternary Research 44, 341–354.

Svensmark, H., Calder, N., 2007. The chilling stars: A new theory of climate change, Icon Books, Allen and Unwin Pty Ltd, p. 246.

Svensmark, H., Friis-Christensen, E., 1997. Variation of cosmic ray flux and global cloud cover—a missing link in solar–climate relationships. Journal of Atmospheric and Solar–Terrestrial Physics 59, 1125–1132.

Svensmark, H., Pedersen, J.O., Marsh, N.D., Enghoff, M.B., Uggerhøj, U.I., 2007. Experimental evidence for the role of ions in particle nucleation under atmospheric conditions. Proceedings of the Royal Society A 463, 385–396.

Usoskin, I.G., Mursula, K., Solanki, S.K., Schussler, M., Alanko, K., 2004. Reconstruction of solar activity for the last millenium using ^{10}Be data. Astronomy and Astrophysics 413, 745–751.

Verdon, D.C., Franks, S.W., 2006. Long-term behaviour of ENSO: interactions with the PDO over the past 400 years inferred from paleoclimate records. Geophysical Research Letters 33.

Zhang, Y., Wallace, J.M., Battisti, D., 1997. ENSO-like interdecadal variability. Journal of Climatology 10, 1004–1020.

Chapter 2

Evidence for Synchronous Global Climatic Events: Cosmogenic Exposure Ages of Glaciations

Don J. Easterbrook[*], John Gosse[†], Cody Sherard[‡], Ed Evenson[**] and Robert Finkel[††]

[*] *Department of Geology, Western Washington University, Bellingham, WA 98225, USA,* [†] *Dalhousie University, Halifax, Nova Scotia, Canada B3H 4R2,* [‡] *Department of Geology, Western Washington University, Bellingham, WA 98225, USA,* [**] *Earth and Environmental Sciences Department, Bethlehem, PA 18015, USA,* [††] *Lawrence Livermore National Laboratory, Livermore, CA 94550*

1. INTRODUCTION

Why is knowledge of short-term sensitivity of glaciers to climatic/oceanic changes important? Despite three decades of research on abrupt climate changes, such as the Younger Dryas (YD) event, the geological community is only now arriving at a consensus about its global extent, but has not established an unequivocal cause or a mechanism of global glacial response to rapid climate changes. At present, although Greenland ice cores have allowed the development of highly precise ^{18}O curves, we cannot adequately explain the cause of abrupt onset and ending of global climatic reversals. In view of present global warming, understanding the cause of climate change is critically

Evidence-Based Climate Science. DOI: 10.1016/B978-0-12-385956-3.10002-6

important to human populations—the initiation and cessation of the YD ice age were both completed within a human generation.

Whether or not large magnitude but short-term fluctuations of late Pleistocene ice sheets and alpine glaciers were synchronous globally is still being debated, although evidence is compelling that the YD was a global climatic event. Some researchers continue to argue that the YD did not affect western North America and that alpine glaciers and ice sheets were out of phase at the last glacial maxima (LGM). Others argue that YD cooling did not affect the Southern Hemisphere. At the root of these debates is a lack of precise time control on discrete regional climatic events, incomplete understanding of the causal mechanisms of abrupt climatic change, problems linked to uncertainties of some of the dating methods when comparing chronologies, stratigraphic complexities, and too few dates at critical localities.

2. LATE PLEISTOCENE CLIMATE OSCILLATIONS RECORDED BY GLACIERS

In addition to the evidence for abrupt, late Pleistocene, climatic changes in ice cores, late Pleistocene climate changes have also been recorded in alpine moraines, which are useful indicators of climate change because they record temporal fluctuations in ice volume. However, they do not necessarily record *every* climate change because: (1) moraines only mark an ice-marginal position for a particular maximum, so if a subsequent advance is greater, the earlier evidence is obliterated; (2) not all glaciers form moraines during a particular glacial maximum because the glacier margin may not have been stable long enough at a single ice-marginal position to build a significant moraine; (3) glaciological conditions may force glacier advance or retreat out of phase with expected climate change; and (4) glaciers respond to changes in both temperature and precipitation that may induce a complicated moraine record out of phase with global temperature change. Thus, at any single site, these factors may complicate climatic inferences. However, when a consistent pattern emerges over wide regions, the likelihood that all would have been the result of non-climatic influences is much diminished. Thus, well-dated ice-marginal fluctuations can reveal useful information regarding the timing of major paleoclimate reversals and whether or not the effects of some late Pleistocene climate events were felt globally or only in some regions.

Establishing detailed chronologies of glacier fluctuations is limited by stratigraphic context relating the dated material to the climate record and the accuracy of the dating method. Dating of moraines has been accomplished directly using terrestrial in situ cosmogenic nuclide (TCN) dating or ^{14}C dating of incorporated organic material and indirectly by ^{14}C dating of associated sediments. Measurements of ^{10}Be, ^{26}Al, ^{36}Cl, and ^{3}He from boulders give exposure ages with approximately 1−8% random error (1 s), but uncertainties in the TCN production rates and other systematic errors of these isotope systems diminish the obtainable accuracy. However, identifying the 1,300−1,500-year span of the YD is well within the precision of TCN dating (e.g., Gosse et al., 1995a,b; Ivy-Ochs et al., 1999; Gosse and Phillips, 2001). Likewise, although the internal error of ^{14}C dating is about ±1% (1 s), uncertainty in the δ^{14}C calibration curve at about the time of the YD decreases the total accuracy. One might therefore argue that neither dating method provides sufficient accuracy to distinguish events *within* the YD. However, identifying multiple YD or other late Pleistocene events does not depend on the ability to distinguish double YD moraines solely on the basis of TCN and ^{14}C dating. If *both* moraines of a YD doublet fall within the age range of the YD event, the existence of two moraines will show that the YD at that site consisted of more than one event. A consistent pattern of double YD moraines at many sites indicates that the double moraines are not due to some local glaciological effect. TCN and ^{14}C dating methods provide the necessary level of precision for this purpose.

In this paper, morphologic and chronologic evidence is presented that the YD climatic oscillation that affected glaciers in the western U.S. consisted of two phases, and glaciers in the Swiss Alps, Scandinavia, and New Zealand recorded double YD moraines, suggesting that glaciers in both hemispheres reacted sensitively and virtually simultaneously (Easterbrook, 1994a, 2002, 2003a,b; Easterbrook et al., 2004). Evidence is also presented for the synchronicity of late Pleistocene oscillations of the Cordilleran and Scandinavian ice sheets, which were also in phase with alpine glacier fluctuations. Not all post-LGM cirque moraines are YD and not every cirque contains double YD moraines. In some places, pre-YD (probably Intra-Allerød Cold Period) moraines have also been documented and some YD localities have single or multiple moraines. However, enough double YD moraines exist to show a two-fold YD climate change.

This paper is not intended as an in-depth review of the Younger Dryas—rather it presents evidence that glaciers in both hemispheres advanced twice during the YD, and the timing is matched in the GISP2 core. This suggests that climate was globally connected during the late Pleistocene and implies that climate connections are via means other than oceanic-atmospheric teleconnections. Mercer (1982) suggested that glaciations in the Northern and Southern Hemispheres were synchronous and YD moraines have been found in many areas. Our contribution is the suggestion that *two* global YD advances must mean that glaciers were so sensitive to global climate change that such changes could not have been initiated in one region and then propagated via the oceans globally without introducing significant time lags in the glacial response in different hemispheres.

The essential point of this paper is that such sensitive, global, climate responses cannot be adequately explained by the North Atlantic deep current hypothesis. Our objective is to introduce and provide support for the proposed climatic significance of the widely occurring twin-moraine YD records. We establish a new YD twin-moraine record in the western USA by providing geochronological constraints on late Pleistocene glacial dynamics which ended with the YD readvances. Then we compare this chronology with that of previously dated twin-moraine records. Much discussion has centered around whether or not glaciers in western North America advanced and retreated synchronously or reacted progressively in accordance to theoretical hypotheses related to secondary effects generated by the Laurentide Ice Sheet, calling for progressive westward transgression of glacial climates in the west and asynchronicity of glacial events (Gillespie and Molnar, 1995). Others have suggested an absence of Younger Dryas glacial advances in the western U.S. or asynchroniety of post-LGM events. The data in this paper address these significant questions.

2.1. Sawtooth Mts. Moraine Sequences

2.1.1. Geologic Setting

More than 50 peaks in the Sawtooth Range rise above 10,000 ft (3,300 m), making it among the highest mountain ranges in the northern Rocky Mts. of Idaho. Most of the range consists of granitic rocks of the Idaho batholith, including both Cretaceous and Eocene phases and all of the moraines studied consist largely of granitic material. The contact between the Cretaceous and Eocene granitic rocks occurs in the headwaters of the Redfish Lake drainage so both rock types occur on moraines in the area. The Eocene granite tends to yield much larger erratics than the Cretaceous granite and does not appear to weather as rapidly. All of the moraines in the Redfish Lake area are mantled with large erratics (several meters high) of Eocene granite. Moraines in the valleys south of Redfish Lake consist mostly of Cretaceous granite and erratics are generally considerably smaller. Large erratic boulders suitable for cosmogenic dating are abundant in the Redfish Lake area but are rare on moraines in the southern part of the range. The very large erratics at Redfish Lake may also reflect jointing characteristics of the Eocene granite, allowing big, joint-bounded blocks to be quarried by glaciers.

The Sawtooth Range of central Idaho (Fig. 1) was deeply affected by late Pleistocene alpine glaciation. Valley glaciers from the high mountains extended out onto the piedmont area of the Stanley Basin where they constructed groups of massive moraines that cut across one another in successive glacial advances. These moraines provide an exceptional opportunity for establishing the glacial chronology of this part of the Rocky Mts.

2.1.2. Glacial Geomorphology

The last glacial maximum (LGM) is represented by large, massive moraines rising 330 m above their bases on the valley floors in more than a dozen drainages (Fig. 1). At Redfish Lake, glacial advances emerged from the

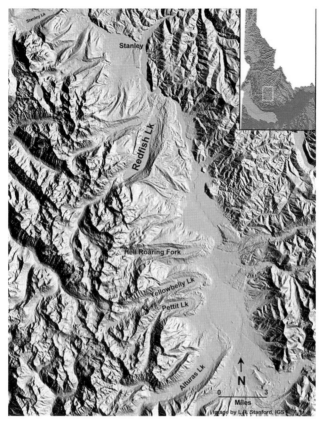

FIGURE 1 DEM of the Sawtooth Range and Stanley Basin. Moraines 300 m high extend from the mountains well into the basin *(Image by L.R. Stanford, IGS)*.

mountains and left moraines at nearly right angles to the mountain front (Figs. 1, 2). Massive late Pinedale moraines cut across the older moraines and extend 10 km across the Stanley Basin, temporarily blocking the Salmon River. During its retreat, this glacier built at least two dozen recessional moraines. Four cirques at progressively higher elevations above Redfish Lake contain moraines that record Younger Dryas, successively-rising, equilibrium line altitudes (ELAs).

The moraine record in the Sawtooth Range and Stanley Basin was first studied in detail by Williams (1961), who established relative moraine chronology based on morphology and relative weathering. He mapped several crosscutting morainal systems, which he called Bull Lake 1, Bull Lake 2, and Pinedale, and established a relative chronology, but had no numerical dates. Breckenridge et al. (1988) discussed moraines and lake sediments in the Stanley Basin.

FIGURE 2   Morainal groups and ^{10}Be ages in the Redfish Lake area *(Image by L.R. Stanford, IGS)*.

Thackray et al. (2004) mapped moraines at Alturas, Pettit, and Yellowbelly Lakes, and Hell Roaring valleys south of Redfish Lake. They recognized two morainal groups, each consisting of several lateral and end moraines, which they distinguished on the basis of soil characteristics and moraine morphometry and concluded that the outer moraines were at least 10 ka older than the inner moraines. Our ^{10}Be dates on these moraines indicate that all of these moraines are late Pinedale and confirm that the prominent cirque moraines indeed fall within the YD chron.

2.1.3. Methods

Samples were collected using a diamond—blade saw to extract uniformly thick (2 cm) specimens from the center of the top of each boulder, minimizing the effects of neutron loss (Masarik and Weiler, 2003) and edge effects (Gosse and Phillips, 2001), and reducing variability in thickness measurements. None of the sampled boulders had peaked or significantly ($>12°$) dipping surfaces. Only boulders greater than 1.5 m high (except one) situated on moraine crests were sampled in order to minimize the possibility of significant snow burial or post-depositional exhumation.

All samples were cleaned of lichen and soil, crushed, ground, and dry sieved. Following Kohl and Nishiizumi (1992), ca. 300 g of the 250—500 μm fraction was leached in 4 l of deionized water with 140 ml of HF acid and 50 ml of HNO_3 acid in ultrasonic tanks for 50 h, during which the grains were rinsed and solution replaced two times. After confirming that greater than 35% of the original quartz mass was dissolved, quartz purity was verified using Al concentrations determined with Quant-EM strips (all samples had <100 mg/ml Al). To achieve a sufficient AMS current for BeO and a $^{10}Be/^9Be$ above 5×10^{-13}, 0.2 mg of ^{10}Be carrier was added to approximately 15—40 g of each dried sample in a 500 ml Teflon digestion vessel. The mixture was dissolved in 20 h in a solution of HF (540 ml), $HClO_4$ (13 ml) and Aqua Regia (50 ml), then evaporated and reduced to a chloride. The Be carrier is a 985 mg/ml solution from a $BeCl_3$ prepared by J. Klein (U. Pennsylvania) using a beryl crystal collected deep within the Homestake Gold mine in South Dakota, USA (long-term average $^{10}Be/^9Be$ is 4×10^{-15}). The Be carrier concentration has been re-measured regularly with ICP.MS, AA, and ICP.OES and found to be within 3% of this value. Be was isolated using ion chromatography (10 ml and 17 ml of anion and cation Biorad® resins) and controlled precipitations with ultrapure ammonia gas, with final ignition to BeO in a furnace. The $^{10}Be/^9Be$ measurements at CAMS-LLNL against KNSTD3110 Be standard yielded precisions of 1.1—2.9%, and most blank corrections were less than 2%; one $^{10}Be/^9Be$ process blank had a high ratio (29×10^{-15}) which caused a 5% adjustment to the measured ratio of sample ID-RL-04-024. Isobaric boron interference corrections were low ($<1\%$).

Following data reduction, the concentrations were corrected for snow cover (0—3% based on boulder height and assumed snow depth of 2 m at $2 \, g \, cm^{-3}$ density for 4 months per year, Gosse and Phillips, 2001), tree cover (0—2% in an open pine forest, Plug et al., 2007), and topographic shielding ($<2\%$; Gosse and Phillips, 2001). No correction for boulder erosion has been made as our sampling strategy was devised to minimize this effect. However, it is possible that erosion may have decreased the apparent ages of LGM and pre-LGM boulders by a few percent. Additional details including ^{10}Be concentrations and AMS error are provided in Sherard (2006). The ages have been calculated using the ^{10}Be online CRONUS-CALCULATOR v. 2.2 (Balco et al., 2008). The ages reported in the text are the average of the three time-dependent geomagnetic field corrected scaling methods. Uncertainty in age at 1σ reflects the AMS error and 10% error in production rate.

2.1.4. Pre-LGM Moraines

2.1.4.1. Elk Meadow Morainal Group

The Elk Meadow morainal group consists mostly of multiple, inset, arcuate, lateral moraines on the east side of the large Redfish Lake morainal complex

FIGURE 3   Elk Meadow and Salmon River moraine groups and [10]Be ages south of Redfish Lake. These moraines are truncated by the high Redfish Lake lateral (Pinedale) in the center of the image *(Image by L.R. Stanford).*

(Fig. 2). A group of younger moraines, the Salmon River morainal group, crosscut the youngest Elk Meadow moraines (Figs. 3, 4).

About half a dozen nested, Elk Meadow moraines are the oldest moraines mapped by Williams (1961) in the Redfish Lake area. He mapped them as Bull Lake 1, based on their weathering characteristics and topographic position relative to the next younger moraines.

Four samples were collected from large granitic boulders at three locations on the moraines near the terminus. The [10]Be ages of the samples were 200 ± 25 ka, 161 ± 20 ka, 107 ± 14 ka, and 105 ± 15 ka.

2.1.4.2. Salmon River Morainal Group

The Salmon River morainal group consists of at least seven nested lateral and end moraines southeast of Redfish Lake (Figs. 3, 4). The Salmon River moraines cut across the upper Elk Meadow moraines (Figs. 3, 4) and are bounded by a pair of long lateral moraines enclosing at least five lateral and end moraines (Fig. 4). A high Pinedale lateral moraine (~300 m above Redfish Lake) cuts across the Salmon River moraines (Fig. 4). Williams (1961) mapped these moraines as Bull Lake 2, based on their weathering characteristics and topographic position relative to the Pinedale Redfish Lake moraines (Fig. 4). The morphology of the moraines is less subdued than the Elk Meadow moraines and the boulder frequency seems unusually high for Bull Lake moraines.

Two samples from moraines of the Salmon River Group were dated at 312 ± 42 ka and 229 ± 30 ka. The overlap in ages between the Elk Meadow

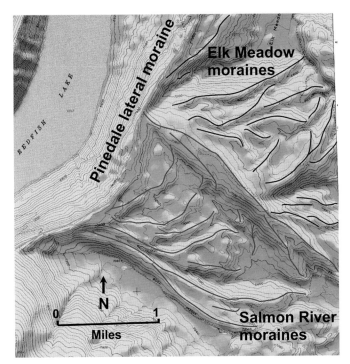

FIGURE 4 Salmon River moraines cutting across Elk Meadow moraines SE of Redfish Lake.

and Salmon River group may be due to a combination of inherited [10]Be and differences in boulder weathering rates. The six boulders dated on the two moraine sequences have a similar distribution to boulders exposure dated at the Bull Lake type locality in Wyoming, and, considering soils and moraine morphology observations are unlikely to have been deposited during the LGM.

2.1.5. LGM Moraines

2.1.5.1. Redfish Lake Morainal Group

High, massive lateral moraines that bound Redfish Lake (Figs. 3, 5) are here named the Redfish Lake morainal group. They were mapped by Williams (1961) as Pinedale moraines on the basis of their morphology, stratigraphic position, and degree of weathering, but he had no numerical chronology for them. The morainal group includes at least two dozen lateral moraines and several end moraines extending over a distance of several kilometers (Fig. 5). The dominant, enclosing lateral moraines are massive, reaching 300 m above lake level and cut across the Salmon River and Elk Meadow morainal groups (Fig. 4). The surface of the moraines is littered with many

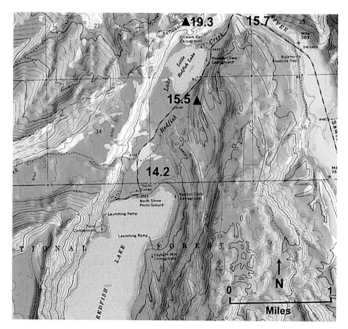

FIGURE 5   The Redfish Lake morainal group. Curved lines indicate moraine crests.

very large (several meters high) granitic boulders, mostly pink granite derived from the Eocene phase of the Idaho batholith, well suited to cosmogenic exposure dating.

The Pinedale LGM at Redfish Lake was dated at 18.8 ± 2.3 ka from a boulder on the terminal moraine and dates of 20.3 ± 2.5 ka and 18.1 ± 2.2 ka from the highest lateral moraines. An age of 13.6 ± 1.7 ka was obtained from the highest lateral moraine at Stanley Lake.

Terminal moraines at Alturas Lake (Fig. 2) were first mapped as Pinedale by Williams (1961). Thackray et al. (2004) distinguished two groups of moraines there, based on moraine morphology and relative weathering. They inferred that the older set of moraines is at least 10 ka older than the younger moraines and that the age is probably considerably older, prior to 27,000 years B.P. We attempted to locate boulders suitable for TCN dating on the outer Alturas and Pettit Lake moraines, but boulders are rare because the granitic source rock is the Cretaceous portion of the Idaho batholith, which does not generate boulders as large as does the Eocene granite at Redfish Lake. Only two boulders met our sampling criteria, both on the outer Alturas Lake moraine. Sample ID-RL-04-033 is from a boulder on the second oldest Alturas Lake moraine (Fig. 1). The ^{10}Be ages of the boulders are 17.5 ± 2.1 ka and 18.4 ± 2.3 ka. These dates indicate that the ages of the Busterback Ranch moraines inferred by Thackery et al. (2004) are incorrect and that all of the moraines at Alturas Lake are late Pinedale.

About two dozen nested lateral moraines mark the systematic retreat of Pinedale ice from the Redfish Lake LGM (Fig. 5). Six [10]Be dates from these moraines range from 16.1 ka near the terminus to 14.6 ka at Redfish Lake (Table 1, Fig. 5). Other LGM recessional positions at Redfish Lake but higher in elevation have dates of 15.1 ka, 16.2 ka, and 16.7 ka. The innermost recessional moraine at Lake Stanley was dated at 13.6 ± 1.7 ka.

2.1.6. Younger Dryas Moraines

2.1.6.1. Bench Lakes Morainal Group

Bench Lakes comprise five paternoster lakes that occur at successively higher elevations in the mountains on the north side of Redfish Lake (Figs. 2, 6). Four of the lakes are rimmed by moraines representing sequentially rising ELAs.

The [10]Be ages of three boulders from a moraine at the distal end of Fourth Bench Lake (Fig. 6) ranged from 11.8 ± 1.4 ka to 12.2 ± 1.5 ka. A boulder on a moraine at Third Bench Lake about 100 m lower (Fig. 6) was dated at 11.9 ± 1.4 ka. Although all four ages fall within the Younger Dryas, at least two phases of moraine building took place during the YD, the younger of which occurred when ELAs were ~100 m higher. No boulders meeting our systematic sampling strategy were present on the other (higher) moraines.

2.1.7. Summary of Sawtooth Mts. Late Pleistocene Ice Dynamics

The moraine maps and TCN chronology support the following conclusions:

1. The oldest moraines in the region appear equivalent to the Bull Lake moraine deposits at their type locality, and may represent an advance in OIS-6 if we assume some of the older TCN apparent ages were due to inheritance, or a combination of OIS-6 and older glaciations as proposed by Phillips et al. (1998).
2. No evidence for an OIS-4 glaciation exists on the basis of the exposure ages, so the LGM advance was more extensive than OIS-4 in the Sawtooth Range.
3. The timing of initial retreat from the last Pinedale maximum was between about 16.2 ka and 15.8 ± 1 ka.
4. Late Pleistocene retreat rate in the piedmont was 2 m/year. No other significant ice-marginal positions occur between the piedmont and Bench Lakes and on this basis we interpret the Bench Lakes moraines to represent a significant climate signal, not merely a series of recessional moraines, although of course we cannot preclude the latter.

2.1.8. Correlations

While it is tempting to correlate the Sawtooth Moraines to other moraine records of ice sheet and ice caps, we recognize that large ice masses may not be in

TABLE 1 Summary of ^{10}Be Exposure Ages for the Sawtooth Mountains Moraines

Age (ka ± 1s)	Sample number	Description	Latitude (°)	Longitude (°)	Elevation (m)
Younger Dryas Dates from Cirque Moraines at Bench Lakes					
11.8 ± 1.4	ID-RL-04-017	Post-LGM Pinedale moraine, Bench Lk	N44 06.653	W114 58.073	2,532
12.0 ± 1.5	ID-RL-04-018	Post-LGM Pinedale moraine, Bench Lk	N44 06.685	W114 58.042	2,518
12.2 ± 1.5	ID-RL-04-016	Post-LGM Pinedale moraine, Bench Lk	N44 06.694	W114 57.970	2,508
11.9 ± 1.4	ID-RL-04-020	Post-LGM Pinedale moraine, Bench Lk	N44 06.862	W114 57.740	2,424
Recessional Moraines from LGM at Redfish Lake					
14.6 ± 1.8	ID-RL-04-023	Inner moraine	N44 08.888	W114 54.667	2,014
15.1 ± 1.8	ID-RL-04-015	Lateral moraine outside Bench Lks	N44 06.704	W114 57.347	2,450
15.8 ± 1.9	ID-RL-04-024	Recessional moraine	N44 09.181	W114 54.407	2,022
15.8 ± 1.9	ID-RL-04-026	Recessional moraine	N44 09.585	W114 54.267	2,013
15.8 ± 1.9	ID-RL-04-025	Recessional moraine	N44 09.223	W114 54.408	2,003
16.1 ± 2.0	ID-RL-04-012	Moraine near terminus	N44 09.931	W114 53.888	2,011
16.7 ± 2.0	ID-RL-04-039	Right lateral moraine	N44 06.012	W114 55.164	2,271
16.2 ± 2.0	ID-RL-04-014	Lateral moraine outside Bench Lks	N44 06.932	W114 57.113	2,386

Recessional Moraine from LGM at Stanley Lake					
13.6 ± 1.7	ID-SL-04-035	Recessional moraine at Stanley Lake	NN44 15.071	W115 03.492	2,002
Redfish Lake LGM					
18.1 ± 2.2	ID-RL-04-040	Highest right lateral moraine	N44 05.962	W114 55.180	2,279
18.8 ± 2.3	ID-RL-04-013	Terminal moraine	N44 09.969	W114 54.819	2,003
20.3 ± 2.5	ID-RL-04-038	Highest right lateral moraine	N44 06.316	W114 55.042	2,311
Alturas Lake LGM					
17.5 ± 2.1	ID-AL-04-034	Alturas Lake terminal moraine	N43 57.206	W114 50.630	2,153
18.4 ± 2.3	ID-AL-04-033	Alturas Lake terminal moraine	N43 57.299	W114 50.707	2,151
Stanley Lake LGM					
20.3 ± 2.5	ID-SL-04-049	Highest left lateral moraine			
South Redfish Bull Lake 2 Moraines					
229 ± 30	ID-RL-04-045	Salmon River moraine	N44 05.395	W114 54.370	2,183
312 ± 42	ID-RL-04-043	Salmon River moraine	N44 05.460	W114 55.262	2,226
South Redfish Bull Lake 1 Moraines					
105 ± 13	ID-RL-04-028	Elk Meadow moraine	N44 08.009	W114 53.200	2,069
107 ± 14	ID-RL-04-027	Elk Meadow moraine	N44 08.433	W114 53.498	2,077
161 ± 20	ID-RL-04-031	Elk Meadow moraine	N44 05.787	W114 52.308	2,048
200 ± 25	ID-RL-04-029	Elk Meadow moraine	N44 06.530	W114 52.558	2,100

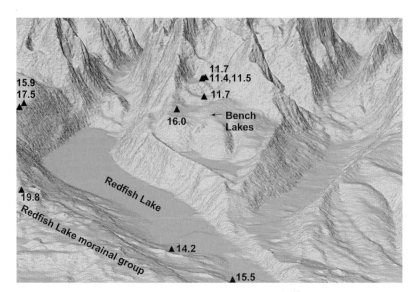

FIGURE 6　DEM of Redfish Lake moraines and [10]Be ages.

equilibrium with climate everywhere along their margins. However, smaller alpine glaciations are demonstrably sensitive to regional precipitation and temperature changes and such correlations can be instructive. We first compare the new Sawtooth chronology with LGM records in the region and globally. Then we compare the YD moraine chronologies. We recognize that there have been several important changes in the ages since the time that these previous results were published. For instance, radiocarbon calibrations have matured, and the production rates, scaling methods, and half-life of cosmogenic isotopes have changed. However, as demonstrated by Heyman et al. (2011), the changes in the TCN ages have not been greater than 10% in most locations. Furthermore, we are uncertain of the actual snow and tree shielding corrections that are necessary at many locations, or if they were calculated in a fashion consistent with our ages. Furthermore, for [10]Be ages, we are unable to re-standardize the original [10]Be/[9]Be measurements for the new half-life in cases where the authors have not provided the AMS standard used. Therefore, for the following discussion we use the originally published ages. Even with a 20% uncertainty to account for these differences, certain correlations are still apparent.

2.2. Global LGM Moraines

2.2.1. Cordilleran Ice Sheet

The Cordilleran Ice Sheet advanced across the Canadian border into the U.S. 18−21,000 ^{14}C years B.P., retreated, then advanced to its LGM terminal position 14−15,000 ^{14}C years B.P. (Fig. 7) (Easterbrook, 1969, 1992,

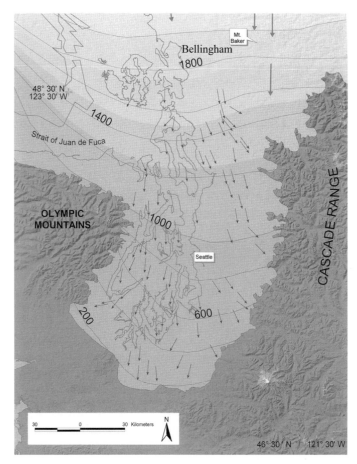

FIGURE 7   The Cordilleran Ice Sheet at its LGM during the Fraser glaciation (Kovanen and Easterbrook, 2002).

2003a,b,c,d,e, 2007; Easterbrook et al., 2003; Easterbrook et al., 2007; Mullineaux et al., 1965; Porter and Swanson, 1998).

2.2.2. Scandinavian Ice Sheet

The Scandinavian Ice Sheet reached its LGM in northern Europe about 20–22,000 years ago. Deglaciation began about 17,500 years B.P. and ended at the close of the Younger Dryas (Tschudi et al., 2000).

2.2.3. Wallowa Lake, Cascade Range, Oregon

Wallowa Lake is held in by multiple late Pinedale moraines (Figs. 8, 9). Two glacial phases are represented—older moraines giving ^{10}Be ages of 20–22,000 ka and younger inset moraines dated at 15–17,000 ka.

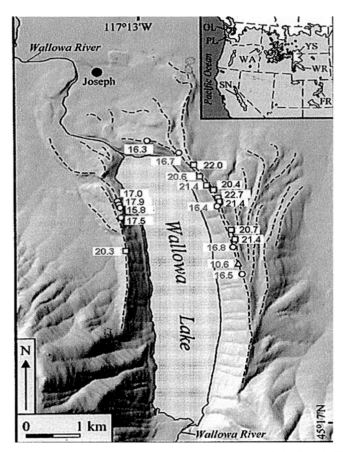

FIGURE 8 Late Pleistocene moraines at Wallowa Lake, Oregon (Licciardi et al., 2004).

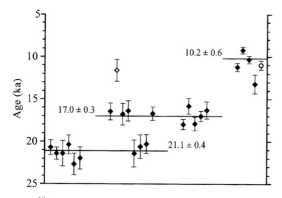

FIGURE 9 [10]Be ages from Wallowa Lake moraines (Licciardi et al., 2004).

2.2.4. Bloody Canyon, Sierra Nevada, California

The LGM at Bloody Canyon consists of two moraines, Tioga 2 (early LGM) and Tioga 3 (late LGM). The less extensive, latest Pleistocene glacial readvances (Tioga 4) have been dated at 13–15 ka.

[10]Be ages range from the Tioga 3 moraine, range from 16.2 ka to 19.7 ka, with a mean value of 17.8 ± 1.5 ka. These ages are consistent with earlier [36]Cl data from the Bloody Canyon Tioga 3 moraines with a mean 16.4 ± 1.4 ka, and with [36]Cl ages from the Tioga 3 moraine at nearby Bishop Creek (mean 17.7 ± 0.7 ka) (Phillips et al., 1996).

[36]Cl ages (ka)

Tioga 3, Bloody Canyon	
BC90-5	14.5
BC86-1	17.4
BC86-3	16.1
BC86-5	17.5
Tioga 3, Bishop Creek	
BPCR91-4	17.3
BPCR90-73	16.8
BPCR90-74	17.3
BPCR90-75	18.9
BPCR91-1	17.2
BPCR91-3	18.1
BPCR90-22	18.3
BPCR90-24	16.8
BPCR90-25	18.2
BPCR90-26	18.1

2.2.5. Fremont Lake, Wind River Range, Wyoming

The classic late Pinedale moraines at Fremont Lake in the Wind River Range, Wyoming were [10]Be-dated by Gosse et al. (1995a,b). Dates from the outer moraines range 18.5–22 ka and dates from inner moraines range from 14.3 ka to 17.7 ka with an average of 17.6 ± 0.8 ka (Phillips et al., 1997). These dates correlate well with [36]Cl-dated, late Pinedale moraines in southern and northern Colorado (Benson et al., 2005) (Fig. 10).

2.2.6. Yellowstone National Park, Wyoming

[10]Be ages from late Pinedale moraines of the Yellowstone ice cap range from 18.6 ka to 14.4 ka with a mean age of 16.5 ± 0.4 ka (Licciardi et al., 2001) (Fig. 11).

2.2.7. Northern Switzerland

During the LGM, the Rhone piedmont glacier expanded into the northern foreland of the Swiss Alps. The oldest [10]Be age of four boulders on the

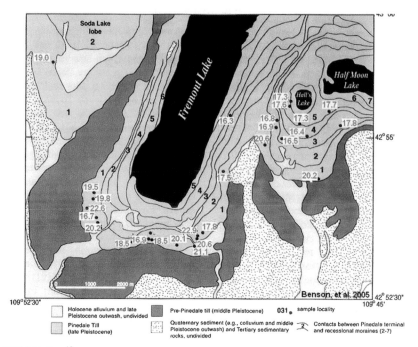

FIGURE 10   ¹⁰Be ages from boulders on late Pinedale moraines at Fremont Lake, Wyoming (Gosse et al., 1995a,b).

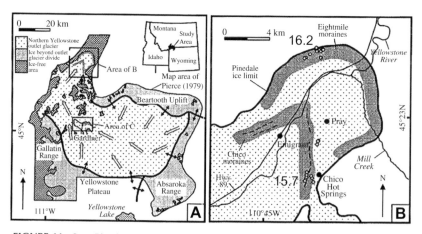

FIGURE 11   Late Pinedale moraines of the Yellowstone ice cap (from Licciardi et al., 2001).

outermost terminal moraine is 20.8 ± 0.9 ka and the mean is 19.3 ± 1.8 ka. A limiting date of $17,700$ [14]C cal years B.P. has been obtained from the inner moraines.

2.2.8. New Zealand

LGM moraines bounding Lake Pukaki in the Southern Alps of New Zealand range in age from 19.3 to 14.1 (Schaefer et al., 2006; Easterbrook et al., 2004) (Fig. 12).

2.2.9. Australia

Moraines in Snowy Mountains of the Tasmanian highlands of SE Australia consist of a distinct outermost LGM moraine and recessional moraines

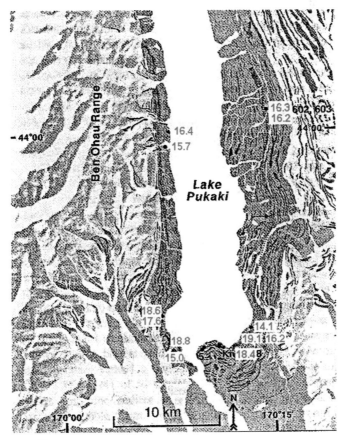

FIGURE 12 [10]Be dates from moraines at Lake Pukaki, New Zealand *(Schaefer et al., 2006).*

(Barrows et al., 2001, 2002). Nineteen [10]Be ages range from 15.0 ka to 19.1 ka, with a mean of 16.8 ± 1.3 kyr.

2.2.10. Lago Buenos Aires, Andes Mts., Argentina

Two sets of LGM moraines (Fenix II and Fenix I) surround Lago Buenos Aires in Patagonia (Kaplan et al., 2004). The oldest boulder age on Fenix II (outer) moraines is 20.4 ± 0.7 ka. [10]Be ages of 16.0 ± 0.4 ka and 18.8 ± 1.5 ka were obtained for two boulders on the Fenix I moraine

2.3. Global Younger Dryas Moraines

2.3.1. Multiple Younger Dryas Moraines

Despite early evidence for a late-glacial readvance of the CIS in western North America (Armstrong, 1960; Easterbrook, 1963; Armstrong et al., 1965), the apparent absence of pollen evidence for the YD in North America led to a generally-held belief that the YD climatic event did not affect North America. However, in recent years, effects of the YD climatic changes have been recognized at a number of localities in the Pacific Northwest (Easterbrook, 1994a,b, 2002, 2003a,b; Easterbrook and Kovanen, 1998; Kovanen and Easterbrook, 2001, 2002; Kovanen, 2002; Kovanen and Slaymaker, 2005; Licciardi et al., 2004), in the Rocky Mts. (Gosse et al., 1995a,b; Easterbrook et al., 2004), and in California (Owen et al., 2003). Planktonic faunal records from the Pacific Northwest and Alaska confirm that the YD event did indeed affect western North America. Alkenone SST estimates from marine cores west off Vancouver Island indicate a temperature drop of ~3 °C during the YD (Kienast and McKay, 2001). Cool-water foraminifera, suggesting YD cooling, have been found on the British Columbia shelf and in the Santa Barbara Basin. Cooling during the YD is also shown from pollen records in SW British Columbia, NW Washington, Oregon, and SE Alaska.

2.3.2. Cordilleran Ice Sheet

Morphologic, stratigraphic, and chronologic evidence of multiple moraines associated with oscillations of the remnants of the Cordilleran Ice Sheet (CIS) in the Fraser Lowland of British Columbia and Washington has revealed multiple, post-LGM fluctuations of the CIS (Kovanen and Easterbrook, 2002). The chronology of the ice margin fluctuations and timing of ice retreat during the Sumas Stade (Fig. 13) has been bracketed by 70 radiocarbon dates and tied to morphologic and stratigraphic evidence. Two of these, fall within the YD. The CIS chronology, which closely matches that of the GISP2 and GRIP ice cores from Greenland, and sea surface temperatures in the north Pacific (Kienast and McKay, 2001) also compares well with the chronology of post-LGM alpine moraines in the western U.S.

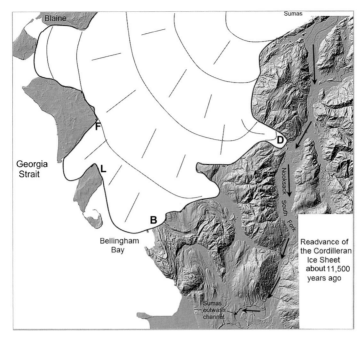

FIGURE 13 Reconstruction of the extent of the oldest of the Sumas glacial readvances in the Fraser Lowland of Washington about 11,500 years ago.

2.3.3. Laurentide Ice Sheet

The Laurentide Ice Sheet readvanced during the YD and built moraines in SW Canada (Grant and King, 1984; Stea and Mott, 1986, 1989). These moraines have been ^{14}C dated as YD in age.

2.3.4. Scandinavian Ice Sheet

Multiple YD moraines of the Scandinavian Ice Sheet (Fig. 14) have long been documented and a vast literature exists. The Scandinavian Ice Sheet readvanced during the YD and built two extensive Salpausselka end moraines across southern Finland, the central Swedish moraines, and the Ra moraines of southwestern Norway. [14]C dates suggest an age of ~10,700 ^{14}C years B.P. for the outer Salpausselka moraine and ~10,200 ^{14}C years B.P. for the inner moraine, very similar to the Cordilleran and Laurentide Ice Sheets.

Thus, all three major Pleistocene ice sheets experienced double moraine-building episodes. Apparently, at least two significant climatic changes that occurred in the Northern Hemisphere during the YD were synchronous.

2.3.5. Younger Dryas in the Western U.S.

The extent to which the climate change of the YD was recorded in North America has long been contentious. Failure to find well-defined YD changes in the pollen

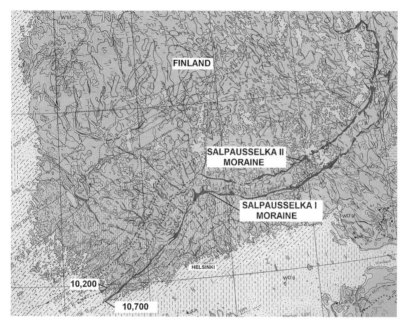

FIGURE 14   Double Younger Dryas moraines of the Scandinavian Ice Sheet.

record led to early beliefs that the YD did not affect North America. In recent years, with the advent of cosmogenic dating, more and more evidence for YD moraines indicates that the YD did indeed affect North American glaciers.

Dated YD moraines occur in the Wind River Range at Titcomb Basin and Temple Lake, and many similar, but more poorly dated, moraines occur throughout the Rocky Mts. What is apparent from these examples of YD moraines is that not only was the YD climatic event recorded by alpine glaciers in western North America, but in many places double moraines record a dual YD climatic change.

2.3.5.1. Wind River Range, Wyoming

Cirque glaciers expanded twice during the YD at Titcomb Lakes in the Wind River Range, WY. Erratics on moraines and glaciated bedrock (Fig. 15) ~33 km upvalley from LGM moraines at Freemont Lake, Wyoming, Pinedale have been [10]Be dated between 12.3 and 10.6 [10]Be years B.P. (Gosse et al., 1995a,b). Nine of ten boulder cosmogenic exposure ages of the inner of the two moraines plot between these dates. The Titcomb Lakes moraines (Fig. 15) have been correlated with the Temple Lake moraines in the Wind River Range to the southwest where multiple late Pleistocene moraines occur (Miller and Birkeland, 1974; Zielinski and Davis, 1987; Davis, 1988; Davis and Osborn, 1987) (Fig. 15).

FIGURE 15   Double Younger Dryas moraines at Titcomb Lakes, WY.

2.3.5.2. Sawtooth Range, Idaho

Cirque moraines at multiple elevations in the Sawtooth Range of Idaho also record two YD climatic events. Bench Lakes, north of Redfish Lake, consist of five tarns at successively higher elevations, representing sequentially rising YD ELAs. Moraines rim four of the lakes. The [10]Be ages of three boulders from a moraine at the distal end of Fourth Bench Lake range from 11.7 ± 0.6 ka to 11.4 ± 0.5 ka. A boulder on a moraine at Third Bench Lake about 100 m lower was dated at 11.7 ± 0.6 ka. Thus, at least two phases of moraine building took place during the YD, the younger of which occurred when ELAs were ~100 m higher than the preceding phase.

2.3.5.3. North Cascades, Washington

Distinctive moraines and ice-contact deposits derived from local sources in the North Cascades and carried to their depositional sites by post-LGM alpine valley glaciers 23–45 km long have been recognized (Kovanen and Easterbrook, 2001; Easterbrook et al., 2004). Soon after 12 ka ^{14}C years B.P., the Nooksack Middle Fork alpine glacier retreated upvalley and then built a moraine containing logs dated at $10,680 \pm 70$ to $10,500 \pm 70$ [14]C years B.P. (Fig. 16) (Kovanen and Easterbrook, 2001; Kovanen and Slaymaker, 2005), establishing their YD age. In the Nooksack North Fork, outwash contains charcoal layers dated at $10,603 \pm 69$ and $10,788 \pm 77$.

2.3.5.4. Mt. Rainier, Washington

In the Cascade Range near Mt. Rainier, Crandell and Miller (1974) mapped cirque moraines designated as McNeeley I (outer moraine) and McNeeley II

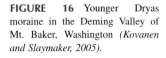

FIGURE 16 Younger Dryas moraine in the Deming Valley of Mt. Baker, Washington *(Kovanen and Slaymaker, 2005)*.

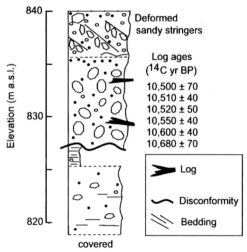

(inner moraine). The McNeeley I moraines were overlain by Mt. Rainier R ash dated at 8750 ± 280 ^{14}C years B.P., suggesting that the moraines were probably late Pleistocene. Heine (1998) cored bogs and lakes associated with the moraines and concluded that McNeeley I moraines were older than the YD, and that McNeeley II moraines were post-YD. However, lakes just upvalley from McNeeley I moraines became ice-free shortly before 11,000 ^{14}C years B.P. and are floored with clastic sediment deposited between 11,090 and 10,150 ^{14}C years ago, the source of which must have been upvalley (i.e., at the site of the McNeeley II moraines), strongly suggesting that the McNeeley II moraines were built during the YD when the basin behind the moraines was filled with ice and meltwater was flowing from there into the lakes behind the McNeeley I moraines. Cores from lakes behind the McNeeley II moraine have *no* sediment deposited between 10 and 11,000 ^{14}C years B.P., confirming that the basin was filled with ice until sometime after 10,000 ^{14}C years B.P. Thus, the McNeeley II moraines were deposited during the YD. Whether or not a second YD moraine exists in these basins remains unclear as field reconnaissance has shown the presence of additional moraines not mapped by Heine.

2.3.5.5. Icicle Creek, Cascade Range, Washington

Double, post-LGM moraines occur at the junctions of Eight-mile and Rat Creeks with Icicle Creek about 12 km upvalley from LGM moraines at Leavenworth, WA (Page, 1939; Porter, 1976; Long, 1989). Boulders at Eight-mile Creek were ^{10}Be dated at 12.6 ± 0.6 ka and 12.3 ± 1.1 ka and boulders on the Rat Creek moraines were dated at 11.3 ± 0.7 ka and 11.9 ± 0.6 ka. All of the dates from the inner and outer moraines fall within the YD (Fig. 17).

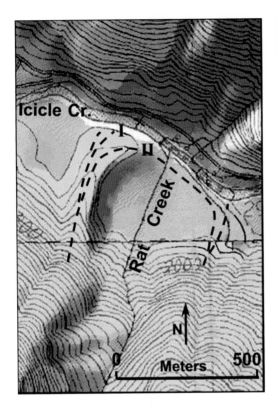

FIGURE 17   Double Younger Dryas moraines at Icicle Cr., North Cascades, WA.

^{36}Cl dates of the Icicle Creek moraines by Porter and Swanson (2008) yielded ages ranging from 11 ka to 17 ka with averages of 13,575 and 13,145 years for the inner and outer moraines respectively. These ages are about 1,500 to 2,000 years older than the ^{10}Be ages, probably as a result of a ~2,000 year error in the ^{36}Cl production rate used (Swanson and Caffee, 2001; Easterbrook, 2003a).

Multiple, as-yet-undated, post-LGM moraines occur at Snoqualmie Pass in the North Cascades. Half a dozen closely spaced moraines at Snoqualmie Pass well upvalley from the LGM record several periods of glacial retreat and stillstand. No direct numerical ages have been published for these moraines, although a ^{14}C date of 11,050 ± 50 ^{14}C years B.P. was obtained from wood at the contact of basal peat on gravel downvalley from the youngest moraines (Porter, 1976).

2.3.5.6. San Bernadino Mts., California

Three sets of moraines occur in the San Bernadino Mts. of southern California. ^{10}Be of the outermost LGM moraines ranges from 18 ka to 20 ka and ages from the inner moraines range from 15 ka to 16 ka (Owen et al., 2003) (Fig. 18).

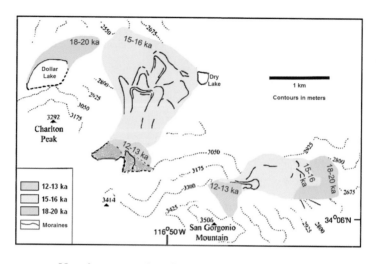

Moraine ages, San Bernadino Mts.

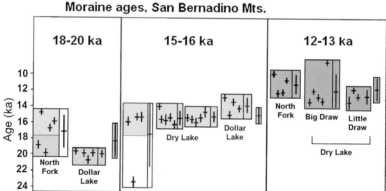

FIGURE 18　LGM and YD moraines in the San Bernadino Mts., CA *(Modified from Owen et al., 2003).*

[10]Be dates from moraines less than 1 km from the cirques were dated 12–13 ka and were considered by Owen et al. (2003) to be broadly correlative with the Younger Dryas.

2.3.6. Swiss Alps

At Julier Pass near St. Moritz, Switzerland, a complex moraine system contains two main morainal ridges. The outer moraine has been dated by [10]Be, [26]Al, and [36]Cl at 11.75 ka and the inner moraine at 10.47 ka (Fig. 19) (Ivy-Ochs et al., 1996, 1999; Kerschner et al., 1999). At Maloja Pass, less than 10 km from Julier Pass, a bog just inside the outermost of three Egesen moraines yielded a ^{14}C age of 10,700 ^{14}C years B.P. (Heitz et al., 1982).

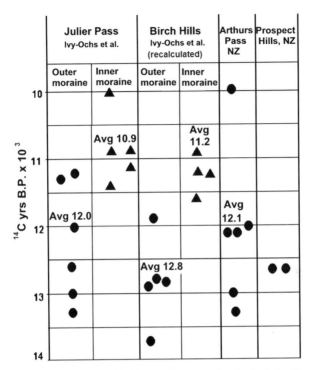

FIGURE 19 Ages of double Younger Dryas moraines in the Swiss Alps.

2.3.7. Scotland

Among the first multiple YD moraines to be recognized were the Loch Lomond moraines of the Scottish Highlands (Sissons, 1974, 1979, 1980; Ballantyne, 1989, 2002; Ballantyne et al., 1998; Benn and Ballantyne, 2000, 2005; Benn et al., 1992; Bennett and Boulton, 1993; Rose et al., 1998). Alpine glaciers and icefields in Britain readvanced or re-formed during the YD and built extensive moraines at the glacier margins. The largest YD icefield at this time was the Scottish Highland glacier complex, but smaller alpine glaciers occurred in the Hebrides and Cairngorms of Scotland (Sissons, 1980), in the English Lake District, and in Ireland.

The Loch Lomond moraines consist of single or several moraines, sometimes multiple, nested, recessional moraines. Radiocarbon dates constrain the age of the Loch Lomond moraines between 12.9 and 11.5 cal years. B.P. Although the Loch Lomond YD doesn't show a consistent pattern of double moraines, the multiple moraines are consistent with multiple phases of YD global climate.

2.3.8. Southern Alps, New Zealand

The YD double-moraine pattern is also found in the Southern Alps of New Zealand at Arthur's Pass and at Birch Hills along Lake Pukaki ~40 km upvalley

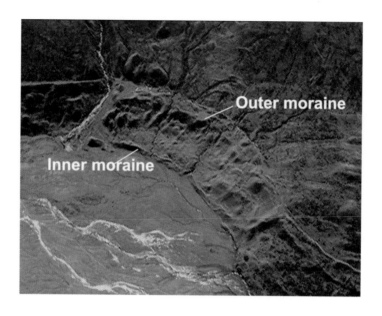

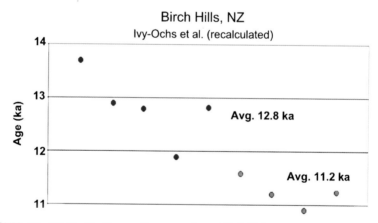

FIGURE 20   (A) Double Younger Dryas moraine at Birch Hills, New Zealand. (B) Ages of Younger Dryas moraines in the Southern Alps of New Zealand.

from the LGM moraine. Five [10]Be dates from the outermost Birch Hills moraine (Figs. 20, 21) average 12.8 ka (Fig. 20) and four [10]Be dates from the inner moraine average 11.2 ka (recalculated ages by Ivy-Ochs, personal communication). Another pair of YD moraines occurs at Arthur's Pass where the mean [10]Be age of the distal moraine is 11.8 ka and of the proximal moraine

FIGURE 21   Younger Dryas Birch Hill moraines, Lake Pukaki, New Zealand.

FIGURE 22   Younger Dryas moraines at Prospect Hill, New Zealand.

is 11.4 ka (Ivy-Ochs et al., 1999). A morainal complex at Prospect Hills in the Arrowsmith Range (Burrows, 1975) yielded ^{10}Be dates of 12.7 and 12.8 years B.P. (Easterbrook, 2002) (Fig. 22).

On the west coast of South Island, wood in the Waiho Loop moraine, deposited by the Franz Josef Glacier about 20 km behind the LGM moraine, has been dated at 11,200 ^{14}C years B.P. (Fig. 14) (Mercer, 1982, 1988; Denton and Hendy, 1994).

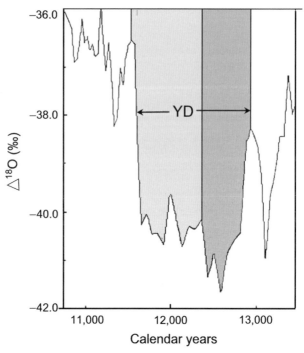

FIGURE 23 Double Younger Dryas event recorded in the GISP2 ice core (Grootes and Stuiver, 1997; Stuiver and Grootes, 2000).

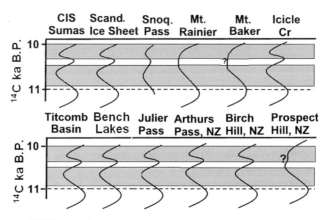

FIGURE 24 Localities having double Younger Dryas moraines.

3. CONCLUSIONS

The multiple nature of LGM and YD moraines in widely separated areas in both hemispheres suggests a common, global, climatic cause. From the evidence presented above, the following conclusions are reached:

1. The LGM in both the Northern and Southern Hemispheres is characterized by two phases, ~20−24 ka and ~14−17 ka with multiple recessional moraines.
2. The YD is characterized by two distinct moraines in widely separated parts of both the Northern and Southern Hemispheres and in the Pacific and Atlantic regions, indicating that the YD consisted of more than a single climatic event.
3. The twin YD response occurred virtually simultaneously globally.
4. Both ice sheets and alpine glaciers were sensitive to the dual YD phases.
5. The GISP2 ice core shows two peaks within the YD that appear to match the morainal record (Fig. 23).
6. The global synchronicity of the late Pleistocene deglaciation and twin YD phases indicates a global atmospheric cause.
7. The absence of a time lag between the N and S Hemispheres glacial fluctuations precludes an ocean cause.
8. The sensitivity and synchronicity of worldwide glacial events with no apparent time lag between hemispheres mean that abrupt climatic changes such as the YD were caused by virtually instantaneous global changes.
9. The lack of a lag time in glacial events between hemispheres infers that changes in the North Atlantic deep water circulation were not cause the Younger Dryas cooling.

These synchronous glacial fluctuations in the western U.S., Europe, and New Zealand, i.e., in both hemispheres and on both sides of North America (Fig. 24), suggest that several abrupt, global, simultaneous, climatic changes occurred during the late Pleistocene. Such changes cannot be explained by changes in the North Atlantic Deep Current alone because of the contemporaneity of glacial responses in both hemispheres with no time lag between hemispheres. The global sensitivity of the double YD suggests a common global cause, rather than an oceanic event that was propagated across the equator. Both ^{14}C and [10]Be production rates in the upper atmosphere changed during the YD, raising the possibility of changes in incoming radiation.

ACKNOWLEDGMENTS

We wish to thank Louden Stanford for creating the DEMs used in this paper. JCG was supported by an NSERC-Discovery Grant.

REFERENCES

Armstrong, J.A., 1960. Surficial geology of the Sumas map area, British Columbia. Geological Survey of Canada, Paper 92, G/1, p. 27.

Armstrong, J.A., Crandell, D.R., Easterbrook, D.J., Noble, J., 1965. Pleistocene stratigraphy and chronology in southwestern British Columbia and northwestern Washington. Geological Society of America Bulletin 76, 321−330.

Balco, G., Stone, J., Lifton, N., Dunai, T., 2008. A simple, internally consistent, and easily accessible means of calculating surface exposure ages and erosion rates from Be-10 and Al-26 measurements. Quaternary Geochronology 3, 174−195.

Ballantyne, C.K., 1989. The Loch Lomond readvance on the Isle of Skye, Scotland: glacier reconstruction and palaeoclimatic implications. Journal of Quaternary Science 4, 95−108.

Ballantyne, C.K., 2002. The Loch Lomond Readvance on the Isle of Mull, Scotland: glacier reconstruction and palaeoclimatic implications. Journal of Quaternary Science 17, 759−771.

Ballantyne, C.K., McCarroll, D., Nesje, A., Dahl, S.O., Stone, J.O., 1998. The last ice sheet in North-west Scotland: reconstruction and implications. Quaternary Science Reviews 17, 1149−1184.

Barrows, T.T., Stone, J.O., Fifield, L.K., Cresswell, R.G., 2001. Late Pleistocene glaciation of the Kosciuszko Massif, Snowy Mountains, Australia: Quaternary Research 55, 179−189.

Barrows, T.T., Stone, J.O., Fifield, L.K., Cresswell, R.G., 2002. The timing of the last glacial maximum in Australia. Quaternary Science Reviews 21, 159−173.

Benn, D.I., Ballantyne, C., 2000. Classic Landforms of the Isle of Skye. Geographical Association, Sheffield, p. 56.

Benn, D.I., Ballantyne, C.K., 2005. Palaeoclimatic inferences from reconstructed Loch Lomond readvance glaciers, West Drumochter Hills, Scotland. Journal of Quaternary Science 20, 577−592.

Benn, D.I., Lowe, J.J., Walker, M.J.C., 1992. Glacier response to climatic change during the Loch Lomond Stadial and early Flandrian: geomorphological and palynological evidence from the Isle of Skye, Scotland. Journal of Quaternary Science 7, 125−144.

Bennett, M.R., Boulton, G.S., 1993. Deglaciation of the Younger Dryas or Loch Lomond Stadial ice-field in the northern Highlands, Scotland. Journal of Quaternary Science 8, 133−145.

Benson, L., Madole, R., Landis, G., Gosse, J., 2005. New data for late Pleistocene Pinedale alpine glaciation from southwestern Colorado. Quaternary Science Reviews 24, 49−65.

Breckenridge, R.M., Stanford, L.R., Cotter, J.F.P., Bloomfield, J.M., Evenson, E.B., 1988. Field guides to the Quaternary geology of central Idaho: Part B, Glacial geology of the Stanley Basin. In: Lin, P.K., Hackett, W.R. (Eds.), Guidebook to the Geology of Central and Southern Idaho, Idaho Geological Survey Bulletin, 27, pp. 209−221.

Burrows, C.J., 1975. Late Pleistocene and Holocene moraines of the Cameron Valley, Arrowsmith Range, Canterbury, New Zealand. Arctic and Alpine Research 7.

Crandell, D.R., Miller, R.D., 1974. Quaternary stratigraphy and extent of glaciation in the Mount Rainier region, Washington. U.S. Geological Survey Professional Paper 450-D, p. 59.

Davis, P.T., 1988. Holocene glacier fluctuations in the American Cordillera. Quaternary Science Reviews 7, 129−157.

Davis, P.T., Osborn, G., 1987. Age of pre-Neoglacial cirque moraines in the central North American Cordillera. Geographie Physique et Quaternaire 41, 365−375.

Denton, G.H., Hendy, C.H., 1994. Younger Dryas age advance of Franz Josef Glacier in the Southern Alps of New Zealand. Science 264, 1434—1437.

Easterbrook, D.J., 1963. Late Pleistocene glacial events and relative sea-level changes in the northern Puget Lowland, Washington. Geological Society of America Bulletin 74, 1465—1483.

Easterbrook, D.J., 1992. Advance and retreat of Cordilleran Ice Sheets in Washington, USA. Géographie physique et Quaternaire 46, 51—68.

Easterbrook, D.J., 1994a, Evidence for abrupt late Wisconsin climatic changes during deglaciation of the Cordilleran Ice Sheet in Washington: Geological Society of America Abstracts with Program.

Easterbrook, D.J., 1994b. Stratigraphy and chronology of early to late Pleistocene glacial and interglacial sediments in the Puget Lowland, Washington. In: Swanson, D.A., Haugerud, R.A. (Eds.), Geologic Field Trips in the Pacific Northwest. Geological Society of America 1J23—1J38.

Easterbrook, D.J., 2002. Implications of Younger Dryas glacial fluctuations in the western U.S., New Zealand, and Europe. Geological Society of America, Abstracts with Program 130.

Easterbrook, D.J., 2003a. Global, double Younger Dryas glacial fluctuations in ice sheets and alpine glaciers: XVI INQUA Congress. Geological Society of America, Abstracts with Program 73.

Easterbrook, D.J., 2003b. Synchronicity and sensitivity of alpine and continental glaciers to abrupt, global, climate changes during the Younger Dryas. Geological Society of America, Abstracts with Program 35, 350.

Easterbrook, D.J., 2003c. Cordilleran Ice Sheet glaciation of the Puget Lowland and Columbia Plateau and alpine glaciation of the North Cascade Range, Washington. INQUA 2003 Field Guide Volume. In: Easterbrook, D.J. (Ed.), Quaternary Geology of the United States. Desert Research Institute, Reno, NV, pp. 265—286.

Easterbrook, D.J., 2003d. Cordilleran Ice Sheet glaciation of the Puget Lowland and Columbia Plateau and alpine glaciation of the North Cascade Range, Washington. In: Swanson, T.W. (Ed.), Western Cordillera and Adjacent Areas, Field Guide, pp. 137—157.

Easterbrook, D.J., 2003e. Cordilleran Ice Sheet Glaciation of the Puget Lowland and Columbia Plateau and Alpine Glaciation of the North Cascade Range, Washington. Field Guide 4. Geological Society of America, 137—157.

Easterbrook, D.J., 2007. Late Pleistocene and Holocene glacial fluctuations: implications for the cause of abrupt global climate changes. Geological Society of America, Abstracts with Programs 39, 594.

Easterbrook, D.J., Evenson, E., Gosse, J.C., Ivy-Ochs, S., Kovanen, D.J., Sherard, C.A., 2004. Synchronous, global, late Pleistocene ice sheet and alpine glacial fluctuations. Geological Society of America, Abstracts with Program 36, 344.

Easterbrook, D.J., Kovanen, D.J., 1998. Pre-Younger Dryas resurgence of the southwestern margin of the Cordilleran Ice Sheet, British Columbia, Canada: Comments. Boreas 27, 229—230.

Easterbrook, D.J., Kovanen, D.J., Slaymaker, O., 2007. New developments in Late Pleistocene and Holocene glaciation and volcanism in the Fraser Lowland and North Cascades, Washington. In: Stelling, P., Tucker, D.S. (Eds.), Geological Society of America Field Guide, 9, pp. 36—51.

Easterbrook, D.J., Pierce, K., Gosse, J., Gillespie, A., Evenson, E., Hamblin, K., 2003. Quaternary Geology of the Western United States, INQUA 2003. Field Guide Volume. Desert Research Institute, Reno, NV, pp. 19—79.

Gillespie, A.R., Molnar, P., 1995. Asynchronism of maximum advances in of mountain and continental glaciations. Reviews of Geophysics 33, 311−364.

Gosse, J.C., Evenson, E.B., Klein, J., Lawn, B., Middleton, R., 1995a. Precise cosmogenic ^{10}Be measurements in western North America: support for a global Younger Dryas cooling event. Geology 23, 877−880.

Gosse, J., Evenson, J.B., Klein, J., Lawn, B., Middleton, R., 1995b. Beryllium-10 dating of the duration and retreat of the last Pinedale glacial sequence. Science 268 (5215), 1329−1333.

Gosse, J.C., Phillips, F.M., 2001. Terrestrial in situ cosmogenic nuclides: theory and application. Quaternary Science Reviews 20, 1475−1560.

Grant, D.R., King, L.H., 1984. A Stratigraphic Framework for the Quaternary History of the Atlantic Provinces, Canada, vol. 84-10. Geological Survey of Canada, pp. 173−191.

Grootes, P.M., Stuiver, M., 1997. Oxygen 18/16 variability in Greenland snow and ice with 10^{-3}- to 10^5-year time resolution. Journal of Geophysical Research 102, 26455−26470.

Heine, J.T., 1998. Extent, timing, and climatic implications of glacier advances Mount Rainier, Washington, U.S.A. at the Pleistocene/Holocene transition. Quaternary Science Reviews 17, 1139−1148.

Heitz, A., Punchakunnel, P., Zoller, P., 1982. NonEnVegetations-, klima- and gletchergeschichte des Obergrengadins. In: Gamper, M. (Ed.), Physische Geographie: Beitrage zur Quartarfoschung in der Schweize. Geographishes Institut der Universitat Zurich, pp. 157−170.

Heyman, J., Stroeven, A.P., Harbor, J.M., Caffee, M.W., 2011. Too young or too old: Evaluating cosmogenic exposure dating based on an analysis of compiled boulder exposure ages: Earth and Planetary Science Letters 302, 71−80.

Ivy-Ochs, S., Schlüchter, C., Kubik, P.W., Denton, G.H., 1999. Moraine exposure dates imply synchronous Younger Dryas glacier advance in the European Alps and in the Southern Alps of New Zealand. Geografiska Annaler 81A, 313−323.

Ivy-Ochs, S., Schlüchter, C., Kubik, P.W., Synal, H.-A., Beer, J., Kerschner, H., 1996. The exposure age of an Egesen moraine at Julier Pass measured with ^{10}Be, ^{26}Al and ^{36}Cl. Eclogae Geologicae Helvetiae 89, 1049−1063.

Kerschner, H., Ivy-Ochs, S., Schlüchter, C., 1999. Paleoclimatic interpretation of the early late-glacial glacier in the Gschnitz Valley, Central Alps, Austria. Annals of Glaciology 28, 135−140.

Kienast, S.S., McKay, J.L., 2001. Sea surface temperatures in the subarctic Northeast Pacific reflect millennial-scale climate oscillations during the last 16 kyrs. Geophysical Research Letters 28, 1563−1566.

Kohl, C.P., Nishiizumi, K., 1992. Chemical isolation of quartz for measurement of in-situ-produced cosmogenic nuclides. Geochemica et Cosmochemica Acta 56, 3583−3587.

Kovanen, D.J., 2002. Morphologic and stratigraphic evidence for Allerød and Younger Dryas age glacier fluctuations of the Cordilleran Ice Sheet, British Columbia, Canada and Northwest Washington, USA. Boreas 31, 163−184.

Kovanen, D.J., Easterbrook, D.J., 2001. Late Pleistocene, post-Vashon, alpine glaciation of the Nooksack drainage, North Cascades, Washington. Geological Society of America Bulletin 113, 274−288.

Kovanen, D.J., Easterbrook, D.J., 2002. Extent and timing of Allerød and Younger Dryas age (ca. 12,500−10,000 ^{14}C yr BP) oscillations of the Cordilleran Ice Sheet in the Fraser Lowland, Western North America. Quaternary Research 57, 208−224.

Kovanen, D.J., Slaymaker, O., 2005. Fluctuations of the Deming glacier and theoretical equilibrium line elevations during the late Pleistocene and early Holocene on Mt. Baker, Washington, USA. Boreas.

Licciardi, J.M., Clark, P.U., Brook, E.J., Elmore, D., Sharma, P., 2004. Variable responses of western U.S. glaciers during the last deglaciation. Geology 32, 81–84.

Licciardi, J.M., Clark, P.U., Brook, E.J., Pieerce, K.L., Kurz, M.D., Elmore, D., Sharma, P., 2001. Cosmogenic ^{3}He and ^{10}Be chronologies of the late Pinedale northern Yellowstone ice cap, Montana, USA. Geology 29, 1095–1098.

Long, W.A., 1989. A probable sixth Leavenworth glacial substage in the Icicle-Chiwaukum Creeks area, North Cascades Range, Washington. Northwest Science 63, 96–103.

Masarik, J., Weiler, R., 2003. Production rates of cosmogenic nuclides in boulders. Earth and Planetary Science Letters 216, 201–208.

Mercer, J.H., 1982. Simultaneous climatic change in both hemispheres and similar bipolar interglacial warming: evidence and implications. Geophysical Monograph 29, 307–313.

Mercer, J.H., 1988. The age of the Waiho Loop terminal moraine, Franz Josef Glacier, Westland, New Zealand. New Zealand Journal of Geology and Geophysics 31, 95–99.

Miller, C.D., Birkeland, P.W., 1974. Probable pre-Neoglacial age for the type Temple Lake moraine, Wyoming: Discussion and additional relative-age data. Arctic and Alpine Research 6, 301–306.

Mullineaux, D.R., Waldron, H.H., Rubin, M., 1965. Stratigraphy and chronology of late interglacial and early Vashon time in the Seattle area, Washington. 1194-O. U.S. Geological Survey Bulletin, 1–10.

Owen, L.A., Finkel, R.C., Minnich, R.A., Perez, A.E., 2003. Extreme southwestern margin of late Quaternary glaciation in North America: timing and controls. Geology 31 (8), 729–732.

Page, B.M., 1939. Multiple alpine glaciation in the Leavenworth area, Washington. Journal of Geology 47, 785–815.

Phillips, F.M., Zreda, M.G., Benson, L.V., Plummer, M.A., Elmore, D., Sharma, P., 1996. Chronology for fluctuations in late Pleistocene Sierra Nevada glaciers: Science 274, 749–751.

Phillips, F.M., Zreda, M.G., Gosse, J.C., Klein, J., Evenson, E.B., Hall, R.D., Chadwick, O.A., Sharma, P., 1997. Cosmogenic ^{36}Cl and ^{10}Be ages of Quaternary glacial and fluvial deposits of the Wind River Range, Wyoming. Geological Society of America Bulletin 109, 1453–1463.

Plug, L., Gosse, J., West, J., Bigley, R., 2007. Attenuation of cosmic ray flux in temperate forest. Journal of Geophysical Research 112, F02022. doi:10.1029/2006JF000668.

Porter, S.C., 1976. Pleistocene glaciation in the southern part of the North Cascade Range, Washington: Geological Society of America Bulletin 87, 61–75.

Porter, S.C., Swanson, T.W., 1998. Radiocarbon age constraints on rates of advance and retreat of the Puget lobe of the Cordilleran Ice Sheet during the last glaciation. Quaternary Research 50, 205–213.

Porter, S.C., Swanson, T.W., 2008. Surface exposure ages and paleoclimatic environment of Middle and icicle Late Pleistocene glacier advances, northeastern Cascade Range, Washington: American Journal of Science 308, 130–166.

Rose, J., Lowe, J.J., Switsur, R., 1998. A radiocarbon date on plant detritus beneath till from the type area of the Loch Lomond readvance. Scotland. Journal of Geology 24, 113–124.

Schaefer, J.M., Denton, G.H., Barrell, D.J.A., Ivy-Ochs, S., Kubik, P.W., Anderson, B.J., Phillips, F.M., Lowell, T.V., Schluchter, C., 2006. Near-synchronous interhemispheric termination of the last glacial maximum in mid-latitudes. Science 312, 1510–1513.

Sherard, C., 2006. Regional correlations of late Pleistocene climatic changes based on cosmogenic nuclei exposure dating of moraine in Idaho. M.Sc. Thesis, Western Washington University, Bellingham. 91.

Sissons, J.B., 1974. A late-glacial ice-cap in the Central Grampians, Scotland. Transactions of Institute of British Geographers 62, 95–114.

Sissons, J.B., 1979. Palaeoclimatic inferences from former glaciers in Scotland and the Lake District. Nature 278, 518–521.

Sissons, J.B., 1980. The Loch Lomond advance in the Lake District, northern England, 71. Transactions of the Royal Society of Edinburgh, U.K. 13–27.

Stea, R.R., Mott, R.J., 1986. Late-glacial climatic oscillation in Atlantic Canada equivalent to the Allerod/Younger Dryas event. Nature 323, 247–250.

Stea, R.R., Mott, R.J., 1989. Deglaciation environments and evidence for glaciers of Younger Drays age in Nova Scotia, Canada. Boreas 18, 169–187.

Stuiver, M., Grootes, P.M., 2000. GISP2 oxygen isotope ratios. Quaternary Research 54/3.

Thackray, G.D., Lundeen, K.A., Borgert, J.A., 2004. Latest Pleistocene alpine glacier advances in the Sawtooth Mountains, Idaho, USA: reflections of midlatitude moisture transport at the close of the last glaciation. Geology 32 (3), 225–228.

Tschudi, S., Ivy-Ochs, S., Schlüchter, C., Kubik, P.W., Raino, H., 2000. [10]Be dating of Younger Dryas Salpausselkä I Formation in Finland. Boreas 29, 287–293.

Williams, P.L., 1961. Glacial geology of the Stanley Basin: Moscow, Idaho Bureau of Mines and Geology. Pamphlet 123, 29.

Zielinski, G.A., Davis, P.T., 1987. Late Pleistocene age of the Type Temple Lake Moraine. Wind River Range, Wyoming, U.S.A. Géographie Physique et Quaternaire, XLI, 397–401.

Part II

Temperature Measurements

Chapter 3

A Critical Look at Surface Temperature Records

Joseph D'Aleo

CCM, AMS Fellow, 18 Glen Drive, Hudson, NH 03051, USA

Chapter Outline

1. INTRODUCTION

Although warming from 1979 to 1998 is well supported, major questions exist about long-term trends. Climategate inspired investigations suggest global surface-station data are seriously compromised. The data suffer significant contamination by urbanization and other local factors such as land-use/

Evidence-Based Climate Science. DOI: 10.1016/B978-0-12-385956-3.10003-8

land-cover changes and instrument siting that does not meet government standards. There was a major station dropout, which occurred suddenly around 1990 and a significant increase in missing monthly data in the stations that remained. There are also uncertainties in ocean temperatures; no small issue, as oceans cover 71% of the Earth's surface.

These factors all lead to significant uncertainty and in most cases a tendency for overestimation of century-scale temperature trends. Indeed, numerous peer-reviewed papers cataloged here have estimated that these local issues with the observing networks may account for **30%, 50% or more** of the warming shown since 1880. After the data with all its issues are collected, further adjustments are made, each producing more warming,

"[W]hen data conflicts with models, a small coterie of scientists can be counted upon to modify the data" to agree with models' projections," says MIT meteorologist Dr. Richard Lindzen.

In this paper, we look at some of the issues in depth and the recommendations made for a reassessment of global temperatures necessary to make sensible policy decisions.

2. THE GLOBAL DATA CENTERS

Five organizations publish global temperature data. Two — Remote Sensing Systems (RSS) and the University of Alabama at Huntsville (UAH) — are satellite data sets. The three terrestrial data sets provided by the institutions — NOAA's National Climatic Data Center (NCDC), NASA's Goddard Institute for Space Studies (GISS/ GISTEMP), and the University of East Anglia's Climatic Research Unit (CRU) — all depend on data supplied by surface stations administered and disseminated by NOAA under the management of the National Climatic Data Center in Asheville, North Carolina. The Global Historical Climatology Network (GHCN) is the most commonly cited measure of global surface temperature for the last 100 years.

Around 1990, NOAA/NCDC's GHCN data set lost more than three-quarters of the climate measuring stations around the world. A study by Willmott et al. (1991) calculated a +0.2C bias in the global average owing to pre-1990 station closures. Douglas Hoyt had estimated approximately the same value in 2001 due to station closures around 1990. A number of station closures can be attributed to Cold-War era military base closures, such as the DEW Line (The Distant Early Warning Line) in Canada and its counterpart in Russia.

The world's surface observing network had reached its golden era in the 1960s to 1980s, with more than 6,000 stations providing valuable climate information. Now, there are fewer than 1,500 remaining.

It is a fact that the three data centers each performed some final adjustments to the gathered data before producing their own final analysis. These adjustments are frequent and often poorly documented. The result was almost always to produce an enhanced warming even for stations which had a cooling trend in the raw data. The metadata, the information about precise location, station moves and equipment changes were not well documented and shown frequently to be in error which complicates the assignment to proper grid boxes and make the efforts of the only organization that attempts to adjust for urbanization, NASA GISS problematic.

As stated here relative to Hansen et al. (2001),[1] "The problem [accuracy of the latitude/longitude coordinates in the metadata] is, as they say, "even worse than we thought." One of the consumers of GHCN metadata is of course GISTEMP, and the implications of imprecise latitude/longitude for GISTEMP are now considerably greater, following the change in January 2010 to use of satellite-observed night light radiance to classify stations as rural or urban throughout the world, rather than just in the contiguous United States as was the case previously. As about a fifth of all GHCN stations changed classification as a result, this is certainly not a minor change."

Among some major players in the global temperature analyses, there is even disagreement about what the surface air temperature really is. (See "The Elusive Absolute Surface Air Temperature (SAT)" by Dr. James Hansen.[2] Essex et al. questioned whether a global temperature existed here.[3])

Satellites measurements of the lower troposphere (around 600 mb) are clearly the better alternative. They provide full coverage and are not contaminated by local factors. Even NOAA had assumed satellites would be the future solution for climate monitoring. Some have claimed satellite measurements are subject to error. RSS and UAH in 2005 jointly agreed[4] that there was a small net cold bias of 0.03 °C in their satellite-measured temperatures, and corrected the data for this small bias. In contrast, the traditional surface-station data we will show suffer from many warm biases that are orders of magnitude greater in size than the satellite data, yet that fact is often ignored by consumers of the data.

Some argue that satellites measure the lower atmosphere and that this is not the surface. This difference is real but it is irrelevant (CCSP[5]). The lower troposphere around 600 mb was chosen because it was above the mixing level and so with polar orbiters the issues of the diurnal variations are eliminated. Also there is a high correlation between temperatures in the lower to middle troposphere and the surface.

1. http://oneillp.wordpress.com/2010/03/13/ghcn-metadata/.

2. http://data.giss.nasa.gov/gistemp/abs_temp.html.

3. http://www.uoguelph.ca/~rmckitri/research/globaltemp/globaltemp.html.

4. http://www.marshall.org/article.php?id=312.

5. http://www.climatescience.gov/Library/sap/sap1-1/finalreport/.

Anomalies from satellite data and surface-station data have been increasing in the last 3 decades. When the satellites were first launched, their temperature readings were in better agreement with the surface-station data. There has been increasing divergence over time which can be seen below (derived from Klotzbach et al., 2009). In the first plot, we see the temperature anomalies as computed from the satellites and assessed by UAH and RSS and the station-based land surface anomalies from NOAA NCDC (Fig. 1).

The divergence is made clearer when the data are scaled such that the difference in 1979 is zero (Fig. 2).

The Klotzbach paper finds that the divergence between surface and lower-tropospheric trends is consistent with evidence of a warm bias in the surface temperature record but not in the satellite data.

Klotzbach et al. described an 'amplification' factor for the lower tropo-sphere as suggested by Santer et al. (2005) and Santer et al. (2008) due to greenhouse gas trapping relative to the warming at the surface. Santer refers to the effect as "tropospheric amplification of surface warming." This effect is a characteristic of all of the models used in the UNIPCC and the USGRCP "ensemble" of models by Karl et al. (2006) which was the source for Karl et al. (2009) which in turn was relied upon by EPA in its recent Endangerment Finding. (Federal Register/Vol. 74, No. 239/Tuesday, December 15, 2009/Rules and Regulations at 66510.)

As Dr. John Christy, keeper of the UAH satellite data set describes it, "The amplification factor is a direct calculation from model simulations that show over 30-year periods that the upper air warms at a faster rate than the surface — generally 1.2 times faster for global averages. This is the so-called "lapse rate feedback" in which the lapse rate seeks to move toward the moist adiabat as the

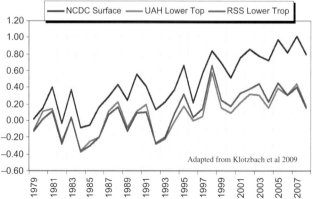

Annual Land Surface vs Equivalent Lower Troposphere Anomalies

FIGURE 1   Annual land surface anomalies compared to UAH and RSS lower-tropospheric temperature anomalies since 1979 (sources: NOAA and Klotzbach).

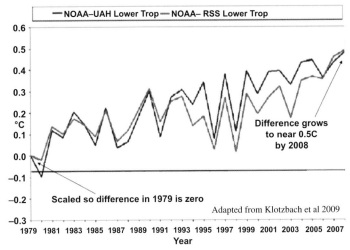

FIGURE 2   NOAA annual land temperatures minus annual UAH lower troposphere (blue line) and NOAA annual land temperatures minus annual RSS lower troposphere (green line) over the period from 1979 to 2008.

surface temperature rises. In models, the convective adjustment is quite rigid, so this vertical response in models is forced to happen. The real world is much less rigid and has ways to allow heat to escape rather than be retained as models show." This latter effect has been documented by Chou and Lindzen (2005) and Lindzen and Choi (2009).

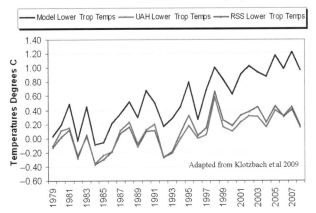

FIGURE 3   Model amplification-based forecast lower troposphere (blue line) and actual UAH (green line) and RSS lower troposphere (purple line) over the period from 1979 to 2008.

The amplification factor was calculated from the mean and median of the 19 GCMs that were in the CCSP SAP 1.1 report (Karl et al., 2006). A fuller discussion of how the amplification factor was calculated is available in the Klotzbach paper.[6]

The ensemble model forecast curve (upper curve) in Fig. 3 was calculated by multiplying the NOAA NCDC surface temperature for each year by the amplification factor, and thus is the model projected tropospheric temperature. The lower curves are the actual UAH and RSS lower-tropospheric satellite temperatures.

This strongly suggests that instead of atmospheric warming from greenhouse effects dominating, surface-based warming very likely due to uncorrected urbanization and land-use contamination is the biggest change. Since these surface changes are not fully adjusted for, trends from the surface networks are not reliable.

3. THE GOLDEN AGE OF SURFACE OBSERVATION

In this era of ever-improving technology and data systems, one would assume that measurements would be constantly improving. This is not the case with the global station observing network. The Golden Age of Observing was several decades ago. It is gone.

The Hadley Centre's Climate Research Unit (CRU) at East Anglia University is responsible for the CRU global data. NOAA's NCDC, in Asheville, NC, is the source of the Global Historical Climate Network (GHCN) and of the U.S. Historical Climate Network (USHCN). These two data sets are relied upon by NASA's GISS in New York City and by Hadley/CRU in England.

All three have experienced degradation in data quality in recent years.

Ian "Harry" Harris, a programmer at the Climate Research Unit, kept extensive notes of the defects he had found in the data and computer programs that the CRU uses in the compilation of its global mean surface temperature anomaly data set. These notes, some 15,000 lines in length, were stored in the text file labeled "Harry_Read_Me.txt", which was among the data released by the whistle-blower with the Climategate emails. This is just one of his comments:

"[The] hopeless state of their (CRU) database. No uniform data integrity, it's just a catalogue of issues that continues to grow as they're found…I am very sorry to report that the rest of the databases seem to be in nearly as poor a state as Australia was. There are hundreds if not thousands of pairs of dummy stations, one with no WMO and one with, usually overlapping and with the same station name and very similar coordinates.

---

6. http://pielkeclimatesci.files.wordpress.com/2009/11/r-345.pdf.

I know it could be old and new stations, but why such large overlaps if that's the case? Aarrggghhh! There truly is no end in sight.

"This whole project is SUCH A MESS. No wonder I needed therapy!!

"I am seriously close to giving up, again. The history of this is so complex that I can't get far enough into it before by head hurts and I have to stop. Each parameter has a tortuous history of manual and semi-automated interventions that I simply cannot just go back to early versions and run the updateprog. I could be throwing away all kinds of corrections - to lat/lons, to WMOs (yes!), and more. So what the hell can I do about all these duplicate stations?"

According to Phil Jones, former director of the Climatic Research Unit (CRU), 'there is some truth' to the charge that he failed to update and organize the raw data supporting the CRU temperature data set, on which the IPCC relies in its reports to make temperature projections and that at least some of the original raw data were lost. This should raise questions about the quality of global data.

In the following email, CRU's Director at the time, Dr. Phil Jones, acknowledges that CRU mirrors the NOAA data:

"Almost all the data we have in the CRU archive is exactly the same as in the GHCN archive used by the NOAA National Climatic Data Center."

In the Russell inquiry into CRU's role in Climategate, they estimated at least 90% of the data were the same. Steve McIntyre's analysis showed 95.6% concordance. NASA uses the GHCN as the main data source for the NASA GISS data.

Dr. Roger Pielke Sr. in this post[7] on the three data sets notes:

"The differences between the three global surface temperatures that occur are a result of the analysis methodology as used by each of the three groups. They are not "completely independent." Each of the three surface temperature analysis suffer from unresolved uncertainties and biases as we documented, for example, in our peer reviewed paper[8]"

Dr. Richard Anthes, President of the University Corporation for Atmospheric Research, in testimony to Congress[9] in March 2009, noted:

"The present federal agency paradigm with respect to NASA and NOAA is obsolete and nearly dysfunctional, in spite of best efforts by both agencies."

---

7. http://pielkeclimatesci.wordpress.com/2009/11/25/an-erroneous-statement-made-by-phil-jones-to-the-media-on-the-independence-of-the-global-surface-temperature-trend-analyses-of-cru-giss-and-ncdc/.

8. http://pielkeclimatesci.files.wordpress.com/2009/10/r-321.pdf.

9. http://www.ucar.edu/oga/pdf/Anthes%20CJS%20testimony%203-19-09.pdf.

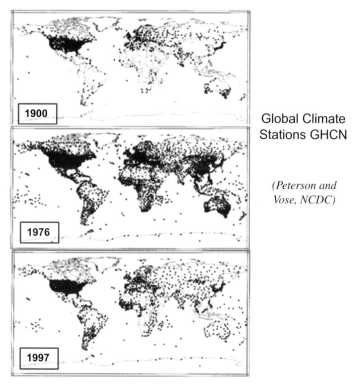

1900

1976

1997

Global Climate
Stations GHCN

*(Peterson and
Vose, NCDC)*

FIGURE 4 Stations in 1900, 1976, and 1997 used in the global GHCN database (*sources: Peterson and Vose NCDC, 1997*).

4. VANISHING STATIONS

More than 6,000 stations were in the NOAA database for the mid-1970s, but just 1,500 or less are used today. NOAA claims the real-time network includes 1,200 stations with 200–300 stations added after several months and included in the annual numbers. NOAA is said to be adding additional U.S. stations now that USHCN v2 is available, which will inflate this number, but make it disproportionately U.S.

There was a major disappearance of recording stations in the late 1980s to the early 1990s. Figure 4 compares the number of global stations in 1900, the 1970s, and 1997, showing the increase and then decrease (Peterson and Vose[10]).

Dr. Kenji Matsuura and Dr. Cort J. Willmott at the University of Delaware have prepared this animation.[11] See the lights go out in 1990, especially in Asia.

10. http://www.ncdc.noaa.gov/oa/climate/ghcn-monthly/images/ghcn_temp_overview.pdf.

11. http://climate.geog.udel.edu/~climate/html_pages/Ghcn2_images/air_loc.mpg.

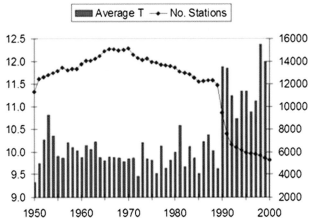

FIGURE 5   Plot of the number of total station ID's in each year since 1950 and the average temperatures of the stations in the given year.

The following chart[12] of all GHCN stations and the average annual temperature show the drop focused around 1990. In this plot, those stations with multiple locations over time are given separate numbers, which inflates the total number. While a straight average is not meaningful for global temperature calculation (because areas with more stations would have higher weighting), it illustrates that the disappearance of so many stations in an uneven fashion may have introduced a distribution bias (Fig. 5).

As can be seen in the figure, the straight average of all global stations does not fluctuate much until 1990, at which point the average temperature jumps up. This observational bias can influence the calculation of area-weighted averages to some extent. As previously noted, a study by Willmott, Robeson, and Feddema ("Influence of Spatially Variable Instrument Networks on Climatic Averages", 1991) calculated a +0.2C bias in the global average owing to pre-1990 station closures. Others have attempted experiments (Mosher, Grant, Lilligren) that purport to show this does not necessarily translate into a warm bias given the 'anomaly method' (using anomalies or departures from normal base period values instead of the actual temperatures). The effect may not be definitively known until a full data reconstruction can take place.

Global databases all compile data into latitude/longitude-based grid boxes and calculate temperatures inside the boxes using data from the stations within them or use the closest stations (weighted by distance) in nearby boxes.

The use of anomalies instead of mean temperatures greatly improve the chances of filling in some of the smaller holes (empty grid boxes) or not producing significant differences in areas where the station density is high, they

---

12. http://www.uoguelph.ca/~rmckitri/research/nvst.html.

can't be relied on to accurately estimate anomalies in the many large data sparse areas (Canada, Greenland, Brazil, Africa, parts of Russia). To fill in these areas requires NOAA and NASA to reach out as far as 1200 km

There are 8,000 grid boxes globally (land and sea). If the Earth is 71% ocean, approximately 2,320 grid boxes would be over land (actual number will vary as some grid boxes will overlap or may just touch the coast).

With 1,200 stations in the real-time GHCN network that would be enough to have 51.7% of the land boxes with a station. However, since stations tend to cluster, that number is smaller. Our calculation is that that number is around 44% or 1,026 land grid boxes without a station.

For data in empty boxes, GHCN will look to surrounding areas as far away as 1,200 km (in other words using Atlanta, GA to estimate a monthly or annual anomaly in Chicago, IL, Birmingham Al to estimate New York City, Los Angeles to estimate Jackson Hole, WY).

Certainly an isolated vacant grid box surrounded by boxes with data in them may be able to yield a reasonably representative extrapolated anomaly value from the surrounding data.

But in data sparse regions, such as is much of the Southern Hemisphere, when you have to extrapolate from more than one grid box away you are increasing the data uncertainty. If you bias it towards having to look towards more urbanized or airport regions or lower elevation coastal locations as E.M. Smith has detected, you are added potential warm bias to uncertainty. This has been the case in the north in the large countries bordering on the arctic (Russia and Canada) where the greatest warming is shown in the data analyses but also in Brazil where fast growing cities are used to estimate anomalies in the Amazon.

To ascertain whether a net bias exists, E.M. Smith has conducted first an analysis of mean temperatures for whatever stations existed by country or continent/sub continent. He then applied a dT method[13] which is a variation of 'First Differences' as a means of examining temperature data anomalies independent of actual temperature. dT/year is the "average of the changes of temperature, month now vs. the same month that last had valid data, for each year". He then does a running total of those changes, or the total change, the "Delta Temperature" to date. He is doing this for every country (see footnote 14). His next step will be to attempt to splice/blend the data into the grids.

Even then uncertainty will remain that only more complete data set usage would improve. The following graphic powerfully illustrates this was a factor even before the major dropout. Brohan (2006) showed the degree of uncertainty in surface temperature sampling errors for 1969 (here for CRUTEM3). The degree of uncertainty exceeds the total global warming signal (Fig. 6).

---

13. http://chiefio.wordpress.com/2010/02/28/last-delta-t-an-experimental-approach/.

14. http://chiefio.wordpress.com/.

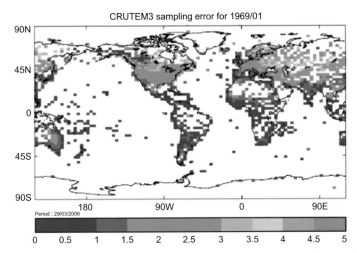

FIGURE 6   Temperature anomaly sampling errors (C) for January 1969 on the HadCM3 atmosphere grid (*source: Brohan et al., 2006 here*).

5. SEE FOR YOURSELF — THE DATA IS A MESS

Look for yourself following these directions using the window into the NOAA, GHCN data provided by NASA GISS.[15] Point to any location on the world map (say, central Canada). You will see a list of stations and approximate populations. Locations with less than 10,000 people are assumed to be rural (even though Oke has shown a town of 1,000 can have an urban warming bias of 2.2C).

You will see that the stations have a highly variable range of years with data. Try to find a few stations where the data extend to the current year. If you find some, you will likely see gaps in the graphs. To see how incomplete the data set is for that station, click in the bottom left of the graph *Download monthly data as text*. For many, many stations you will see the data set in a monthly tabular form has many missing data months mostly after 1990 (designated by 999.9) (Fig. 7).

These facts suggest that the golden age of observations was in the 1950s to 1980s. Data sites before then were more scattered and did not take data at standardized times of day. After the 1980s, the network suffered from loss of stations and missing monthly data. To fill in these large holes, data were extrapolated from greater distances away.

Indeed this is more than just Russia. Forty percent of GHCN v2 stations have at least one missing month (Fig. 8).

This is concentrated in the winter months as analyst Verity Jones has shown here.[16]

---

15. http://data.giss.nasa.gov/gistemp/station_data/.

16. http://diggingintheclay.blogspot.com/2010/03/of-missing-temperatures-and-filled-in.html.

SVERDLOVSK, RUSSIA

	Jan	Feb	Mar	Apr	May	Jun	Jul	Aug	Sep	Oct	Nov	Dec
1987	-18.6	-8.3	-5.5	0.9	14.9	20	19.3	16.2	8.6	2.1	-11.2	-11.5
1988	-12.7	-9.6	-2.6	4.8	10.3	19.5	22.6	18.3	9.7	3.9	-7	-9.9
1989	-15.1	-9.6	-0.3	1.5	11.9	21.4	22.8	14.2	10.2	2.8	-4.2	-10.2
1990	-14.2	-6.8	0.1	5.3	9.6	999.9	19	16.4	999.9	1.2	-5	-8.3
1991	999.9	-12.6	-7.1	9.2	15.5	999.9	17.6	13.6	10.6	7.2	-3.6	-13.6
1992	-12	-9.2	-4.1	1.8	10.1	14	16.3	13.8	10.9	2.1	-5.3	999.9
1993	-8.9	-11.8	-6	3.2	10.8	17.9	18.6	16.5	5.8	2.4	-13.2	-9.9
1994	-10.2	-17.2	999.9	999.9	11.4	999.9	999.9	14.9	11	6.6	-7	-14
1995	-10.2	-4.2	-0.6	10.7	13	17	999.9	16.9	999.9	3.9	-3.7	-12.7
1996	-14.1	-11.2	-3.9	0.6	12.2	19.1	19.2	999.9	7.1	1.9	-2.3	-10.2
1997	-18.4	-9.4	-2.1	6.2	12	16.7	15.9	14	11.2	5.9	-7.3	-14.7
1998	-11	-14.8	-3.3	-1.4	11.8	18.5	21.6	17.6	8.3	3.5	-12.7	-7.1
1999	-12.6	-7.8	-8.6	4.8	9	15.1	20.2	15.6	9.3	7	-10.4	-6.4
2000	-12.9	-6.9	-1.7	7.2	8.3	19.1	20.5	999.9	8.9	2.3	-6.5	-12.2
2001	-12.1	-14.9	-3.4	6.8	13	14.6	17.9	999.9	10.7	0.6	-4.6	-12.3
2002	-9.2	-4.2	-1	3	9.3	14	19	13.1	11.1	2.1	-3.7	-18.5

FIGURE 7 The monthly average temperatures in degrees Celsius 1987 to 2002. 999.9 values are missing months. These require estimation from surrounding sites.

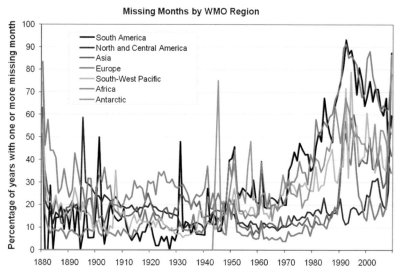

FIGURE 8 Quantification of missing months in annual station data. *(Analysis and graph: Andrew Chantrill.)*

As Verity Jones notes *"Much of the warming signal in the global average data can be traced to winter warming (lows are not as low). If we now have a series of cooler years, particularly cooler winter months with lower lows, my concern is that missing months, particularly winter months could lead to a warm bias."*

NOAA tells us that by 2020, we will have as much data for the 1990s and 2000s as we had in the 1960s and 1970s. We are told that other private sources have been able to assemble more complete data sets in near real time (example: WeatherSource). Why can't our government with a budget far greater than these private sources do the same or better? This question has been asked by others in foreign nations.

6. STATION DROPOUT WAS NOT TOTALLY RANDOM

6.1. Canada

After 1990, just one thermometer remains in the database for everything north of the 65th parallel. That station is Eureka, which has been described as "The Garden Spot of the Arctic" thanks to the flora and fauna abundant around the Eureka area, more so than anywhere else in the High Arctic. Winters are frigid but summers are slightly warmer than at other places in the Canadian Arctic.

NOAA GHCN used only 35 of the 600 Canadian stations in 2009, down from 47 in 2008. A case study by Tim Ball confirmed Environment Canada claims that weather data are available elsewhere from airports across Canada and indeed hourly readings can be found on the internet for many places in Canada (and Russia) not included in the global databases. Environment Canada reported in the National Post,[17] that there are 1,400 stations in Canada with 100 north of the Arctic Circle, where NOAA uses just one. See E.M. Smith's analysis in footnote 18.

Verity Jones plotted the stations from the full network rural, semi-rural and urban for Canada and the northern United States both in 1975 and again in 2009. She also marked with diamonds the stations used in the given year. Notice the good coverage in 1975 and very poor, virtually all in the south in 2009. Notice the lack of station coverage in the higher latitude Canadian region and arctic in 2009 (Fig. 9).

6.2. New Zealand and Australia

Smith found that in New Zealand the only stations remaining had the words "water" or "warm" in the descriptor code. Some 84% of the sites are at airports, with the highest percentage in southern cold latitudes.

In Australia, Torok et al. (2001),[19] observed that in European and North American cities urban—rural temperature differences scale linearly with the logarithms of city populations. They also learned that Australian city heat islands are generally smaller than those in European cities of similar size,

17. http://www.nationalpost.com/news/story.html?id=2465231#ixzz0dY7ZaoIN.

18. http://chiefio.wordpress.com/2009/11/13/ghcn-oh-canada-rockies-we-dont-need-no-rockies/.

19. http://www.co2science.org/articles/V5/N20/C3.php.

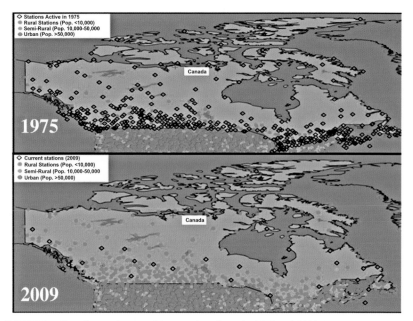

FIGURE 9 Canadian stations used in annual analyses in 1975 and 2009 *(source: Verity Jones from GHCN).*

which in turn are smaller than those in North American cities. The regression lines for all three continents converge in the vicinity of a population of 1,000 people, where the urban–rural temperature difference is approximately $2.2 \pm 0.2\ °C$, essentially the same as what Oke (1973) had reported two decades earlier.

Smith finds the Australian dropout[20] was mainly among higher latitude, cooler stations after 1990, with the percentage of city airports increasing to 71%, further enhancing apparent warming. The trend in "island Pacific without Australia and without New Zealand" is dead flat. The Pacific Ocean islands are NOT participating in "global" warming. Changes of thermometers in Australia and New Zealand are the source of any change.

6.3. Turkey

Turkey had one of the densest networks of stations of any country. E.M. Smith calculated anomaly process similar to First Differences. Then dT is the running total of those changes, or the total change, the "Delta Temperature" to date. Note the step up after 1990 cumulative change in temperature and the change per year for Turkey.[21]

---

20. http://chiefio.wordpress.com/2009/10/23/gistemp-aussy-fair-go-and-far-gone/.

21. http://chiefio.wordpress.com/2010/03/10/lets-talk-turkey/.

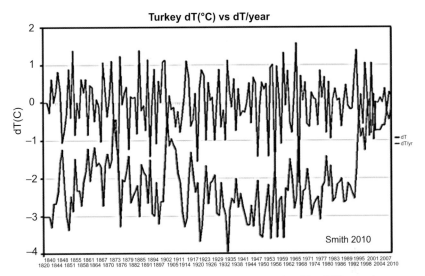

FIGURE 10   Smith analysis of Turkey temperatures using 'First Differences'.

His dT method[22] is a variation of 'First Differences' as a means of examining temperature data anomalies independent of actual temperature. dT/year is the "average of the changes of temperature, month now vs. the same month that last had valid data, for each year" (Fig. 10).

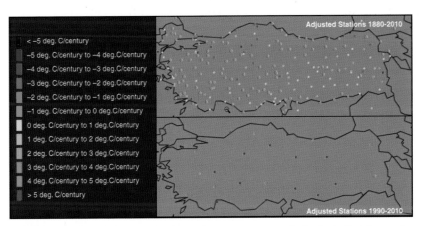

FIGURE 11   Verity Jones maps showing station temperature trends for (top) all stations active during 1880—2010 and (bottom) for stations active after 1990. The result is that Turkey is shown to be warming when the data shows cooling.

22. http://chiefio.wordpress.com/2010/02/28/last-delta-t-an-experimental-approach/.

Despite that apparent warming, the Turkish Met Service finds evidence for cooling. This peer-reviewed paper: Murat Turke, Utku M. Sumer, Gonul Kilic, State Meteorological Service, Department of Research, Climate Change Unit, 06120 Kalaba-Ankara, Turkey which concludes:

"Considering the results of the statistical tests applied to the 71 individual stations data, it could be concluded that annual mean temperatures are generally dominated by a cooling tendency in Turkey." See in Verity Jones website Digging in the Clay *here[23], the dropout of stations from nearly 250 to 39 leaving behind warming stations. 25 of the 39 stations are shown as the other stations did not have complete enough data to determine a reliable trend (less than 10 years without missing months) (Fig. 11).*

7. INSTRUMENT CHANGES AND SITING

The World Meteorological Organization (WMO), a specialized agency of the United Nations,[24] grew out of the International Meteorological Organization (IMO), which was founded in 1873. Established in 1950, the WMO became the specialized agency of the United Nations (in 1951) for meteorology, weather, climate, operational hydrology, and related geophysical sciences.

According to the WMO's own criteria, followed by the NOAA's National Weather Service, temperature sensors should be located on the instrument tower at 1.5 m (5 feet) above the surface of the ground. The tower should be on flat, horizontal ground surrounded by a clear surface, over grass or low vegetation kept less than 4 inches high. The tower should be at least 100 m (110 yards) from tall trees, or artificial heating or reflecting surfaces, such as buildings, concrete surfaces, and parking lots.

Very few stations meet these criteria.

8. ALONG COMES 'MODERNIZATION'

The modernization of weather stations in the United States replaced many human observers with instruments that initially had major errors, or had "warm biases" (HO-83) or were designed for aviation and were not suitable for precise climate trend detection [Automates Surface Observing Systems (ASOS) and the Automated Weather Observing System (AWOS)]. Also, the new instrumentation was increasingly installed on unsuitable sites that did not meet the WMO's criteria.

During recent decades there has been a migration away from old instruments read by trained observers. These instruments were generally in shelters that were properly located over grassy surfaces and away from obstacles to ventilation and heat sources.

---

23. http://diggingintheclay.blogspot.com/2010/03/no-more-cold-turkey.html.

24. http://www.unsystem.org/en/frames.alphabetic.index.en.htm#w.

FIGURE 12   USHCN climate station in Bainbridge, GA, showing the MMTS pole sensor in the foreground near the parking space, building, and air conditioner heat exchanger, with the older Stevenson Screen in the background located in the grassy area (surfacestations.org).

Today we have many more automated sensors (The MMTS) located on poles cabled to the electronic display in the observer's home or office or at airports near the runway where the primary mission is aviation safety.

The installers of the MMTS instruments were often equipped with nothing more than a shovel. They were on a tight schedule and with little budget. They often encountered paved driveways or roads between the old sites and the buildings. They were in many cases forced to settle for installing the instruments close to the buildings, violating the government specifications in this or other ways (Fig. 12).

Pielke and Davey (2005) found a majority of stations, including climate stations in eastern Colorado, did not meet WMO requirements for proper siting.

They extensively documented poor siting and land-use change issues in numerous peer-reviewed papers, many summarized in the landmark paper Unresolved issues with the assessment of multi-decadal global land surface temperature trends[25] (2007).

In a volunteer survey project, Anthony Watts and his more than 650 volunteers www.surfacestations.org found that over 900 of the first 1,067

---

25. http://pielkeclimatesci.files.wordpress.com/2009/10/r-321.pdf.

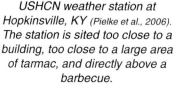

USHCN weather station at Hopkinsville, KY *(Pielke et al., 2006)*. The station is sited too close to a building, too close to a large area of tarmac, and directly above a barbecue.

USHCN station at Tucson, AZ, in a parking lot on pavement. *(Photo by Warren Meyer, courtesy of surfacestations.org.)*

FIGURE 13   USHCN siting issues at Hopkington, KY and Tucson, AZ.

stations surveyed in the 1,221 station U.S. climate network did not come close to meeting the specifications. Only about 3% met the ideal specification for siting. They found stations located next to the exhaust fans of air conditioning units, surrounded by asphalt parking lots and roads, on blistering-hot rooftops, and near sidewalks and buildings that absorb and radiate heat. They found 68 stations located at wastewater treatment plants, where the process of waste digestion causes temperatures to be higher than in surrounding areas. In fact, they found that 90% of the stations fail to meet the National Weather Service's own siting requirements that stations must be 30 m (about 100 feet) or more away from an artificial heating or reflecting source.

The average warm bias for inappropriately-sited stations exceeded 1 °C using the National Weather Service's own criteria, with which the vast majority of stations did not comply.

A report from last spring with some of the earlier findings can be found in footnote 26. Some examples from these sources (Figs. 13, 14):

As of October 25, 2009, 1,067 of the 1,221 stations (87.4%) had been evaluated by the surfacestations.org volunteers and evaluated using the Climate Reference Network (CRN) criteria.[27] 90% were sited in ways that result in errors exceeding 1 °C according to the CRN handbook.

This siting issue remains true even by the older "100 foot rule" criteria for COOP stations, specified by NOAA[28] for the U.S. Cooperative Observer

---

26. http://wattsupwiththat.files.wordpress.com/2009/05/surfacestationsreport_spring09.pdf.

27. http://www1.ncdc.noaa.gov/pub/data/uscrn/documentation/program/X030FullDocumentD0.pdf.

28. http://www.nws.noaa.gov/om/coop/standard.htm.

Numerous sensors are located at waste treatment plants. An infrared image of the scene shows the output of heat from the waste treatment beds right next to the sensor.
(Photos by Anthony Watts, surfacestations.org.)

FIGURE 14 One of many waste treatment plants serving as stations in USHCN.

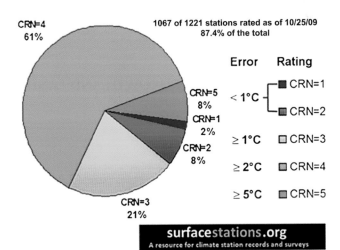

FIGURE 15 Surfacestations.org quality rating by stations for 1,067 U.S. climate stations as of 10/25/2009. Only 10% meet minimal CRN ranking (CRN 1 or 2).

network where they specify "The sensor should be at least 100 feet (~30 m) from any paved or concrete surface (Fig. 15)."

Dr. Vincent Gray, IPPC Reviewer for AR1 through IV published on some issues related to temperature measurements.[29]

In 2008, Joe D'Aleo asked NCDC's Tom Karl about the problems with siting and about the plans for a higher quality Climate Reference Network (CRN — at that time called NERON). He said he had presented a case for a more complete CRN network to NOAA but NOAA said it was unnecessary *because they had satellite monitoring*. The Climate Reference Network was capped at 114 stations and would not provide meaningful trend assessment for about 10 years. NOAA has since reconsidered and now plans to upgrade about 1,000 climate stations, but meaningful results will be even further in the future.

In monthly press releases no satellite measurements are ever mentioned, although NOAA claimed that was the future of observations.

9. ADJUSTMENTS NOT MADE, OR MADE BADLY

The Climategate whistle-blower proved what those of us dealing with data for decades already knew. The data were not merely degrading in quantity and quality: they were also being manipulated. This is done by a variety of post measurement processing methods and algorithms. The IPCC and the scientists supporting it have worked to remove the pesky Medieval Warm Period, the Little Ice Age, and the period emailer Tom Wigley referred to as the "warm 1940s blip". There are no adjustments in NOAA and Hadley data for urban contamination. The adjustments and non-adjustments instead increased the warmth in the recent warm cycle that ended in 2001 and/or inexplicably cooled many locations in the early record, both of which augmented the apparent trend.

10. HEAT FROM POPULATION GROWTH AND LAND-USE CHANGES

10.1. Urban Heat Island

Weather data from cities as collected by meteorological stations are indisputably contaminated by urban heat-island bias and land-use changes. This contamination has to be removed or adjusted for in order to accurately identify true background climatic changes or trends. In cities, vertical walls, steel and concrete absorb the sun's heat and are slow to cool at night. More and more of the world is urbanized (population increased from 1.5 B in 1900 to 6.8 B in 2010).

---

29. http://icecap.us/images/uploads/Gray.pdf.

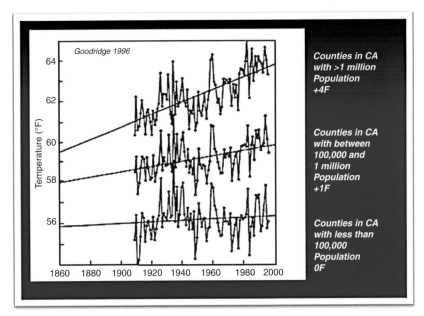

FIGURE 16 Jim Goodrich analysis of warming in California counties by population 1910–1995.

The urban heat-island effect occurs not only for big cities but also for towns. Oke (who won the 2008 American Meteorological Society's Helmut Landsberg award for his pioneer work on urbanization the effect of urbanization on local microclimates) had a formula for the warming that is tied to population. Oke (1973) found that the urban heat-island (in °C) increases according to the formula:

Urban heat-island warming $= 0.317 \ln P$, where $P =$ population.

Thus a village with a population of 10 has a warm bias of 0.73 °C. A village with 100 has a warm bias of 1.46 °C, and a town with a population of 1,000 people has a warm bias of 2.2 °C. A large city with a million people has a warm bias of 4.4 °C.

Goodrich (1996) showed the importance of urbanization to temperatures in his study of California counties in 1996. He found for counties with a million or more population the warming from 1910 to 1995 was 4 °F, for counties with 100,000 to 1 million it was 1 °F, and for counties with less than 100,000 there was no change (0.1 °F) (Fig. 16).

11. U.S. CLIMATE DATA

Compared to the GHCN global database, the USHCN database is more stable (Fig. 17).

GHCN US versus Rest of World (ROW)

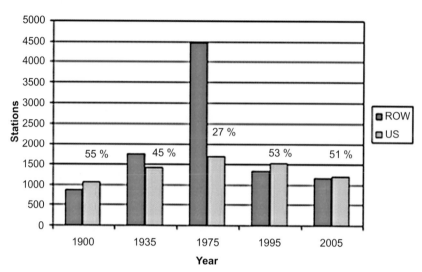

FIGURE 17   Comparison of number of GHCN temperature stations in the U.S. vs. rest of the world (ROW). *http://www.appinsys.com/GlobalWarming/ClimateData.htm*

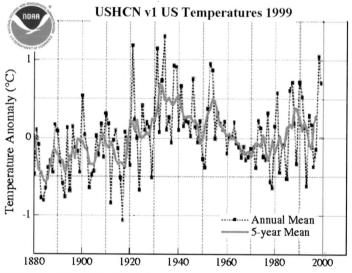

FIGURE 18   NOAA NCDC USHCN version 1 annual U.S. temperatures as of 1999.

When first implemented in 1990 as version 1, USHCN employed 1,221 stations across the United States. In 1999, NASA's James Hansen published this graph of USHCN v.1 annual mean temperature (Fig. 18):

Hansen correctly noted:

"The US has warmed during the past century, but the warming hardly exceeds year-to-year variability. Indeed, in the US the warmest decade was the 1930s and the warmest year was 1934."

USHCN was generally accepted as the world's best database of temperatures. The stations were the most continuous and stable and had adjustments made for time of observation, urbanization, known station moves or land-use changes around sites, as well as instrumentation changes.

Note how well the original USHCN agreed with the state record high temperatures.

12. U.S. STATE HEAT RECORDS SUGGEST RECENT DECADES ARE NOT THE WARMEST

The 1930s were, by far, the hottest period for the time-frame. In absolute terms the 1930s had a much higher frequency of maximum temperature extremes than

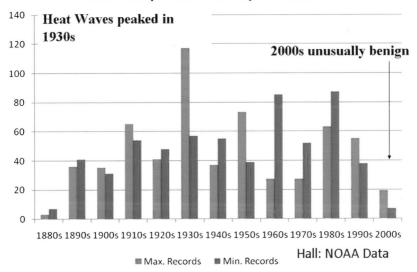

U.S. State Maximum and Minimum Monthly Records by Decade

Heat Waves peaked in 1930s

2000s unusually benign

1880s 1890s 1900s 1910s 1920s 1930s 1940s 1950s 1960s 1970s 1980s 1990s 2000s

■ Max. Records   ■ Min. Records   Hall: NOAA Data

FIGURE 19 United States all-time monthly record lows and highs by decade. Compiled by Hall from NOAA NCDC data.

the 1990s or 2000s or the combination of the last two decades. This was shown by Bruce Hall and Dr. Richard Keen,[30] also covering Canada (Fig. 19).

NCDC's Tom Karl (1988) employed an urban adjustment scheme for the first USHCN database (released in 1990). He noted that the national climate network formerly consisted of predominantly rural or small towns with populations below 25,000 (as of 1980 census) and yet that an urban heat-island effect was clearly evident.

Tom Karl et al.'s adjustments were smaller than Oke had found (0.22 °C annually on a town of 10,000 and 1.81 °C on a city of 1 million and 3.73 °C for a city of 5 million). Karl observed that in smaller towns and rural areas the net urban heat-island contamination was relatively small, but that significant anomalies showed up in rapidly growing population centers.

13. MAJOR CHANGES TO USHCN IN 2007

NOAA had to constantly explain why their global data sets which had no such adjustment was showing warming and the U.S., not so much. NOAA began reducing the UHI around 2000 (noticed by state climatologists and seen in this analysis of New York City's Central Park data here http://icecap.us/index.php/ go/new-and-cool/central_park_temperatures_still_a_mystery/) and then in USHCN version 2 released for the U.S. stations in 2009, the urban heat-island adjustment was eliminated which resulted in an increase of 0.3 °F in warming trend since the 1930s. See animating GIF here http://stevengoddard.files.word-press.com/2010/12/1998uschanges.gif.

In 2007 the NCDC, in its version 2 of USHCN, inexplicably removed the Karl urban heat-island adjustment and substituted a change-point algorithm that looks for sudden shifts (discontinuities). This is best suited for finding site moves or local land-use changes (like paving a road or building next to sensors or shelters), but not the slow ramp up of temperature characteristic of a growing town or city (Fig. 20).

David Easterling, Chief of the Scientific Services Division at NOAA in one of the NASA FOIA emails noted: "One other fly in the ointment, we have a new adjustment scheme for USHCN (V2) that appears to **adjust out** some, if not most, of the "local" trend that includes land-use change and urban warming."

The difference between the old and new is shown here. Note the significant post-1995 warming and mid-20th-century cooling owing to de-urbanization of the database (Fig. 21).

The change can be seen clearly in this animation[31] and in 'blink charts for Wisconsin[32] and Illinois.[33] Here are two example stations with USHCN version

30. http://icecap.us/index.php/go/new-and-cool/more_critique_of_ncar_cherry_picking_temperature_ record_ study/.

31. http://climate-skeptic.typepad.com/.a/6a00e54eeb9dc18834010535ef5d49970b-pi.

32. http://www.rockyhigh66.org/stuff/USHCN_revisions_wisconsin.htm.

33. http://www.rockyhigh66.org/stuff/USHCN_revisions.htm.

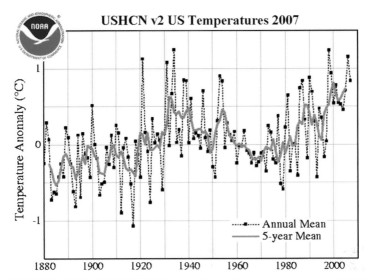

FIGURE 20   NOAA NCDC USHCN version 2 annual mean temperatures as of 2007.

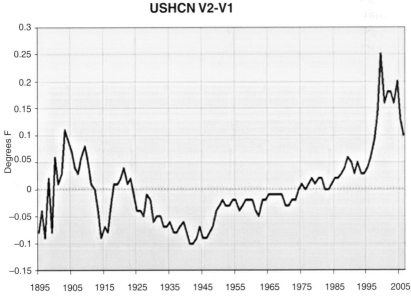

FIGURE 21   NOAA NCDC USHCN version 2 minus version 1 annual mean temperatures.

USHCN v1 Versus v2

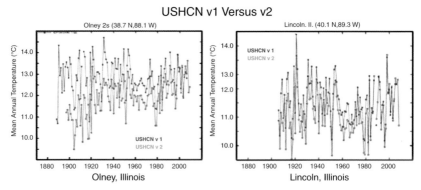

FIGURE 22   NOAA USJCN version 1 vs. version 2 for Olney and Lincoln Illinois.

1 and version 2 superimposed (thanks to Mike McMillan). Notice the clear tendency to cool off the early record and leave the current levels near recently reported levels or increase them. The net result is either reduced cooling or enhanced warming not found in the raw data (Fig. 22).

The new algorithms are supposed to correct for urbanization and changes in siting and instrumentation by recognizing sudden shifts in the temperatures (Fig. 23).

It should catch the kind of change shown above in Tahoe City, CA (Fig. 23).

It is unlikely to catch the slow warming associated with the growth of cities and towns over many years, as in Sacramento, CA, in figure 24 above.

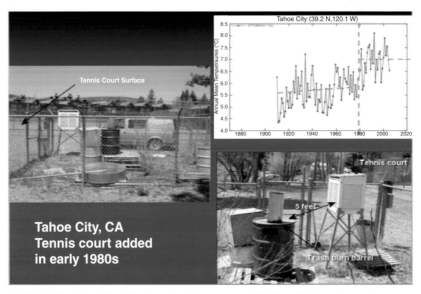

FIGURE 23   Tahoe City, CA data and photos courtesy of Anthony Watts, surfacestations.org.

Sacramento Urban growth and warming will not be seen

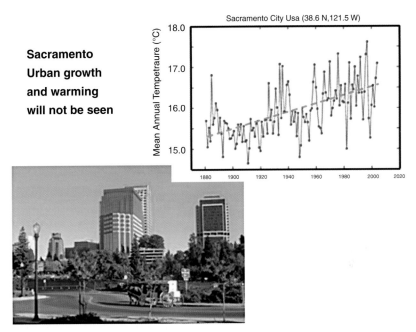

FIGURE 24   Sacramento, CA data and photos courtesy of Anthony Watts, surfacestations.org.

In a conversation during Anthony Watts invited presentation about the surface stations projects to NCDC, on 4/24/2008, he was briefed on USHCN2's algorithms and how they operated by Matt Menne, lead author of the USHCN2 project. While Mr. Watts noted improvements in the algorithm can catch some previously undetected events like undocumented station moves, he also noted that the USHCN2 algorithm had no provision for long-term filtering of signals that can be induced by gradual local urbanization, or by long-term changes in the siting environment, such as weathering/coloring of shelters, or wind blocking due to growth of shrubbery/trees.

When Mr. Menne was asked by Mr. Watts if this lack of detection of such long-term changes was in fact a weakness of the USHCN algorithm, he replied "Yes, that is correct". Essentially USHCN2 is a short period filter only, and cannot account for long-term changes to the temperature record, such as UHI, making such signals indistinguishable from the climate-change signal that is sought.

See some other examples of urban vs. nearby rural.[34] Doug Hoyt, who worked at NOAA, NCAR, Sacramento Peak Observatory, the World Radiation Center, Research and Data Systems, and Raytheon where he was a Senior

---

34. http://www.appinsys.com/GlobalWarming/GW_Part3_UrbanHeat.htm.

NASA (GISS) US48 Adjusments

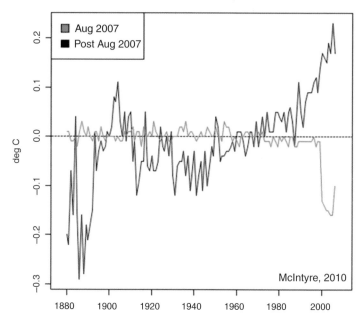

FIGURE 25   Adjustments to U.S. data August 2007 and then post-August 2007 (*source: McIntyre, 2010*).

Scientist did this analysis[35] of the urban heat island. Read beyond the references for interesting further thoughts (Fig. 25).

Even before the version 2 shown above, Balling and Idso (2002)[36] found that the adjustments being made to the raw USHCN temperature data were "producing a statistically significant, but spurious, warming trend" that "approximates the widely-publicized 0.50 °C increase in global temperatures over the past century". There was actually a linear trend of progressive cooling of older dates between 1930 and 1995.

"It would thus appear that in this particular case of "adjustments," the *cure* was much worse than the *disease*. In fact, it would appear that the cure may actually *be* the disease."

It should be noted even with the changes to the USHCN, the correlations with CO_2 are intermittent with just 44 years warming while CO_2 increased and 62 years cooling even as CO_2 rose, not a convincing story for greenhouse CO_2 climate dominance at least with the U.S. data, even with all its warts, generally accepted as the most complete and stable data sets in the world (Fig. 26).

---

35. http://www.warwickhughes.com/hoyt/uhi.htm.

36. http://www.co2science.org/articles/V12/N50/C1.php.

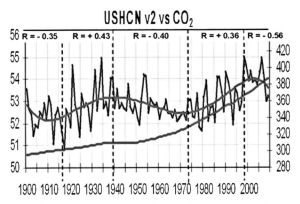

USHCN v2 vs CO$_2$

FIGURE 26   USHCN version 2 annual temperatures vs. ERSL CO$_2$ annual concentrations ppm. Pearson coefficient shown for the warming and cooling intervals.

14. HADLEY AND NOAA

No real urbanization adjustment is made for either NOAA's or CRU's global data. Jones et al. (1990: Hadley/CRU) concluded that urban heat-island bias in gridded data could be capped at 0.05 °C/century. Jones used data by Wang which Keenan[37] has shown was fabricated. Peterson et al. (1998) agreed with the conclusions of Jones, Easterling et al. (1997) that urban effects on 20th century globally and hemispherically-averaged land air temperature time-series do not exceed about 0.05 °C from 1900 to 1990.

Peterson (2003) and Parker (2006) argue urban adjustment is not really necessary. Yet Oke (1973) showed a town of 1,000 could produce a 2.2 °C (3.4 °F warming). The UK Met Office (UKMO) has said[38] future heat waves could be especially deadly in urban areas, where the temperatures could be 9 °C or more above today's, according to the Met Office's Vicky Pope. NASA summer land surface temperature of cities in the Northeast were an average of 7−9 °C (13−16 °F) warmer than surrounding rural areas over a three year period, new NASA research shows. It appears, the warmers want to have it both ways. They argue that the urban heat-island effect is insignificant, but also argue future heat waves will be most severe in the urban areas. This is especially incongruous given that greenhouse theory has the warming greatest in winters and at night.

The most recent exposition of CRU methodology is Brohan et al. (2006), which included an allowance of 0.1 °C/century for urban heat-island effects in the uncertainty but did not describe any adjustment to the reported average temperature. To make an urbanization assessment for all the stations used in the

---

37. http://www.informath.org/WCWF07a.pdf.

38. http://icecap.us/index.php/go/joes-blog/cities_to_sizzle_as_islands_of_heat/.

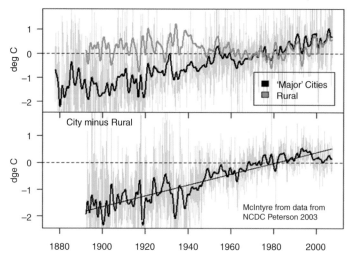

FIGURE 27 GHCN version 2 annual temperatures for stations identified by Peterson in 2003, separated by rural and major cities with the city minus rural (McIntyre, 2007).

HadCRUT data set would require suitable metadata (population, siting, location, instrumentation, etc.) for each station for the whole period since 1850. No such complete metadata are available.

The homepage for the NOAA temperature index[39] cites Smith and Reynolds (2005) as authority. Smith and Reynolds in turn state that they use the same procedure as CRU: i.e., they make an allowance in the error-bars but do not **correct** the temperature index itself. The population of the world went from 1.5 to 6.7 billion in the 20th century, yet NOAA and CRU ignore population growth in the database with only a 0.05−0.1 °C uncertainty adjustment.

Steve McIntyre challenged Peterson (2003), who had said, "Contrary to generally accepted wisdom, no statistically significant impact of urbanization could be found in annual temperatures",[40] by showing that the difference between urban and rural temperatures for Peterson's station set was 0.7 °C and between temperatures in large cities and rural areas 2 °C. He has done the same for Parker (2006) (Fig. 27).[41]

Runnalls and Oke (2006) concluded that:

"Gradual changes in the immediate environment over time, such as vegetation growth or encroachment by built features such as paths, roads, runways, fences, parking lots, and buildings into the vicinity of the instrument site, typically lead to trends in the series.

39. http://www.ncdc.noaa.gov/oa/climate/research/anomalies/anomalies.html.

40. http://climateaudit.org/2007/08/04/1859/.

41. http://climateaudit.org/2007/06/14/parker-2006-an-urban-myth/.

"Distinct régime transitions can be caused by seemingly minor instrument relocations (such as from one side of the airport to another or even within the same instrument enclosure) or due to vegetation clearance. This contradicts the view that only substantial station moves involving significant changes in elevation and/or exposure are detectable in temperature data."

Numerous other peer-reviewed papers and other studies have found that the lack of adequate urban heat-island and local land-use change adjustments could account for up to half of all apparent warming in the terrestrial temperature record since 1900.

Siberia is one of the areas of greatest apparent warming in the record. Besides station dropout and a 10-fold increase in missing monthly data, numerous problems exist with prior temperatures in the Soviet era. City and town temperatures determined allocations for funds and fuel from the Supreme Soviet, so it is believed that cold temperatures were exaggerated in the past. This exaggeration in turn led to an apparent warming when more honest measurements began to be made. Anthony Watts has found that in many Russian towns and cities uninsulated heating pipes[42] are in the open. Any sensors near these pipes would be affected. The pipes also contribute more waste heat to the city over a wide area.

The physical discomfort and danger to observers in extreme environments led to some estimations or fabrications being made in place of real observations, especially in the brutal Siberian winter. See this report.[43] This was said to be true also in Canada along the DEW Line where radars were set up to detect incoming Soviet bombers during the Cold War.

McKitrick and Michaels (2004) gathered weather station records from 93 countries and regressed the spatial pattern of trends on a matrix of local climatic variables and socioeconomic indicators such as income, education, and energy use. Some of the non-climatic variables yielded significant coefficients, indicating a significant contamination of the temperature record by non-climatic influences, including poor data quality.

The two authors repeated the analysis on the IPCC gridded data covering the same locations. They found that approximately the same coefficients emerged. Though the discrepancies were smaller, many individual indicators remained significant. On this basis they were able to rule out the hypothesis that there are no significant non-climatic biases in the data. Both de Laat and Maurellis and McKitrick and Michaels concluded that non-climatic influences add up to a substantial warming bias in measured mean global surface temperature trends.

Ren et al. (2007), in the abstract of a paper on the urban heat-island effect in China, published in *Geophysical Research Letters*, noted that "annual and

---

42. http://wattsupwiththat.com/2008/11/15/giss-noaa-ghcn-and-the-odd-russian-temperature-anomaly-its-all-pipes.

43. http://wattsupwiththat.com/2008/07/17/fabricating-temperatures-on-the-dew-line/.

seasonal urbanization-induced warming for the two periods at Beijing and Wuhan stations is also generally significant, with the annual urban warming accounting for about 65−80% of the overall warming in 1961−2000 and about 40−61% of the overall warming in 1981−2000."

This result, along with the previous mentioned research results, indicates a need to pay more attention to the urbanization-induced bias that appears to exist in the current surface air temperature records.

Numerous recent studies show the effects of urban anthropogenic warming on local and regional temperatures in many diverse, even remote, locations. Jáuregui et al. (2005) discussed the UHI in Mexico, Torok et al. (2001) in southeast Australian cities. Block et al. (2004) showed effects across central Europe. Zhou et al. (2004) and He et al. (2005) across China, Velazquez-Lozada et al. (2006) across San Juan, Puerto Rico, and Hinkel et al. (2003) even in the village of Barrow, Alaska. In all cases, the warming was greatest at night and in higher latitudes, chiefly in winter.

Kalnay and Cai (2003) found regional differences in U.S. data but overall very little change and if anything a slight decrease in daily maximum temperatures for two separate 20-year periods (1980−1999 and 1960−1979), and a slight increase in night-time readings. They found these changes consistent with both urbanization and land-use changes from irrigation and agriculture.

Christy et al. (2006) showed that temperature trends in California's Central Valley had significant nocturnal warming and daytime cooling over the period of record. The conclusion is that, as a result of increases in irrigated land, daytime temperatures are suppressed owing to evaporative cooling and night-time temperatures are warmed in part owing to increased heat capacity from water in soils and vegetation. Mahmood et al. (2006b) also found similar results for irrigated and non-irrigated areas of the Northern Great Plains.

Two Dutch meteorologists, Jos de Laat and Ahilleas Maurellis, showed in 2006 that climate models predict there should be no correlation between the spatial pattern of warming in climate data and the spatial pattern of industrial development. But they found that this correlation does exist and is statistically significant. They also concluded it adds a large upward bias to the measured global warming trend.

Ross McKitrick and Patrick Michaels in 2007 showed a strong correlation between urbanization indicators and the "urban adjusted" temperatures and that the adjustments are inadequate. Their conclusion: "Fully correcting the surface temperature data for non-climatic effects reduce the estimated 1980−2002 global average temperature trend over land by about half."

As Pielke (2007) also notes:

"Changnon and Kunkel (2006) examined discontinuities in the weather records for Urbana, Illinois, a site with exceptional metadata and concurrent records when

important changes occurred. They identified a cooling of 0.17 °C caused by a non-standard height shelter of 3 m from 1898 to 1948. After that there was a gradual warming of 0.9 °C as the University of Illinois campus grew around the site from 1900 to 1983. This was followed by an immediate 0.8 °C cooling when the site moved 2.2 km to a more rural setting in 1984. A 0.3 °C cooling took place with a shift in 1988 to Maximum-Minimum Temperature systems, which now represent over 60% of all USHCN stations. The experience at the Urbana site reflects the kind of subtle changes described by Runnalls and Oke (2006) and underscores the challenge of making adjustments to a gradually changing site."

A 2008 paper[44] by Hadley's Jones et al., has shown a considerable contamination in China, amounting to 1 °C/century. This is an order of magnitude greater than the amount previously assumed (0.05—0.1 °C/century uncertainty).

In a 2009 article,[45] Brian Stone of Georgia Tech wrote:

"Across the US as a whole, approximately 50 percent of the warming that has occurred since 1950 is due to land use changes (usually in the form of clearing forest for crops or cities) rather than to the emission of greenhouse gases. Most large US cities, including Atlanta, are warming at more than twice the rate of the planet as a whole. This is a rate that is mostly attributable to land use change."

In a paper posted on SPPI,[46] Dr. Edward Long summarized his findings as follows: both raw and adjusted data from the NCDC has been examined for a selected Contiguous U.S. set of rural and urban stations, 48 each or one per State. The raw data provides 0.13 and 0.79 °C/century temperature increase for the rural and urban environments (Figs. 28, 29).

One would expect the urban would be adjusted to match the uncontaminated rural data. Instead the rural is adjusted to look more like the urban with the warming since 1895 increased over half a degree from just 0.13 °C to 0.64 °C while the urban trend decreased an insignificant 0.02 °C (Fig. 30).

The adjusted data provide 0.64 and 0.77 °C/century respectively. Comparison of the adjusted data for the rural set to that of the raw data shows a systematic treatment that causes the rural adjusted set's temperature rate of increase to be five-fold more than that of the raw data. This suggests the consequence of the NCDC's protocol for adjusting the data is to cause historical data to take on the time-line characteristics of urban data. The consequence intended or not, is to report a false rate of temperature increase for the Contiguous U.S.

44. http://www.warwickhughes.com/blog/?p=204.

45. http://www.gatech.edu/newsroom/release.html?nid=47354.

46. http://scienceandpublicpolicy.org/images/stories/papers/originals/Rate_of_Temp_Change_Raw_and_Adjusted_NCDC_Data.pdf.

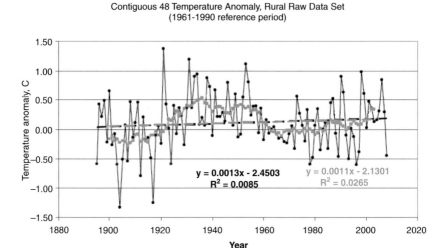

FIGURE 28   Edward long analysis of rural raw stations for the lower 48 states, USHCN version 2. Note the very small trend 0.12 °C/century in this data set and at the significant peak in the 1930s.

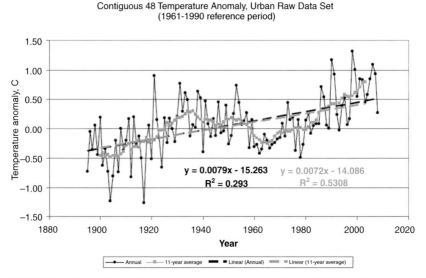

FIGURE 29   Edward Long urban annual temperatures and trend from USHCN version 2 annual temperatures for the lower 48 states, Note the trend of 0.79 °C for this data set with the 1930 peak but with the second recent peak higher.

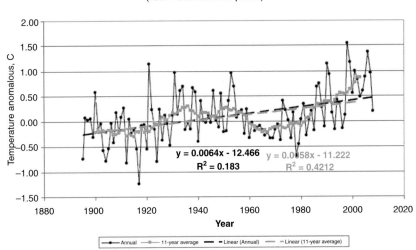

FIGURE 30 Edward Long plot of adjusted rural annual temperatures. Note the trend has increased to 0.64 °C for this data set.

15. FINAL ADJUSTMENTS — HOMOGENIZATION

Dr. William Briggs in a 5 part series on the NOAA/NASA process of homogenization on his blog[47] noted the following:

"At a loosely determined geographical spot over time, the data instrumentation might have changed, the locations of instruments could be different, there could be more than one source of data, or there could be other changes. The main point is that there are lots of pieces of data that some desire to stitch together to make one whole.

Why?

I mean that seriously. Why stitch the data together when it is perfectly useful if it is kept separate? By stitching, you introduce error, and if you aren't careful to carry that error forward, the end result will be that you are far too certain of yourself. And that condition - unwarranted certainty - is where we find ourselves today."

It has been said by NCDC in Menne et al. *"On the reliability of the U.S. surface temperature record"* (in press) and in the June 2009[48] "Talking Points: related to *Is the U.S. Surface Temperature Record Reliable?*" that

---

47. http://wmbriggs.com/blog/?p=1459.

48. www.ncdc.noaa.gov/oa/about/response-v2.pdf.

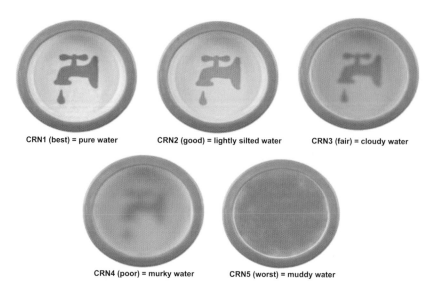

CRN1 (best) = pure water   CRN2 (good) = lightly silted water   CRN3 (fair) = cloudy water

CRN4 (poor) = murky water   CRN5 (worst) = muddy water

FIGURE 31   Simple visual table of CRN station quality ratings and what they might look like as water pollution turbidity levels, rated as 1−5 from best to worst turbidity (Watts).

station siting errors do not matter. However, the way NCDC conducted the analysis gives a false impression because of the homogenization process used.

Here's a way to visualize the homogenization process. Think of it like measuring water pollution. Here's a simple visual table of CRN station quality ratings and what they might look like as water pollution turbidity levels, rated as 1−5 from best to worst turbidity (Fig. 31):

In homogenization the data is weighted against the nearby neighbors within a radius. And so a station might start out as a "1" data wise, might end up getting polluted with the data of nearby stations and end up as a new value, say weighted at "2.5". Even single stations can affect many other stations in the GISS and NOAA data homogenization methods carried out on U.S. surface temperature data (Fig. 32).[49,50]

In the map above, applying a homogenization smoothing, weighting stations by distance nearby the stations with question marks, what would you imagine the values (of turbidity) of them would be? And, how close would these two values be for the east coast station in question and the west coast station in question? Each would be closer to a smoothed center average value based on the neighboring stations.

49. http://wattsupwiththat.com/2009/07/20/and-now-the-most-influential-station-in-the-giss-record-is/.

50. http://wattsupwiththat.com/2008/09/23/adjusting-pristine-data/.

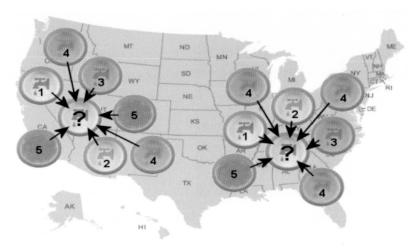

FIGURE 32   In homogenization the data is weighted against the nearby neighbors within a radius. And so a station might start out as a "1" data wise, might end up getting polluted with the data of nearby stations and end up as a new value, say weighted at "2.5".

Essentially, NCDC is comparing *homogenized data* to *homogenized data*, and thus there would *not* likely be any large difference between "good" and "bad" stations in that data. All the differences have been smoothed out by homogenization (pollution) from neighboring stations!

The best way to compare the effect of siting between groups of stations is to use the "raw" data, before it has passed through the multitude of adjustments that NCDC performs. However NCDC is apparently using homogenized data. So instead of comparing apples and oranges (poor sited vs. well sited stations) they essentially just compare apples (Granny Smith vs. Golden Delicious) of which there is little visual difference beyond a slight color change.

They cite 60 years of data in the graph they present, ignoring the warmer 1930s. They also use an early and incomplete surfacestations.org data set that was never intended for analysis in their rush to rebut the issues raised. However, our survey most certainly cannot account for changes to the station locations or station siting quality any further back than about 30 years. By NCDC's own admission, (see Quality Control of pre-1948 Cooperative Observer Network Data[51]) they have little or no metadata posted on station siting much further back than about 1948 on their MMS meta-database. Clearly, siting quality is dynamic over time.

The other issue about siting that NCDC does not address is that it is a significant contributor to extreme temperature records. By NOAA's own admission in PCU6 − Unit No. 2 Factors Affecting the Accuracy and

---

51. http://ams.confex.com/ams/pdfpapers/68379.pdf.

*Baltimore
USHCN station
circa 1990's*
(photo courtesy NOAA)

FIGURE 33   Baltimore USHCN rooftop station circa 1999.

Continuity of Climate Observations[52] such siting issues as the rooftop weather station in Baltimore contributed many erroneous high temperature records, so many in fact that the station had to be closed.

NOAA wrote about the Baltimore station:

"A combination of the rooftop and downtown urban siting explain the regular occurrence of extremely warm temperatures. Compared to nearby ground-level instruments and nearby airports and surrounding COOPs, it is clear that a strong warm bias exists, partially because of the rooftop location.

Maximum and minimum temperatures are elevated, especially in the summer. The number of 80 plus minimum temperatures during the one-year of data overlap was 13 on the roof and zero at three surrounding LCD airports, the close by ground-based inner Baltimore harbor site, and all 10 COOPs in the same NCDC climate zone. Eighty-degree minimum are luckily, an extremely rare occurrence in the mid-Atlantic region at standard ground-based stations, urban or otherwise."

Clearly, siting does matter, and siting errors have contributed to the temperature records of the United States, and likely the world GHCN network. Catching such issues isn't always as easy as NOAA demonstrated in Baltimore (Fig. 33).

There is even some evidence that the change-point algorithm does not catch some site changes it should catch and that homogenization doesn't help. Take, for example, Lampasas, Texas, as identified by Anthony Watts (Fig. 34).

The site at Lampasas, TX, moved close to a building and a street from a more appropriate grassy site after 2001. Note even with the GISS "homogeneity" adjustment (red) applied to the NOAA adjusted data, this artificial

---

52. http://www.weather.gov/om/csd/pds/PCU6/IC6_2/tutorial1/PCU6-Unit2.pdf.

FIGURE 34   Lampasas, Texas relocated station (Photograph by Julie K. Stacy surfacestations.org).

warming remains although the old data (blue) is cooled to accentuate warming even further (Fig. 35).

The net result is to make the recent warm cycle maximum more important relative to the earlier maximum in the 1930s, and note the sudden warm blip after the station move remains.

Other examples (and there are many, many such examples) include (Fig. 36):

Adjustments to the raw data are responsible for the New Zealand warming trend shown by NIWA, the National Institute of Water and Atmospheric

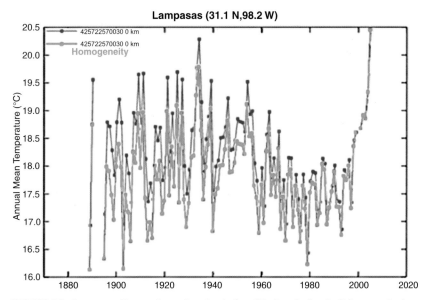

FIGURE 35   Lampasas, Texas relocated station before (blue) and after (red) homogenization. Note the cooling of the old data but no correction for the station move in 2001.

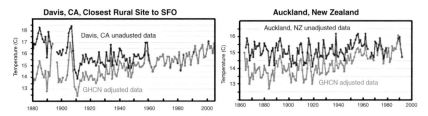

FIGURE 36   GHCN raw vs. adjusted for Davis, CA and Auckland, NZ.

Research (NIWA). New Zealand Climate Science Coalition (NZCSC) publicly called on NIWA http://icecap.us/index.php/go/political-climate/high_court_asked_to_invalidate_niwas_official_nz_temperature_record1/ to admit no valid statistical justification for its claims of a 0.91 °C rise in New Zealand's average temperature last century for the Seven Station Series (7SS) (Fig. 37).

For the globe the final adjusted data set is then used to populate a global grid, interpolating up to 1200 km (745 miles) to grid boxes that had become now vacant by the elimination of stations.

The data are then used for estimating the global average temperature and anomaly and for initializing or validating climate models.

After the Menne et al. (2009) paper, NCDC recognized their position on station siting was untenable and requested $100 million to upgrade the siting of 1,000 climate stations in the 1,220 station network.

NASA/NOAA homogenization process has been shown to significantly alter the trends in many stations where the siting and rural nature suggest the

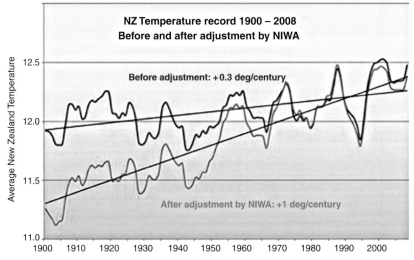

FIGURE 37   NIWA raw vs. adjusted for Seven Sisters Stations (7SS). Adjusted NIWA becomes GHCN raw.

data are reliable. In fact, adjustments account for virtually all the trend in the data (multi-author paper accepted 2011).

16. PROBLEMS WITH SEA SURFACE TEMPERATURE MEASUREMENTS

The world is 71% ocean. The Hadley Centre only trusts data from British merchant ships, mainly plying northern hemisphere routes. Hadley has virtually no data from the southern hemisphere's oceans, which cover four-fifths of the hemisphere's surface. NOAA and NASA use ship data reconstructions. The gradual change from taking water in canvas buckets to taking it from engine intakes introduces uncertainties in temperature measurement. Different sampling levels will make results slightly different. How to adjust for this introduced difference and get a reliable data set has yet to be resolved adequately, especially since the transition occurred over many decades. The chart, taken from Kent (2007), shows how methods of ocean-temperature sampling have changed over the past 40 years (Fig. 38).

We have reanalysis data based on reconstructions from ships, from buoys (which also have problems with changing methodology) and, in recent decades, from satellites. The oceans offer some opportunity for mischief, as the emails released by the Climategate whistle-blower showed clearly.

This report[53] analyzed climate model (Barnet et al., 2001) forecasts of ocean temperatures from 1955 to 2000 vs. actual changes. It found models greatly overstated the warming especially at the surface where the actual change was just about 0.1 °C over that period.

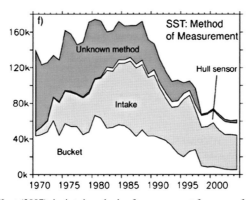

FIGURE 38   Kent (2007) depicted methods of measurement for sea surface temperatures.

---

53. http://www.worldclimatereport.com/archive/previous_issues/vol6/v6n16/feature1.htm#http://
www.world climatereport.com/archive/previous_issues/vol6/v6n16/feature1.htm.

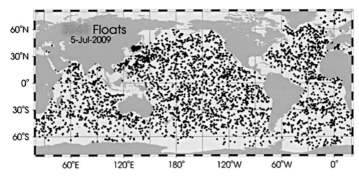

FIGURE 39   ARGO float network as of July 2009.

There is another data set that may better resolve this discrepancy with time, the ARGO buoys. The Argo network[54] which may eventually overcome many of the prior problems, became operational in mid-2003.

Before Argo, starting in the early 1960s, ocean temperatures were measured with bathythermographs (XBTs). They are expendable probes fired into the water by a gun, that transmit data back along a thin wire. They were nearly all launched from ships along the main commercial shipping lanes, so geographical coverage of the world's oceans was poor—for example the huge southern oceans were not monitored. XBTs do not go as deep as Argo floats, and their data are much *less* accurate (Met Office,[55] Argo[56]) (Fig. 39).

Early results showed a cooling, but some issues may exist with the quality control of the early measurements and the strong El Nino in 2009/10 produced a brief pop up now reversing. We believe in the future this data set may give us the best indication of ocean heat content which could be the most robust and reliable indication of climate trends (Pielke, 2008[57]).

17. LONG-TERM TRENDS

Just as the Medieval Warm Period was an obstacle to those trying to suggest that today's temperature is exceptional, and the UN and its supporters tried to abolish it with the "hockey-stick" graph, the warmer temperatures in the 1930s and 1940s were another inconvenient fact that needed to be "fixed".

In each of the databases, the land temperatures from that period were simply adjusted downward, making it look as though the rate of warming in the 20th

54. http://www.argo.ucsd.edu/About_Argo.html.

55. http://www.metoffice.gov.uk/weather/marine/observations/gathering_data/argo.html.

56. http://wwlw.argo.ucsd.edu/Novel_argo.html.

57. http://pielkeclimatesci.files.wordpress.com/2009/10/r-334.pdf

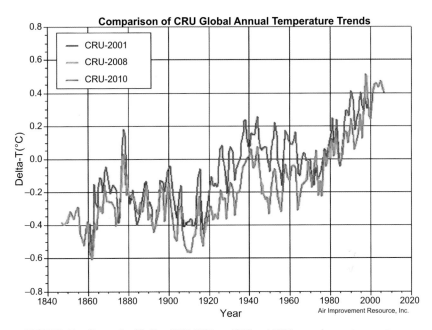

FIGURE 40 Comparing Hadley CRU 2001 vs. 2008 and 2010 annual mean temperatures.

century was higher than it was, and making it look as though today's temperatures were unprecedented in at least 150 years (Figs. 40–44).

Wigley[58] even went so far as to suggest that sea surface temperatures for the period should likewise be "corrected" downward by 0.15C, making the 20th-century warming trend look greater but still plausible. This is obvious data doctoring.

In the Climategate emails, Wigley also noted[59]:

"Land warming since 1980 has been twice the ocean warming — and skeptics might claim that this proves that urban warming is real and important."

NOAA, then, is squarely in the frame. First, the unexplained major station dropout with a bias towards warmth in remaining stations. Next, the removal of the urbanization adjustment and lack of oversight and quality control in the siting of new instrumentation in the United States database degrades what once was the world's best data set, USHCNv1. Then, ignoring a large body of peer review research demonstrating the importance of urbanization and land-use changes NOAA chooses not to include any urban adjustment for the global data set, GHCN.

---

58. http://www.eastangliaemails.com/emails.php?eid=1016&filename=1254108338.txt.

59. http://www.eastangliaemails.com/emails.php?eid=1067&filename=1257546975.txt.

GISS Temperatures Change Yearly			
	2006	*2007*	*2008*	*2009*
1996	−0.18	−0.16	−0.16	−0.06
1997	0.05	0.04	0.04	0.14
1998	1.24	1.24	1.24	1.31
1999	0.94	0.94	0.94	1.07
2000	0.65	0.54	0.54	0.69
2001	0.89	0.78	0.78	0.92
2002	0.67	0.55	0.55	0.69
2003	0.65	0.53	0.53	0.69
2004	0.54	0.46	0.46	0.61
2005	0.99	0.71	0.71	0.92
2006	*	1.15	1.15	1.31
2007	*	*	0.84	0.88
2008	*	*	*	0.12

FIGURE 41 Comparing NASA GISS values for recent years as reported in the year shown. Note the shift down in 2007 after correction for the millennium bug identified by McIntyre and then the shift up again in 2009.

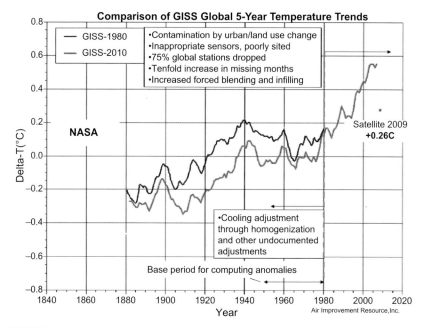

FIGURE 42 Comparing NASA GISS global values from 1980 to 2010. Note the string cooling prior to 1980. Warming post 1980 was due to many issues unaccounted for. Compare to UAH value for 2009.

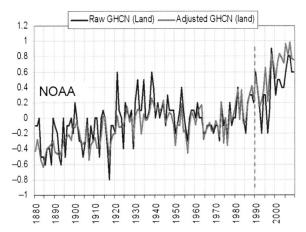

FIGURE 43 Comparing NOAA GHCN version 2 raw vs. adjusted. A similar cooling in the early record and recent warming is clearly shown. Raw data show little warming from peak in 1930s to 1940s to 1990s and 2000s.

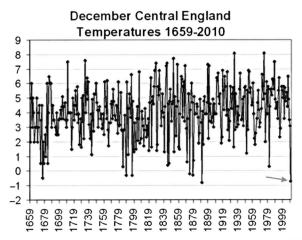

FIGURE 44 Central England Temperatures (CETs) for December from 1659 to 2010 second only 1890.

As shown, these and other changes that have been made alter the historical record and mask cyclical changes that could be readily explained by natural factors like multi-decadal ocean and solar changes.[60]

---

60. http://icecap.us/images/uploads/ATMOSPHERIC_CIRCULATION.doc.

The CRU data have seen changes even in the last decade with a cooling of the early and middle parts of the 20th century and dramatic post 1990 warming when most of the issues discussed emerged (Fig. 40).

The green is the 2001 global temperature plot and the red that in 2010 (with data through 2009).

Is NASA in the clear? No. It works with the same GHCN/USHCN base data (plus the SCAR data from Antarctica). To its credit, as we have shown its U.S. database includes an urban adjustment that is reasonable, but as Steve McIntyre showed[61] uses NASA used population data and adjusted GHCN temperature records for cities in a warming direction as often as they do in a cooling direction. This we have seen is due to very poor metadata from GHCN which GISS uses to match with satellite night light to define a station as urban, suburban or rural.

And their homogenization process and other non-documented final adjustments result in an increase in apparent warming, often by cooling the early record as can be seen in several case studies that follow.

NASA also seems to constantly rehash the surface data. John Goetz[62] showed that 20% of the historical record was **modified 16 times** in the 2½ years ending in 2007. 1998 and 1934 ping pong regularly between first and second warmest year as the fiddling with old data continues.

In 2007, NASA adjusted post-2000 data[63] when Steve McIntyre found a bug in the USHCN data down by 0.12 to 0.15C. Note how the data were adjusted up again in 2009 (USHCN V2.) (see Fig. 41)

Earlier version of NASA data was extracted from an earlier paper by Hansen in 1980 and is compared in the graph below. In 1987 (green on the graph in figure 42), the GISS temperatures were modified down in the middle part of the century from the 1980 version (blue) which enhanced the apparent warming in time for Dr. Hansen's testimony in front of congress.

Cooling before 1980 is dramatic. Warming after 1990 was due to the myriad of issues with the data in this period as we have identified above.

E-mail messages obtained by a Freedom of Information Act request reveal that NASA concluded that its own climate findings were inferior to those maintained by both the University of East Anglia's Climatic Research Unit (CRU) — the scandalized source of the leaked Climategate e-mails — and the National Oceanic and Atmospheric Administration's National Climatic Data Center.

The e-mails from 2007 reveal that when a *USA Today* reporter asked if NASA's data "was more accurate" than other climate-change data sets, NASA's Dr. Reto A. Ruedy replied with an unequivocal no. He said[64] "the National

61. http://icecap.us/images/uploads/US_AND_GLOBAL_TEMP_ISSUES.pdf.

62. http://wattsupwiththat.com/2008/04/08/rewriting-history-time-and-time-again/.

63. http://data.giss.nasa.gov/gistemp/updates/200708.html.

64. http://pajamasmedia.com/files/2010/03/GISS-says-CRU-Better0001.pdf.

Climatic Data Center's procedure of only using the best stations is more accurate," admitting that some of his own procedures led to less accurate readings.

"My recommendation to you is to continue using NCDC's data for the U.S. means and [East Anglia] data for the global means," Ruedy told the reporter.

A similar tale is seen with NOAA GHCN version 2 before and after adjustment (Fig. 43).

The longest history of unaltered data is the Central England Temperature set established during the Little Ice Age in 1659. Note how this past December 2010 was the second coldest December in the record (just 0.1C above 1890). Long-term warming is seen coming out of the LIA but no acceleration upwards can be detected. (Fig. 44)

18. SUMMARY

Climategate has sparked a flurry of examinations of the global data sets not only at CRU, NASA, and NOAA, but in various countries throughout the world. Though the Hadley Centre implied their data were in agreement with other data sets and was thus trustworthy, the truth is that other data centers and the individual countries involved were forced to work with degraded data and appear to be each involved in data manipulation.

Kevin Trenberth, IPCC Lead Author, NCAR and CRU associate said:

"It's very clear we do not have a climate observing system...This may be a shock to many people who assume that we do know adequately what's going on with the climate, but we don't."

Climate change is real. There has been localized warming due to population growth and land-use changes. There are cooling and warming periods that can be shown to correlate well with solar and ocean cycles. You can trust in the data that shows that there has been warming from 1979 to 1998, just as there was warming around 1920 to 1940. But there has been cooling from 1940 to the late 1970s and since 2001. The long-term trend on which this cyclical pattern is superimposed that is exaggerated.

As shown, record highs and rural temperatures in North America show the cyclical pattern but suggest the 1930s to 1940 peak was higher than the recent peak around 1998. Recent ranking was very likely exaggerated by the numerous data issues discussed. Given these data issues and the inconvenient truths in the Climategate emails, the claim that the 2010 was the warmest year and the 2000s was the warmest decade in the record or as some claim in a millennium or two is not credible.

These factors all lead to significant uncertainty and a tendency for over-estimation of century-scale temperature trends. An obvious conclusion from all findings above is that the global databases are seriously flawed and can no longer be trusted to assess climate trends. And, consequently, such surface data should not be used for decision making.

We enthusiastically support Roger Pielke Sr. who, after exchanges with Phil Jones over data sets, called for[65]:

"an inclusive assessment of the surface temperature record of CRU, GISS and NCDC. We need to focus on the science issues. This necessarily should involve all research investigators who are working on this topic, with formal assessments chaired and paneled by mutually agreed to climate scientists who do not have a vested interest in the outcome of the evaluations."

Georgia Tech's Dr. Judith Curry's comments on Roger Pielke Jr.'s blog also support such an effort:

"In my opinion, there needs to be a new independent effort to produce a global historical surface temperature dataset that is transparent and that includes expertise in statistics and computational science...The public has lost confidence in the data sets...Some efforts are underway in the blogosphere to examine the historical land surface data (e.g. such as GHCN), but even the GHCN data base has numerous inadequacies."

Judith is part of the newly announced Berkeley Earth Surface Temperature (BEST) Project which aims to develop an independent analysis of the data from land stations, which would include many more stations than had been considered by the Global Historic Climatology Network. We trust they will include scientists who understand the issues we have raised and will make the reconstructed data sets available for independent review and analysis.

It should be noted that replication is required by the data quality act (DQA) according to the government's own Office of Management and Budget (OMB). Though such an effort can be done locally through tedious research and analysis in the United States, the status of the publicly available global databases (GHCN, GISS, CRU) makes that extremely difficult or impossible currently. Until then, satellite data is the only trustworthy data set.

ACKNOWLEDGMENTS

I wish to thank Anthony Watts who provided invaluable analysis, and considerable constructive feedback and suggestions for this analysis. I wish to also thank Roger Pielke Sr., Steve McIntyre, E.M. Smith and Verity Jones and many others cited in this compilation study for their tireless efforts with regards to issues with temperature measurements.

---

65. http://wattsupwiththat.com/2010/01/14/pielke-senior-correspondence-with-phil-jones-on-klotzbach-et-al/.

REFERENCES

Balling, R.C., Idso, C.D., 2002. Analysis of adjustments to the United States Historical Climatology Network (USHCN) temperature database. Geophysical Research Letters 10.1029/2002GL014825.

Barnett, T.P., Pierce, D.W., Schnur, R., 2001. Detection of anthropogenic climate change in the world's oceans. Science 292, 270–274.

Block, A., Keuler, K., Schaller, E., 2004. Impacts of anthropogenic heat on regional climate patterns. Geophysical Research Letters 31, L12211. doi:10.1029/2004GL019852.

Brohan, P., Kennedy, J.J., Harris, I., Tett, S.F.B., Jones, P.D., 2006. Uncertainty estimates in regional and global observed temperature changes: a new dataset from 1850. Journal of Geophysical Research 111, D12106. doi:10.1029/2005JD006548.

Chou and Lindzen, 2005. Comments on Examination of the Decadal Tropical Mean ERBS Nonscanner Radiation Data for the Iris Hypothesis. Journal of Climate 18, 2123–2127.

Davey, C.A., Pielke Sr, R.A., 2005. Microclimate exposures of surface-based weather stations – implications for the assessment of long-term temperature trends. Bulletin of the American Meteorological Society 86 (4), 497–504.

de Laat, A.T.J., Maurellis, A.N., 2006. Evidence for influence of anthropogenic surface processes on lower tropospheric and surface temperature trends. International Journal of Climatology 26, 897–913.

Essex, Christopher, Andresen, Bjarne, McKitrick, Ross R., 2007. Does a global temperature exist? Journal of Nonequilibrium Thermodynamics 32 No. 1.

Easterling, D.R., Horton, B., Jones, P.D., Peterson, T.C., Karl, T.R., Parker, D.E., Salinger, M.J., Razuvayev, V., Plummer, N., Jamason, P., Folland, C.K., 1997. Maximum and minimum temperature trends for the globe. Science 277, 364–367.

Fall, S., Niyogi, D., Gluhovsky, A., Pielke Sr., R.A., Kalnay, E., Rochon, G., 2009. Impacts of land use land cover on temperature trends over the continental United States: assessment using the North American Regional Reanalysis[66] International Journal of Climatology, accepted.

Gall, R., Young, K., Schotland, R., Schmitz, J., 1992. The recent maximum temperature anomalies in Tucson. Are they real or an instrument problem,. Journal of Climate 5, 657–664.

Goodridge, J.D., 1996. Comments on Regional simulations of Greenhouse warming including natural variability. Bulletin of the American Meteorological Society 77, 1588–1599.

Hansen, J., Ruedy, R., Glascoe, J., Sato, M., 1999. GISS analysis of surface temperature change. Journal of Geophysical Research 104, 30,997–31,022.

Hansen, J., Ruedy, R., Sato, M., Imhoff, M., Lawrence, W., Easterling, D., Peterson, T., Karl, T., 2001. A closer look at United States and global surface temperature change. Journal of Geophysical Research 106 (D20), 23947–23963.

Hinkel, K.M., Nelson, F.E., Klene, S.E., Bell, J.H., 2003. The urban heat island in winter at Barrow, Alaska. International Journal of Climatology 23, 1889–1905. doi: 10.1002/joc.971.

Jáuregui, E., 2005. Possible impact of urbanization on the thermal climate of some large cities in Mexico. Atmosfera 18, 249–252.

Jones, C.G., Young, K.C., 1995. An Investigation of temperature discontinuities introduced by the installation of the HO-83 thermometer. Journal of Climate Volume 8 (Issue 5) (May 1995). p. 1394.

66. http://www.climatesci.org/publications/pdf/R-329.pdf.

Jones, P.D., Groisman, PYa., Coughlan, M., Plummer, N., Wangl, W.C., Karl, T.R., 1990. Assessment of urbanization effects in time series of surface air temperatures over land. Nature 347, 169−172.

Jones, P.D., Lister, D.H., Li, Q., 2008. Urbanization effects in large-scale temperature records, with an emphasis on China. Journal of Geophysical Research 113, D16122. doi:10.1029/2008JD009916.

Kalnay, E., Cai, M., 2003. Impacts of urbanization and land-use change on climate. Nature 423, 528−531.

Karl, T.R., Diaz, H.F., Kukla, G., 1988. Urbanization: its detection and effect in the United States climate record. Journal of Climate 1, 1099−1123.

Karl, T.R., 1995. Critical issues for long-term climate monitoring. Climate Change 31, 185.

Karl, T.R., Hassol, S.J., Miller, C.D., Murray, W.L. (Eds.), 2006. Temperature Trends in the Lower Atmosphere: Steps for Understanding and Reconciling Differences. A Report by the Climate Change Science Program and the Subcommittee on Global Change Research, Washington, DC.

Kent, E.C., Woodruff, S.D., Berry, D.I., 2007. Metadata from WMO Publication No. 47 and an assessment of voluntary observing ship observation heights in ICOADS. Journal of Atmospheric and Oceanic Technology 24, no 2, 214−234.

Klotzbach, P.J., Pielke Sr., R.A., Pielke Jr., R.A., Christy, J.R., McNider, R.T., 2009. An alternative explanation for differential temperature trends at the surface and in the lower troposphere[67]. Journal of Geophysical Research 114, D21102. doi:10.1029/2009JD011841, 2009.

Landsberg, H.E., 1981. The Urban Climate, Academic Press.

Li, Q., et al., 2004. Urban heat island effect on annual mean temperatures during the last 50 years in China. Theoretical and Applied Climatology 79, 165−174.

Lin, X., Pielke Sr., R.A., Hubbard, K.G., Crawford, K.C., Shafer, M.A., Matsui, T., 2007. An examination of 1997−2007 surface layer temperature trends at two heights in Oklahoma[68] Geophysical Research Letters 34, L24705. doi:10.1029/2007GL031652.

Loehle, Craig, 2009. Cooling of the global ocean since 2003. Energy & Environment Vol. 20 (No. 1&2), 101−104 (4).

McKitrick, R.R., Michaels, P.J., 2007. Quantifying the influence of anthropogenic surface processes and inhomogeneities on gridded global climate data. Journal of Geophysical Research 112, D24S09. doi:10.1029/2007JD008465.

McKitrick, R., Michaels, P.J., 2004. A test of corrections for extraneous signals in gridded surface temperature data Climate Research 26(2) pp. 159−173. "Erratum," Climate Research 27 (3), 265−268.

Menne, M.J., Williams, C.N., Palecki, M.A., 2010. On the reliability of the U.S. surface temperature record. Journal of Geophysical Research 115, D11108. doi:10.1029/2009JD013094.

Moberg, D., A, 2003. Hemispheric and large-scale air temperature variations: an extensive revision and update to 2001. Journal of Climate 16, 206−223.

Oke, T.R., 1973. City size and the urban heat island. Atmospheric Environment 7, 769−779.

Parker, D.E., 2004. Climate: large-scale warming is not urban. Nature 432, 290 (18 November 2004); doi:10.1038/432290a.

Peterson, T.C., Vose, R.S., 1997. An overview of the global historical climatology network temperature database. Bulletin of the American Meteorological Society 78, 2837−2849.

---

67. http://sciencepolicy.colorado.edu/admin/publication_files/resource-2792-2009.52.pdf
68. http://www.climatesci.org/publications/pdf/R-333.pdf.

Peterson, T.C., 2003. Assessment of urban versus rural in situ surface temperatures in the contiguous United States: no difference found. Journal of Climate 16 (18), 2941—2959.

Peterson, 2006. Examination of potential biases in air temperature caused by poor station locations. Bulletin of the American Meteorological Society 87, 1073—1089.

Pielke Sr., R.A., 2003. Heat storage within the Earth system[69] Bulletin of the American Meteorological Society 84, 331—335.

Pielke Sr., R.A., Nielsen-Gammon, J., Davey, C., Angel, J., Bliss, O., Doesken, N., Cai, M., Fall, S., Niyogi, D., Gallo, K., Hale, R., Hubbard, K.G., Lin, X., Li, H., Raman, S., 2007. Documentation of uncertainties and biases associated with surface temperature measurement sites for climate change assessment. Bulletin of the American Meteorological Society 88 (6), 913—928.

Pielke Sr., Roger A., 2005: Public Comment on CCSP Report Temperature Trends in the Lower Atmosphere: Steps for Understanding and Reconciling Differences[70] p. 88 including appendices.

Pielke, Sr., R.A, Davey, C., Niyogi, D., Fall, S., Steinweg-Woods, J., Hubbard, K., Lin, X., Cai, M., Lim, Y.-K., Li, H., Nielsen-Gammon, J., Gallo, K., Hale, R., Mahmood, R., Foster, S., McNider, R.T., Blanken, P., 2007. Unresolved issues with the assessment of multi-decadal global land surface temperature trends[71] Journal of Geophysical Research 112, D24S08. doi:10.1029/2006JD008229.

Pielke Sr., R.A., Matsui, T., 2005. Should light wind and windy nights have the same temperature trends at individual levels even if the boundary layer averaged heat content change is the same?[72] Geophysical Research Letters 32 (No. 21), L21813. 10.1029/2005GL024407.

Runnalls, K.E., Oke, T.R., 2006. A technique to detect microclimatic inhomogeneities in historical records of screen-level air temperature. Journal of Climate 19, 959—978.

Santer, B.D., Wigley, T.M.L., Mears, C., Wentz, F.J., Klein, S.A., Seidel, D.J., Taylor, K.E., Thorne, P.W., Wehner, M.F., Gleckler, P.J., Boyle, J.S., Collins, W.D., Dixon, K.W., Doutriaux, C., Free, M., Fu, Q., Hansen, J.E., Jones, G.S., Ruedy, R., Karl, T.R., Lanzante, J.R., Meehl, G.A., Ramaswamy, V., Russell, G., Schmidt, G.A., 2005. Amplification of surface temperature trends and variability in the tropical atmosphere. Science 309, 1551—1556.

Santer, B.D., et al., 2008. Consistency of modeled and observed temperature trends in the tropical troposphere. International Journal of Climatology 28, 1703—1722. doi:10.1002/joc.1756.

Smith, T.M., Reynolds, R.W., 2004. Improved extended reconstruction of SST (1854—1997). Journal of Climate 17, 2466—2477.

Smith, T.M., Reynolds, R.W., 2005. A global merged land air and sea surface temperature reconstruction based on historical observations (1880—1997). Journal of Climate 18, 2021—2036.

Smith, T.A., Reynolds, R.W., Petersen, T.C., Lawrimore, J., 2008. Improvements to NOAA's historical merged land—ocean surface temperature analysis (1880—2006). Journal of Climate vol. 21, 2283—2296.

Torok, S.J., Morris, C.J.G., Skinner, C., Plummer, N., 2001. Urban heat island features of southeast Australian towns. Australian Meteorological Magazine 50, 1—13.

---

69. http://www.climatesci.org/publications/pdf/R-247.pdf.

70. http://www.climatesci.org/publications/pdf/NR-143.pdf.

71. http://www.climatesci.org/publications/pdf/R-321.pdf.

72. http://www.climatesci.org/publications/pdf/R-302.pdf.

Watts, A., 2009. Is the U.S. surface temperature record reliable?[73]. Chicago, IL. The Heartland Institute, ISBN, 10.1-934791-26-6.

Willmott, Robeson, Feddema, Dec. 1991. Influence of spatially variable instrument networks on climatic averages. Geophysical Research Letters vol. 18, No. 12, 2249–2251.

Zhou, L., Dickinson, R., Tian, Y., Fang, J., Qingziang, L., Kaufman, R., Myneni, R., Tucker, C., 2004. Rapid Urbanization warming China's climate faster than other areas. Proceedings of the National Academy of Science June 29, 2004.

---

73. http://wattsupwiththat.files.wordpress.com/2009/05/surfacestationsreport_spring09.pdf.

Chapter 4

2010—The Hottest Year on Record?

Steve Goddard
Real Science Blog

Chapter Outline

Dr. James Hansen of NASA's Goddard Institute for Space Studies (GISS) has announced that 2010 was the "hottest year on record" — by $0.01°$. His claim has been widely touted in the press as strong evidence that the climate is rapidly heating — due to human generated CO_2 emissions. Dr. Hansen has also stated:

"I would not be surprised if most or all groups found that 2010 was tied for the warmest year."

But most groups do not support his claim. The other independent source of surface temperatures HadCRUT shows 2010 cooler than 1998. The graph in Fig. 1 shows the month-to-month differences. Blue represents months where 2010 was cooler than 1998.

The graph in Fig. 2 shows the HadCRUT temperature anomalies for each year since 1998. Last year was not a remarkable year, and was not as warm as 1998.

Similarly, full year satellite temperatures from RSS show 2010 monthly and annual anomalies lower than 1998 (Fig. 3).

Satellite data from UAH also show 2010 slightly cooler than 1998.

Evidence-Based Climate Science. DOI: 10.1016/B978-0-12-385956-3.10004-X

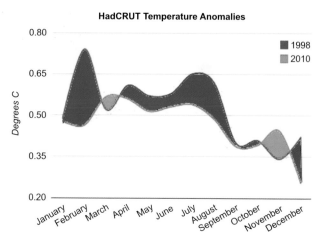

FIGURE 1   HadCRUT monthly temperature anomalies for 1998 and 2010.

The graph in Fig. 4 shows 2010 monthly temperatures for each of the four primary data sets. As you can see, the month-to-month behavior of GISS global temperatures during 2010 was out of kilter with other data sources.

Note that GISS (blue) showed a large temperature spike in March — which was not seen by others. And from July through November, GISS increased sharply while everyone else showed temperatures dropping — due to a near record cold La Nina. La Nina is indicated by ocean temperatures well below normal across much of the Pacific Ocean.

The November spike was followed in December by the largest month-to-month drop in the 130-year GISS record. What sudden change in the climate

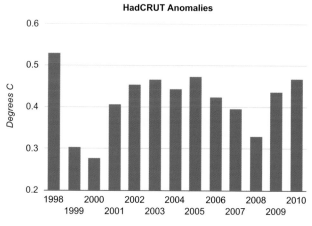

FIGURE 2   HadCRUT annual temperature anomalies for 1998 through 2010.

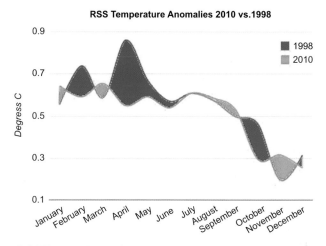

FIGURE 3   RSS monthly temperature anomalies for 1998 and 2010.

could have caused a sharp December drop after 4 months of rise? A plausible explanation is that the August–November reported GISS temperatures were too high, and that December came back more in line with reality.

The NOAA graph (Fig. 5) shows that ocean surface temperatures across much of the Pacific have been the coldest on record since July (yet GISS temperatures rose sharply during that time).

From a physical point of view, it is implausible to have a July–November temperature spike coincident with the rapidly cooling global sea surface temperatures seen in Fig. 6. This is because the oceans make up almost 70% of

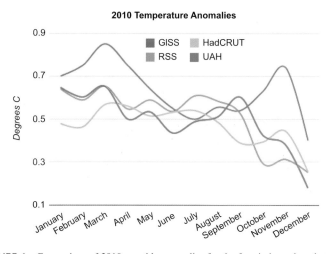

FIGURE 4   Comparison of 2010 monthly anomalies for the four independent data sets.

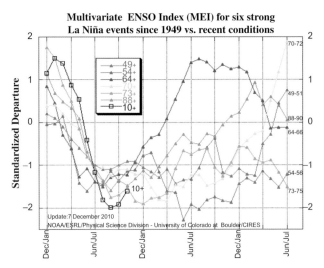

FIGURE 5 Multivariate ENSO index (MEI) for six strong La Nina events since 1949 vs. recent conditions.

the planet's surface, and because sea surface temperatures largely control the temperatures over land.

1. COMPARISONS VS. 1998

The HadCRUT graph (Fig. 7) shows that 2010 was not a remarkable year for temperatures, and was cooler than 1998. In fact, HadCRUT also shows that temperatures have not warmed appreciably (if at all) since 1998.

By contrast, Dr. Hansen claims that temperatures have increased steadily since the mid-1970s.

"Global temperature is rising as fast in the past decade as in the prior two decades"

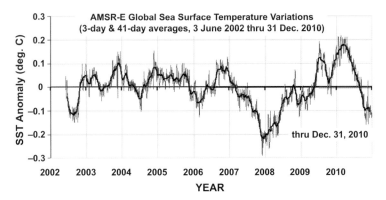

FIGURE 6 Global sea surface temperature variations.

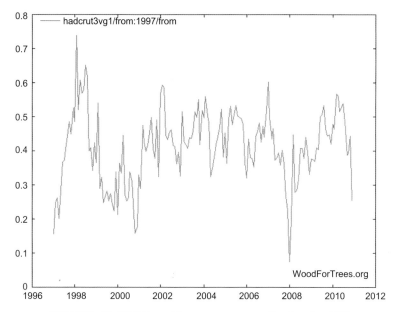

FIGURE 7   HadCRUT monthly temperature anomalies since late 1997.

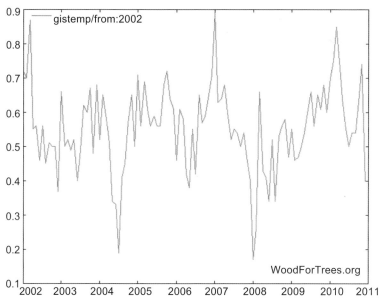

FIGURE 8   GISTEMP monthly temperature anomalies since 2002.

However, even his own data show no significant trend since 2002, and chances are that the GISS January, 2011 anomaly will be one of the lowest of the past decade (Fig. 8).

The UK Met Office explicitly contradicts Dr. Hansen's claim:

"In the last 10 years the rate of warming has decreased."

2. DIVERGENCE FROM OTHER DATA SOURCES

Since the HadCRUT record warm year of 1998, GISS (Fig. 9 − green) has steadily diverged from other data sets, and is now showing temperatures 0.2−0.4° warmer. This is particularly significant because the claimed 2010 record is by a much smaller margin than the discrepancy. Normally scientists will associate a range of errors with their numbers, but Dr. Hansen has avoided mentioning that concept to the press. Rather he has boldly stated that 2010 is the hottest year on record − by 0.01°.

3. WHAT IS GISS DOING WRONG?

The biggest problem is that their warmest regions are in locations where they have little or no data. GISS released the map (Fig. 10) for December 2009 through November 2010, showing large temperature anomalies near the North Pole of 2−6 °C. Those hot red temperatures skew the global average temperature anomaly upwards − by a significant margin.

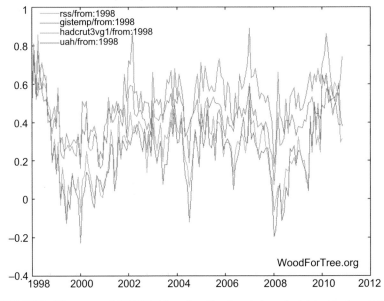

FIGURE 9 Divergence of GISTEMP from the other three independent data sets since 1998.

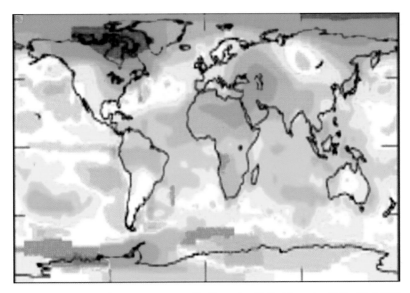

FIGURE 10   GISSTEMP temperature anomalies December 2009—November 2010 with 1,200 km smoothing.

But when we look at where GISS actually has thermometers, we see that they have very few in those "hot" regions. The gray regions in the map shown in Fig. 11 indicate no data.

Temperature anomalies can vary a lot over short distances (note Russia in the map above) yet GISS force fits the entire Arctic to 2—6 °C above normal.

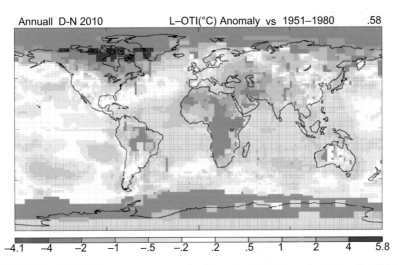

FIGURE 11   GISSTEMP temperature anomalies December 2009—November 2010 with 250 km smoothing.

By averaging in large fabricated numbers, they skew the "global temperature" average upwards — by a significant amount. Remember that the GISS record is only *one hundredth of a degree* warmer than their previous record.

Hansen's claimed precision is much larger than his accuracy. He has no temperature data for more than 10% of the planet. His error bar is probably ±0.3°, yet he claims a record by 0.01°. That mistake would cause him to fail an undergraduate science or engineering class. GISS openly acknowledges that their artificial Arctic data are the causes of the discrepancy with HadCRUT.

"A likely explanation for discrepancy in identification of the warmest year is the fact that the HadCRUT analysis excludes much of the Arctic ... (whereas GISS) estimates temperature anomalies throughout most of the Arctic."

4. WAS 2010 A RECORD HOT YEAR?

There is little evidence to support that. For example, the UK Met Office reported that 2010 was the coldest year in England since 1986. Temperatures there have dropped significantly over the past decade (Fig. 12).

Last winter and the current one have seen record or near record cold across much of the Northern Hemisphere. The Rutgers University graph (Fig. 13) shows that North America had the most extensive snow cover ever recorded last winter. One day in February, all 48 contiguous U.S. states had snow cover. This was due largely to unusually cold temperatures in the deep south.

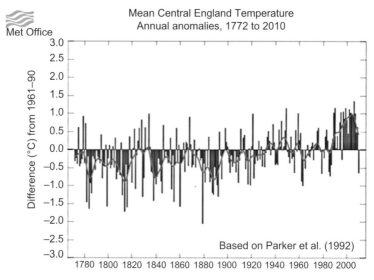

FIGURE 12   Central England temperatures 1772—2010.

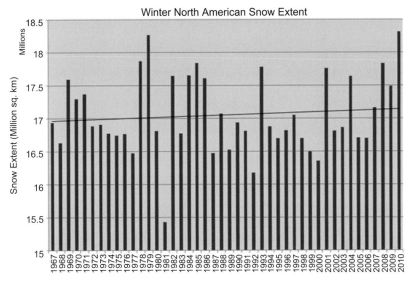

FIGURE 13   Winter North American snow extent (Rutgers University global snow lab).

For the entire Northern Hemisphere, last winter had the second greatest snow extent on record. This was again due to unusually cold temperatures at low latitudes — places like Florida, Europe, and China (Fig. 14).

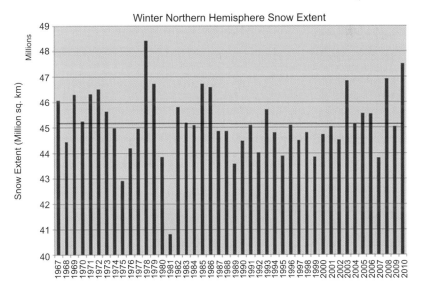

FIGURE 14   Winter Northern Hemisphere snow extent (Rutgers University Global Snow Lab).

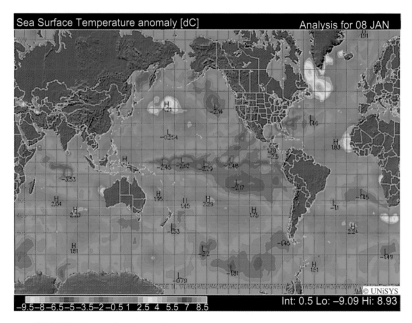

FIGURE 15   UNISYS sea surface temperature (SST) anomalies, January 8, 2011.

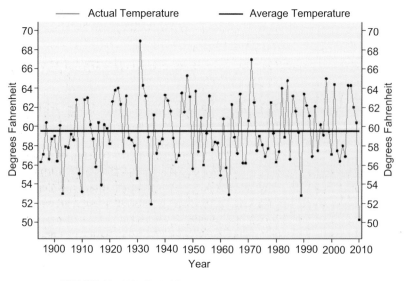

FIGURE 16   NCDC Florida December temperatures 1890–2010.

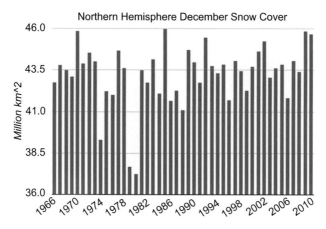

FIGURE 17 Northern Hemisphere December snow cover 1966–2010 (Rutgers University Global Snow Lab).

Like last year, winter 2010–2011 is again bringing unusually cold weather across much of the planet. Sea surface temperatures are running below normal across vast swaths of the ocean. Recent satellite data show that global temperatures are well below the 30-year mean. This is not consistent with "record heat" (Fig. 15).

December 2010 was the second coldest out of 353 years in England and the coldest on record in the UK. Florida also had their coldest December on record in 2010, as shown in Fig. 16.

The current winter has also seen near record snow across much of Asia, Europe, and North America. December had the fourth largest Northern Hemisphere December snow extent ever measured (Fig. 17).

5. WHAT CAUSED THE POSITIVE ANOMALIES IN EARLY 2010?

GISS showed a large spike in temperatures in March. Was this actually due to "hot" weather? The map in Fig. 18 is the RSS satellite temperature anomaly map for March, and it shows some interesting features.

There is an anomalously warm region in Northern Canada and Western Greenland, where temperatures were running about −20 °C, 5° warmer than the normal −25 °C. This was due to a negative Arctic Oscillation, which brings warmer than normal temperatures to portions of Canada — and cold weather to Europe, Russia, the United States, and much of Asia.

Is −20 °C (−4 °F) hot? Not exactly. GISS extrapolated the anomalous "warmth" across the entire Arctic Map (Fig. 19) and reported a large March temperature spike which was inconsistent with other data sets. GISS also somehow missed the cold weather in Europe and much of Russia.

MSU/AMSU Channel TLT Brightness Temperature Anomaly, March, 2010

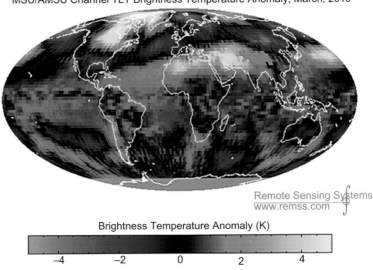

Brightness Temperature Anomaly (K)

FIGURE 18   RSS temperature anomalies, March, 2010.

Once again we see that Hansen's 0.01 record temperature is based on suspect data.

Another important factor in the warmth of 2010 was a strong El Niño, which dominated the first half of the year. This caused a short-term spike in temperatures which should not be interpreted as a trend (Fig. 20).

March 2010   L–OTI(°C) Anomaly vs 1951–1980   .84

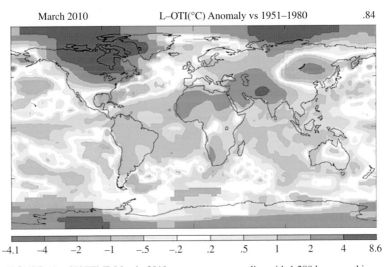

FIGURE 19   GISTEMP March, 2010 temperature anomalies with 1,200 km smoothing.

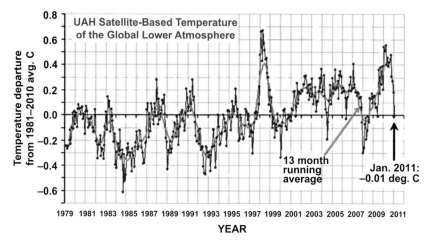

FIGURE 20 UAH monthly temperature anomalies 1978–2010 (http://www.drroyspencer.com/).

6. SOME HISTORY ABOUT DR. JAMES HANSEN

Dr. Hansen has a long history of exaggerated warming forecasts.

*Hansen said that the average U.S. temperature had risen from one to two degrees since 1958 and is **predicted to increase an additional 3 or 4 degrees sometime between 2010 and 2020.** The Press-Courier (Milwaukee) June 11 1986*

"Within 15 years (before 2001)," said Goddard Space Flight Center honcho James Hansen, "global temperatures will rise to a level which hasn't existed on earth for 100,000 years". The News and Courier, June 17th 1986

Pursuing present plans for coal and oil, Hansen found, the climate in the middle of the 21st century "would approach the warmth of the age of the dinosaurs." The Leader-Post, January 9th, 1982

(Hansen's 1988 forecast for Manhattan in 2008) "The West Side Highway [which runs along the Hudson River] will be under water. And there will be tape across the windows across the street because of high winds. And the same birds won't be there. The trees in the median strip will change." Then he said, "There will be more police cars." Why? "Well, you know what happens to crime when the heat goes up."

http://dir.salon.com/books/int/2001/10/23/weather/

[Since Dr. Hansen made that prediction, NOAA tide gauges show less than one half inch rise in sea level near Manhattan.]

"How far can it go? The last time the world was three degrees warmer than today — which is what we expect later this century — sea levels were 25m higher."

http://www.climateimc.org

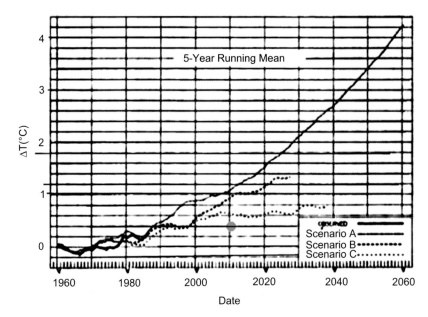

FIGURE 21   Hansen's 1988 temperature forecasts for various emission scenarios (A, B, C).

Temperatures have not come close to what Dr. Hansen formally forecast in 1988 — Fig. 21. The red dot is the GISS December anomaly. (January will likely be even lower.) The top line is where Hansen forecast we should be under a high-emissions scenario. The bottom dashed line is where he forecast we would be if CO_2 emissions were cut dramatically. Yet CO_2 emissions have greatly increased, but temperatures have not.

"Scenario A assumes continued exponential trace gas growth, scenario B assumes a reduced linear growth of trace gases, and scenario C assumes a rapid curtailment of trace gas emissions such that the net climate forcing ceases to increase after the year 2000."

Manhattan is not disappearing under water, but Dr. Hansen continues to project similar bloated forecasts into the future.

7. CONCLUSION

The 2010 record temperature claim is not scientifically supportable for many reasons — rather it is a global warming marketing bullet. 2011 is starting out as a very cold year in the U.S. and across much of the rest of the planet — particularly Asia. Chances are 2011 will be one of the coolest years in recent memory. It will be interesting to see what claims will be made by Hansen this year (Fig. 22).

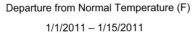

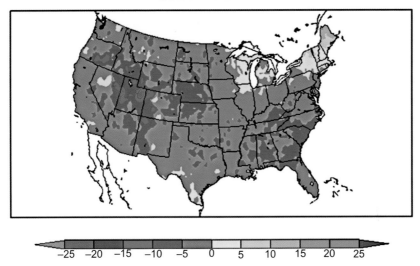

Generated 1/16/2011 at HPRCC using provisional data.   Regional Climate Centers

FIGURE 22   Temperature anomaly map January 1 to January 15, 2011 (High Plains Regional Climate Center).

Dr. Hansen functions in the roles of global warming head coach, cheerleader, referee, and scorekeeper. Temperature measurements need to be made by neutral third parties in the global warming debate.

Part III

The Role of Oceans

Relationship of Multidecadal Global Temperatures to Multidecadal Oceanic Oscillations

Joseph D'Aleo[*] and Don Easterbrook[†]

[*] Icecap US, 18 Glen Drive, Hudson, NH 03051, USA

[†] Department of Geology, Western Washington University, Bellingham, WA 98225

Chapter Outline

1. INTRODUCTION

The sun and ocean undergo regular changes on regular and predictable time frames. Temperatures likewise have exhibited changes that are cyclical. Sir Gilbert Walker was generally recognized as the first to find large-scale

Evidence-Based Climate Science. DOI: 10.1016/B978-0-12-385956-3.10005-1

oscillations in atmospheric variables. As early as 1908, while on a mission to explain why the Indian monsoon sometimes failed, he assembled global surface data and did a thorough correlation analysis.

On careful interpretation of statistical data, Walker and Bliss (1932) were able to identify three pressure oscillations:

1. A flip flop on a big scale between the Pacific Ocean and the Indian Ocean which he called the Southern Oscillation (SO).
2. A second oscillation, on a much smaller scale, between the Azores and Iceland, which he named the North Atlantic Oscillation.
3. A third oscillation between areas of high and low pressure in the North Pacific, which Walker called the North Pacific Oscillation.

Walker further asserted that the SO is the predominant oscillation, which had a tendency to persist for at least 1—2 seasons. He went so far in 1924 as to suggest the SOI had global weather impacts and might be useful in predicting the world's weather. He was ridiculed by the scientific community at the time for these statements. Not until four decades later was the Southern Oscillation recognized as a coupled atmosphere pressure and ocean temperature phenomena (Bjerknes, 1969) and more than two decades further before it was shown to have statistically significant global impacts and could be used to predict global weather/climate, at times many seasons in advance. Walker was clearly a man ahead of his time.

Global temperatures, ocean-based teleconnections, and solar variances interrelate with each other. A team of mathematicians (Tsonis et al., 2003, 2007), led by Dr. Anastasios Tsonis, developed a model suggesting that known cycles of the Earth's oceans—the Pacific Decadal Oscillation, the North Atlantic Oscillation, El Nino (Southern Oscillation), and the North Pacific Oscillation—all tend to synchronize with each other. The theory is based on a branch of mathematics known as Synchronized Chaos. The model predicts the degree of coupling to increase over time, causing the solution to "bifurcate", or split. Then, the synchronization vanishes. The result is a climate shift. Eventually the cycles begin to synchronize again, causing a repeating pattern of warming and cooling, along with sudden changes in the frequency and strength of El Nino events. They show how this has explained the major shifts that have occurred including 1913, 1942, and 1978. These may be in the process of synchronizing once again with its likely impact on climate very different from what has been observed over the last several decades.

2. THE SOUTHERN OSCILLATION INDEX (SOI)

The Southern Oscillation Index (SOI) is the oldest measure of large-scale fluctuations in air pressure occurring between the western and eastern tropical Pacific (i.e., the state of the Southern Oscillation) during El Nino and La Nina episodes (Walker et al., 1932). Traditionally, this index has been calculated based on the differences in air pressure anomaly between Tahiti and Darwin, Australia. In general, smoothed time series of the SOI correspond very well with changes in ocean temperatures across the eastern tropical Pacific. The negative phase of the SOI represents below-normal air pressure at Tahiti and above-normal air pressure at Darwin. Prolonged periods of negative SOI values coincide with abnormally warm ocean waters across the eastern tropical Pacific typical of El Nino episodes. Prolonged periods of positive SOI values coincide with abnormally cold ocean waters across the eastern tropical Pacific typical of La Nina episodes.

As an atmospheric observation-based measure, SOI is subjected not only to underlying ocean temperature anomalies in the Pacific but also intra-seasonal oscillations, like the Madden–Julian Oscillation (MJO). The SOI often shows month-to-month-swings, even if the ocean temperatures remain steady due to these atmospheric waves. This is especially true in weaker El Nino or La Nina events, as well as La Nadas (neutral ENSO) conditions. Indeed, even week-to-week changes can be significant. For that reason, other measures are often preferred.

2.1. Nino 3.4 Anomalies

On February 23, 2005, NOAA announced that the NOAA National Weather Service, the Meteorological Service of Canada and the National Meteorological Service of Mexico reached a consensus on an index and definitions for El Nino and La Nina events (also referred to as the El Nino Southern Oscillation or ENSO). Canada, Mexico, and the United States all experience impacts from El Nino and La Nina.

The index was called the ONI and is defined as a 3-month average of sea surface temperature departures from normal for a critical region of the equatorial Pacific (Niño 3.4 region; 120W–170W, 5N–5S). This region of the tropical Pacific contains what scientists call the "equatorial cold tongue", a band of cool water that extends along the equator from the coast of South America to the central Pacific Ocean. North America's operational definitions for El Nino and La Nina, based on the index, are:

El Nino: A phenomenon in the equatorial Pacific Ocean characterized by a positive sea surface temperature departure from normal (for the 1971–2000 base period) in the Niño 3.4 region greater than or equal in magnitude to 0.5 °C (0.9 °F), averaged over three consecutive months.
La Nina: A phenomenon in the equatorial Pacific Ocean characterized by a negative sea surface temperature departure from normal (for the 1971–2000 base period) in the Niño 3.4 region greater than or equal in magnitude to 0.5 °C (0.9 °F), averaged over three consecutive months.

3. MULTIVARIATE ENSO INDEX (MEI)

Wolter (1987) attempted to combine oceanic and atmospheric variables to track and compare ENSO events. He developed the Multivariate ENSO Index (MEI) using the six main observed variables over the tropical Pacific. These six variables are: sea-level pressure (P), zonal (U), and meridional (V) components of the surface wind, sea surface temperature (S), surface air temperature (A), and total cloudiness fraction of the sky (C).

The MEI is calculated as the first unrotated Principal Component (PC) of all six observed fields combined. This is accomplished by normalizing the total variance of each field first, and then performing the extraction of the first PC on the co-variance matrix of the combined fields (Wolter and Timlin, 1993).

In order to keep the MEI comparable, all seasonal values are standardized with respect to each season and to the 1950–1993 reference period. Negative values of the MEI represent the cold ENSO phase (La Nina) while positive MEI values represent the warm ENSO phase (El Nino). Figure 2 is a plot of the three indices since 2000 (Wolter and Timlin, 1993).

NINO 34 is well correlated with the MEI. The SOI is much more variable month-to-month than the MEI and NINO 34. The MEI and NINO are more reliable determinants of the true state of ENSO, especially in weaker ENSO events.

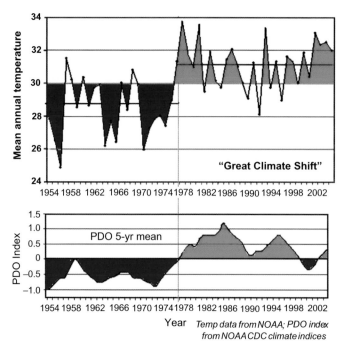

FIGURE 1 Correlation of the Great Pacific Climate Shift and the Pacific Decadal Oscillation.

4. THE PACIFIC DECADAL OSCILLATION (PDO)

The first hint of a Pacific basin-wide cycle was the recognition of a major regime change in the Pacific in 1977 that became to known as the Great Pacific Climate Shift (Fig. 1). Later, this shift was shown to be part of a cyclical regime change with decadal-like ENSO variability (Zhang et al., 1996, 1997; Mantua et al., 1997) and given the name Pacific Decadal Oscillation (PDO) by fisheries scientist Steven Hare (1996) while researching connections between Alaska salmon production cycles and Pacific climate.

Mantua et al. (1997) found the "Pacific Decadal Oscillation" (PDO) is a long-lived El Nino-like pattern of Pacific climate variability. While the two climate oscillations have similar spatial climate fingerprints, they have very different behavior in time. Two main characteristics distinguish PDO from El Nino/Southern Oscillation (ENSO): (1) 20th century PDO "events" persisted for 20-to-30 years, while typical ENSO events persisted for 6—18 months; (2) the climatic fingerprints of the PDO are most visible in the North Pacific/North American sector, while secondary signatures exist in the tropics — the opposite is true for ENSO. Note in Figures 1 and 2 how CO_2 showed no change during this PDO shift, suggesting it was unlikely to have played a role. Figures 3 and 4 show average annual PDO values.

A study by Gershunov and Barnett (1998) showed that the PDO has a modulating effect on the climate patterns resulting from ENSO. The climate signal of El Nino is likely to be stronger when the PDO is highly positive;

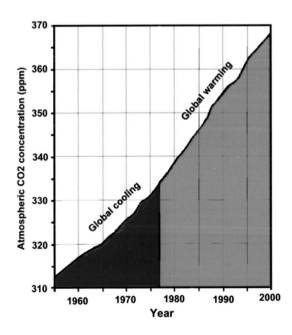

FIGURE 2 Atmospheric CO_2 showed no change across the Great Pacific Shift so could not have been the cause of it.

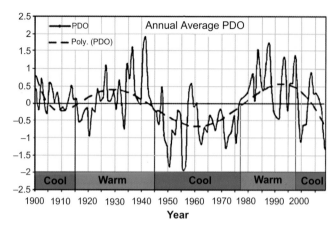

FIGURE 3   Annual average PDO 1900–2009. Note the multidecadal nature of the cycle with a period of approximately 60 years.

conversely the climate signal of La Nina will be stronger when the PDO is highly negative. This does not mean that the PDO physically controls ENSO, but rather that the resulting climate patterns interact with each other. The annual PDO and ENSO (Multivariate ENSO Index) track well since 1950.

5. FREQUENCY AND STRENGTH OF ENSO AND THE PDO

Warm PDOs are characterized by more frequent and stronger El Ninos than La Ninas. Cold PDOs have the opposite tendency. Figure 4 shows how well one

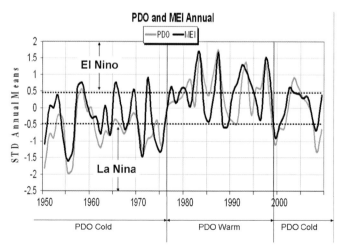

FIGURE 4   Annual average PDO and MEI (Multivariate ENSO Index) from 1950 to 2007 clearly correlate well. Note how the ENSO events amplify or diminish the favored PDO state.

ENSO measure, Wolter's MEI, correlates with the PDO. Mclean et al. (2009) showed that the mean monthly global temperature (GTTA) using the University of Alabama Huntsville MSU temperatures corresponds in general terms with the another ENSO measure, the Southern Oscillation Index (SOI) of 7 months earlier. The SOI is a rough indicator of general atmospheric circulation and thus global climate change.

Temperatures also follow suit (Fig. 5). El Ninos and the warm mode PDOs have similar land-based temperature patterns, as do cold-mode PDOs and La Ninas.

Strong similarity exists between PDO and ENSO ocean basin patterns. Land temperatures also are very similar between the PDO warm modes and El Ninos and the PDO cold modes and La Ninas. Not surprisingly, El Ninos occur more frequently during the PDO warm phase and La Ninas during the PDO cold phase. It maybe that ocean circulation shifts drive it for decades favoring El Ninos which leads to a PDO warm phase or La Ninas and a PDO cold phase (the proverbial chicken and egg), but the 60-year cyclical nature of this cycle is well established (Fig. 6).

About 1947, the PDO (Pacific Decadal Oscillation) switched from its warm mode to its cool mode and global climate cooled from then until 1977, despite the sudden soaring of CO_2 emissions. In 1977, the PDO switched back from its cool mode to its warm mode, initiating what is regarded as 'global warming' from 1977 to 1998 (Fig. 7).

During the past century, global climates have consisted of two cool periods (1880–1915 and 1945–1977) and two warm periods (1915–1945 and 1977–1998). In 1977, the PDO switched abruptly from its cool mode, where it had been since about 1945, into its warm mode and global climate shifted from cool to warm (Miller et al., 1994). This rapid switch from cool to warm has become to known as "The Great Pacific Climatic Shift" (Fig. 1). Atmospheric

Temperature Correlations

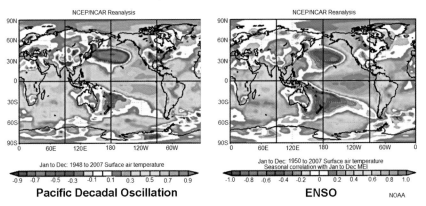

FIGURE 5 PDO and ENSO compared.

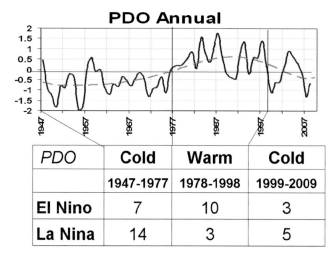

FIGURE 6 Note how during the PDO cold phases, La Nina dominate (14–7 in the 1947–1977 cold phase) and 5–3 in the current, while in the warm phase from 1977 to 1998, the El Ninos had a decided frequency advantage of 10–3.

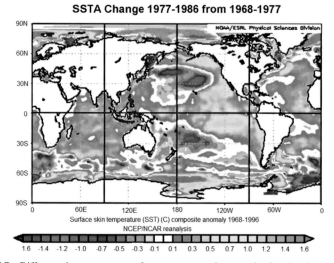

FIGURE 7 Difference in average sea surface temperatures between the decade prior to the GPCS and the decade after the GPCS. Yellow and green colors indicate warming of the NE Pacific off the coast of North America relative to what it had been from 1968 to 1977. Note the cooling in the west central North Pacific.

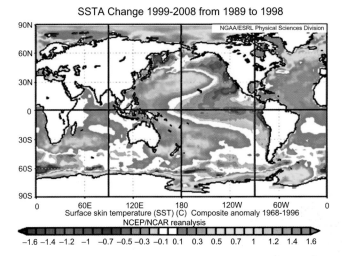

SSTA Change 1999-2008 from 1989 to 1998

FIGURE 8   Sea surface temperature difference image of the decade after the GPCS minus the decade before the GPCS. Note the strong cooling in the eastern Pacific and the warming of the west central North Pacific.

CO_2 showed no unusual changes across this sudden climate shift and was clearly not responsible for it. Similarly, the global warming from ~1915 to ~1945 could not have been caused by increased atmospheric CO_2 because that time preceded the rapid rise of CO_2, and when CO_2 began to increase rapidly after 1945, 30 years of global cooling occurred (1945–1977). The two warm and two cool PDO cycles during the past century (Fig. 3) have periods of about 25–30 years.

The PDO flipped back to the cold mode in 1999. The change can be seen with this sea surface temperature difference image of the decade after the GPCS minus the decade before the GPCS (Fig. 8).

Verdon and Franks (2006) reconstructed the positive and negative phases of PDO back to A.D. 1662 based on tree ring chronologies from Alaska, the Pacific Northwest, and subtropical North America as well as coral fossil from Rarotonga located in the South Pacific. They found evidence for this cyclical behavior over the whole period (Fig. 9).

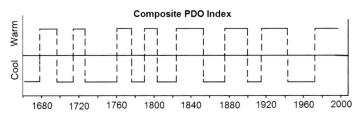

Composite PDO Index

FIGURE 9   Verdon and Franks (2006) reconstructed the PDO back to 1662 showing cyclical behavior over the whole period.

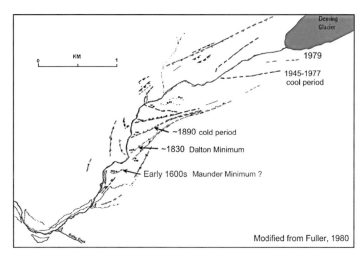

FIGURE 10 Ice marginal deposits (moraines) showing fluctuations of the Deming glacier, Mt. Baker, WA corresponding to climatic warming and cooling in Greenland ice cores.

6. CORRELATION OF THE PDO AND GLACIAL FLUCTUATIONS IN THE PACIFIC NORTHWEST

The ages of moraines downvalley from the present Deming glacier on Mt. Baker (Fuller, 1980; Fuller et al., 1983) match the ages of the cool periods in the Greenland ice core. Because historic glacier fluctuations (Harper, 1993) coincide with global temperature changes and PDO, these earlier glacier fluctuations could also well be due to oscillations of the PDO (Fig. 10).

Glaciers on Mt. Baker, WA show a regular pattern of advance and retreat (Fig. 11) which matches the Pacific Decadal Oscillation (PDO) in the NE Pacific Ocean. The glacier fluctuations are clearly correlated with, and probably driven by, changes in the PDO. An important aspect of this is that the PDO record extends to the about 1900 but the glacial record goes back many years and can be used as a proxy for older climate changes.

7. ENSO VS. TEMPERATURES

Douglass and Christy (2008) compared the NINO 34 region anomalies to the tropical UAH lower troposphere and showed a good agreement, with some departures during periods of strong volcanism. During these volcanic events, high levels of stratospheric sulfate aerosols block incoming solar radiation and produce multi-year cooling of the atmosphere and oceans. A similar comparison of UAH global lower tropospheric data with the MEI Index also shows good agreement, with some departure during periods of major

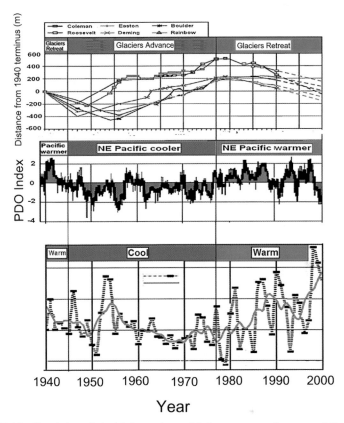

FIGURE 11 Correlation of glacial fluctuations, global temperature changes, and the Pacific Decadal Oscillation.

volcanism in the early 1980s and 1990s. Alaskan temperatures clearly show discontinuities associated with changes in the PDO.

8. THE ATLANTIC MULTIDECADAL OSCILLATION (AMO)

Like the Pacific, the Atlantic exhibits multidecadal tendencies and a characteristic tripole structure (Figs. 12, 13). For a period that averages about 30 years, the Atlantic tends to be in what is called the warm phase with warm temperatures in the tropical North Atlantic and far North Atlantic and relatively cool temperatures in the central (west central). Then the ocean flips into the opposite (cold) phase with cold temperatures in the tropics and far North Atlantic and a warm central ocean. The AMO (Atlantic sea surface temperatures standardized) is the average anomaly standardized from 0 to 70N. The AMO has a period of 60 years maximum to maximum and minimum to minimum.

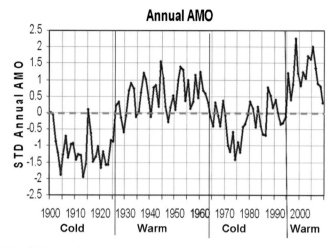

FIGURE 12   AMO annual mean (STD) showing a similar 60−70-year cycle as the PDO but with a lag of about 15 years to the PDO.

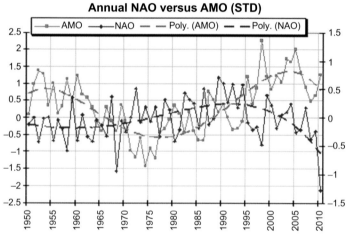

FIGURE 13   Annual Average AMO and NAO compared. Note the inverse relationship with a slight lag of the NAO to the AMO.

9. NORTH ATLANTIC OSCILLATION, THE ARCTIC OSCILLATION, AND THE AMO

The North Atlantic Oscillation (NAO), first found by Walker in the 1920s, is the north−south flip flop of pressures in the eastern and central North Atlantic (Walker and Bliss, 1932). The difference of normalized MSLP anomalies between Lisbon, Portugal, and Stykkisholmur, Iceland has become the widest used NAO index and extends back in time to 1864 (Hurrell, 1995), and to 1821 if Reykjavik is used instead of Stykkisholmur and Gibraltar instead of Lisbon (Jones et al., 1997). Hanna et al. (2003) and Hanna et al. (2006) showed how these cycles in the Atlantic sector play a key role in temperature variations in Greenland and Iceland. Kerr (2000) identified the NAO and AMO (Fig. 13) as key climate pacemakers for large-scale climate variations over the centuries.

The Arctic Oscillation (also known as the Northern Annular Mode Index (NAM)) is defined as the amplitude of the pattern defined by the leading empirical orthogonal function of winter monthly mean NH MSLP anomalies poleward of 20°N (Thompson and Wallace, 1998, 2000). The NAM/Arctic Oscillation (AO) is closely related to the NAO.

Like the PDO, the NAO and AO tend to be predominantly in one mode or the other for decades at a time, although since, like the SOI, it is a measure of atmospheric pressure and subject to transient features, it tends to vary much more from week-to-week and month-to-month. All we can state is that an inverse relationship exists between the AMO and NAO/AO decadal tendencies. When the Atlantic is cold (AMO negative), the AO and NAO tend more often to the positive state, when the Atlantic is warm, on the other hand, the NAO/AO tend to be more often negative. The AMO tri-pole of warmth in the 1960s below was associated with a predominantly negative NAO and AO while the cold phase was associated with a distinctly positive NAO and AO in the 1980s and early 1990s (Figs. 14, 15). A lag of a few years occurs after the flip of the AMO and the tendencies appear to be

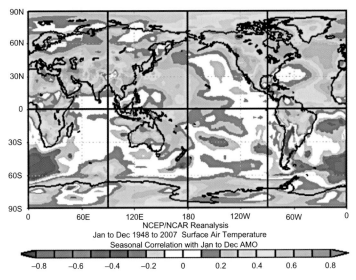

FIGURE 14 Correlation of the AMO with annual surface temperatures.

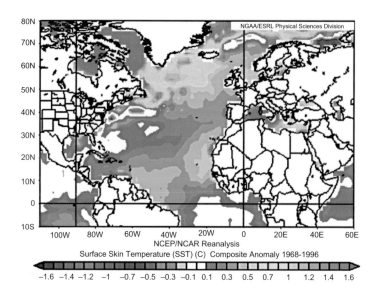

FIGURE 15   Difference in sea surface temperatures 1996–2004 from 1986 to 1995. It shows the evolution to the warm Atlantic Multidecadal Oscillation.

greatest at the end of the cycle. This may relate to timing of the maximum warming or cooling in the North Atlantic part of the AMO or even the PDO/ENSO interactions. The PDO typically leads the AMO by 10–15 years. The relationship is a little more robust for the cold (negative AMO) phase than for the warm (positive) AMO. There tends to be considerable intra-seasonal variability of these indices that relate to other factors (stratospheric warming and cooling events that are correlated with the Quasi-Biennial Oscillation or QBO for example).

10. SYNCHRONIZED DANCE OF THE TELECONNECTIONS

The record of natural climate change and the measured temperature record during the last 150 years gives no reason for alarm about dangerous warming caused by human CO_2 emissions. Predictions based on past warming and cooling cycles over the past 500 years accurately predicted the present cooling phase (Easterbrook, 2001, 2005, 2006a,b, 2007, 2008a,b,c) and the establishment of cool Pacific sea surface temperatures confirms that the present cool phase will persist for several decades.

Latif and his colleagues at Leibniz Institute at Germany's Kiel University predicted the new cooling trend in a paper published in 2009 and warned of it again at an IPCC conference in Geneva in September 2009.

'A significant share of the warming we saw from 1980 to 2000 and at earlier periods in the 20th Century was due to these cycles — perhaps as much as 50 per cent. They have now gone into reverse, so winters like this one will become much more likely. Summers will also probably be cooler, and all this may well last two decades or longer. The extreme retreats that we have seen in glaciers and sea ice will come to a halt. For the time being, global warming has paused, and there may well be some cooling.'

According to Latif and his colleagues (Latif and Barnett, 1994; Latif et al., 2009) this in turn relates to much longer-term shifts — what are known as the Pacific and Atlantic 'multi-decadal oscillations' (MDOs). For Europe, the crucial factor here is the temperature of the water in the middle of the North Atlantic, now several degrees below its average when the world was still warming.

Prof. Anastasios Tsonis, head of the University of Wisconsin Atmospheric Sciences Group, has shown (2007) that these MDOs move together in a synchronized way across the globe, abruptly flipping the world's climate from a 'warm mode' to a 'cold mode' and back again in 20—30-year cycles.

'They amount to massive rearrangements in the dominant patterns of the weather,' he said yesterday, 'and their shifts explain all the major changes in world temperatures during the 20th and 21st Centuries. We have such a change now and can therefore expect 20 or 30 years of cooler temperatures.'

The period from 1915 to 1940 saw a strong warm mode, reflected in rising temperatures, but from 1940 until the late 1970s, the last MDO cold-mode era, the world cooled, despite the fact that carbon dioxide levels in the atmosphere continued to rise. Many of the consequences of the recent warm mode were also observed 90 years ago. For example, in 1922, the Washington Post reported that Greenland's glaciers were fast disappearing, while Arctic seals were 'finding the water too hot'. The Post interviewed Captain Martin Ingebrigsten, who had been sailing the eastern Arctic for 54 years: 'He says that he first noted warmer conditions in 1918, and since that time it has gotten steadily warmer. Where formerly great masses of ice were found, there are now moraines, accumulations of earth and stones. At many points where glaciers formerly extended into the sea they have entirely disappeared. As a result, the shoals of fish that used to live in these waters had vanished, while the sea ice beyond the north coast of Spitsbergen in the Arctic Ocean had melted. Warm Gulf Stream water was still detectable within a few hundred miles of the Pole.'

In contrast, 56% of the surface of the United States was covered by snow. 'That hasn't happened for several decades,' Tsonis pointed out. 'It just isn't true to say this is a blip. We can expect colder winters for quite a while.' He recalled that towards the end of the last cold mode, the world's media were preoccupied by fears of freezing. For example, in 1974, a Time magazine cover story predicted 'Another Ice Age', saying: 'Man may be somewhat responsible — as a result of farming and fuel burning [which is] blocking more and more sunlight from reaching and heating the Earth.'

PDO+AMO vs USHCN V2 Annual Temp

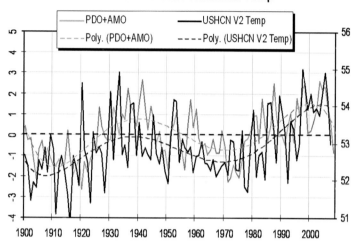

FIGURE 16   NASA GISS version of NCDC USHCN version 2 vs. PDO + AMO. The mutlide-cadal cycles with periods of 60 years match the USHCN warming and cooling cycles. Annual temperatures end at 2007. With an 11-year smoothing of the temperatures and PDO + AMO to remove any effect of the 11-year solar cycles, gives an even better correlation with an r^2 of 0.85.

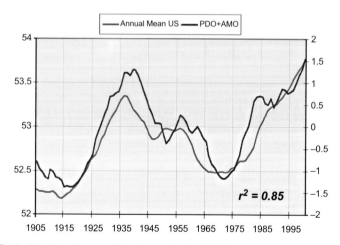

FIGURE 17   With 22 point smoothing, the correlation of U.S. temperatures and the ocean multi-decadal oscillations is clear with an r^2 of 0.85. Figure 18 shows the AMO/PDO regression fit to USHCN version 2. The PDO/AMO works well in predicting temperatures (Fig. 19). Figure 20 shows the difference in U.S. annual mean temperatures for USHCN version 2 minus USHCN version 1.

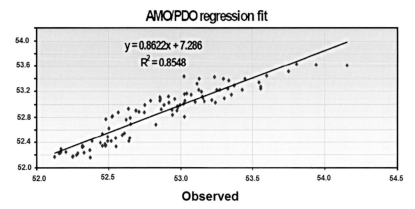

FIGURE 18   The AMO/PDO regression fit to USHCN version 2.

Tsonis observed 'Perhaps we will see talk of an ice age again by the early 2030s, just as the MDOs shift once more and temperatures begin to rise.' Although the two indices (PDO and AMO) are derived in different ways, they both represent a pattern of sea surface temperatures, a tripole with warm in the high latitudes and tropics and colder in between especially west or vice versa. In both cases, the warm modes were characterized by general global warmth and the cold modes with general broad climatic cooling though each with though with regional variations. I normalizing and adding the two indices makes them more comparable. A positive AMO + PDO should correspond to an above normal temperature and the negative below normal. Indeed that is the case for the US temperatures (NCDC USHCN v2) as shown in Fig. 16.

Correlation of U.S. temperatures and the ocean multidecadal oscillations gives an r^2 of 0.85 (Fig. 17). In Figures 18 and 19 the AMO/PDO was used to predict US temperatures using multiple regression approach. The results showed excellent results with some divergence near the end of the period.

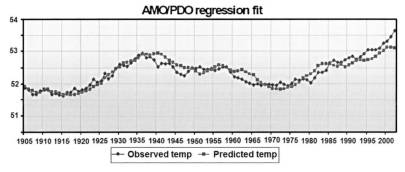

FIGURE 19   Using the PDO/AMO to predict temperatures works well here with some departure after around 2000.

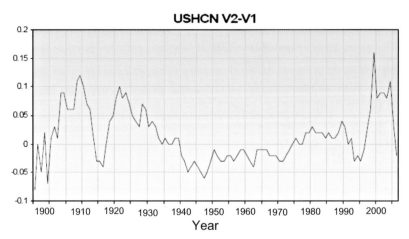

FIGURE 20 The difference in U.S. annual mean temperatures for USHCN version 2 minus USHCN version 1. The elimination of the urbanization adjustment led to a hard-to-explain spike in the 1997−2005 time period.

The plot (Fig. 20) of the difference between version 1 and version 2 suggests the latter as the likely cause. In version 2, the urban adjustment was removed. Note that the upward adjustment of the 1998−2005 temperatures by as much as 0.15 °F is unexplained.

11. SHORT-TERM WARM/COOL CYCLES FROM THE GREENLAND ICE CORE

Variation of oxygen isotopes in ice from Greenland ice cores is a measure of temperature. Most atmospheric oxygen consists of 16O but a small amount consists of 18O, an isotope of oxygen that is somewhat heavier. When water vapor (H_2O) condenses from the atmosphere as snow, it contains a ratio of $^{16}O/^{18}O$ that reflects the temperature at that time. When snow falls on a glacier and is converted to ice, it retains an isotopic 'fingerprint' of the temperature conditions at the time of condensation. Measurement of the $^{16}O/^{18}O$ ratios in glacial ice hundreds or thousands of years old allows reconstruction of past temperature conditions (Stuiver and Grootes, 2000; Stuiver and Brasiunas, 1991, 1992; Grootes and Stuiver, 1997; Stuiver et al., 1995; Grootes et al., 1993). High resolution ice core data show that abrupt climate changes occurred in only a few years (Steffensen et al., 2008). The GISP2 ice core data of Stuiver and Grootes (2000) can be used to reconstruct temperature fluctuations in Greenland over the past 500 years (Fig. 21). Figure 21 shows a number of well-known climatic events. For example, the isotope record shows the Maunder Minimum, the Dalton Minimum, the 1880−1915 cool period, the 1915 to ~1945 warm period, and the ~1945 to 1977 cool period, as well as many other cool and warm periods. Temperatures fluctuated between warm and cool at least 22 times

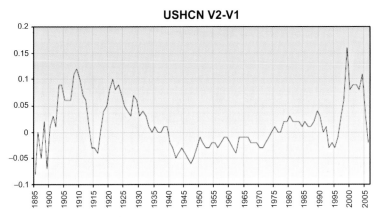

FIGURE 21 Cyclic warming and cooling trends in the past 500 years (plotted from GISP2 data, Stuiver and Grootes, 2000).

between 1480 A.D. and 1950 (Fig. 21). None of the warming periods could have possibly been caused by increased CO_2 because they all preceded rising CO_2.

Only one out of all of the global warming periods in the past 500 years occurred at the same time as rising CO_2 (1977−1998). About 96% of the warm periods in the past 500 years could not possibly have been caused by rise of CO_2. The inescapable conclusion of this is that CO_2 is not the cause of global warming. The Greenland ice core isotope record matches climatic fluctuations recorded in alpine glacier advances and retreats.

12. WHERE ARE WE HEADED DURING THE COMING CENTURY?

The cool phase of PDO is now entrenched. We have shown how the two ocean oscillations drive climate shifts. The PDO leads the way and its effect is later amplified by the AMO. Each of this has occurred in the past century, global temperatures have remained cool for about 30 years.

No statistically significant global warming has taken place since 1998 (Fig. 22), and cooling has occurred during the past several years (Hanna and Cappelen, 2003). A very likely reason for global cooling over the past decade is the switch of the Pacific Ocean from its warm mode (where it has been from 1977 to 1998) to its cool mode in 1999. Each time this has occurred in the past century, global temperatures have remained cool for about 30 years. Thus, the current sea surface temperatures not only explain why we have had stasis or global cooling for the past 10 years, but also should assure that cooler temperatures will continue for several more decades. There will be brief bounces upwards with periodic El Ninos, as we have seen in late 2009 and early 2010, but they will give way to cooling as the favored La Nina states returns. With a net La Nina tendency, the net result should be cooling.

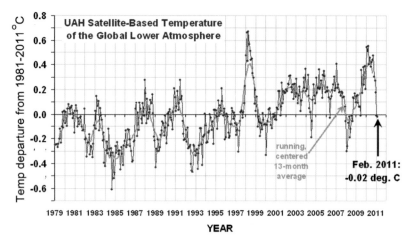

FIGURE 22   UAH globally averaged lower atmospheric temperatures.

12.1. Predictions Based on Past Climate Patterns

The past is the key to understanding the future. Past warming and cooling cycles over the past 500 years were used by Easterbrook (2001, 2005, 2006a,b, 2007, 2008a,b,c) to accurately predict the cooling phase that is now happening. Establishment of cool Pacific sea surface temperatures since 1999 indicates that the cool phase will persist for the next several decades. We can look to past natural climatic cycles as a basis for predicting future climate changes. The climatic fluctuations over the past few hundred years suggest ~30-year climatic cycles of global warming and cooling, on a general warming trend from the Little Ice Age cool period. If the trend continues as it has for the past several centuries, global temperatures for the coming century might look like those in Fig. 23. The

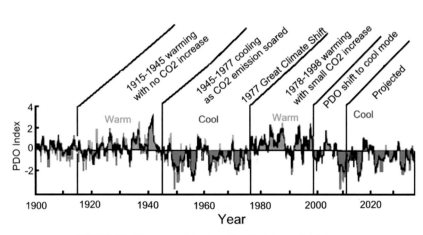

FIGURE 23   Using past behavior of the PDO to predict future.

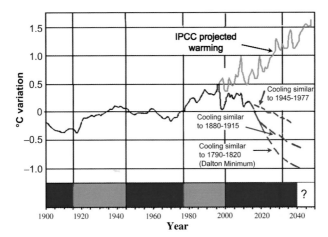

FIGURE 24 projected climate for the century based on climatic patterns over the past 500 years and the switch of the PDO to its cool phase.

left side of Fig. 23 is the warming/cooling history of the past century. The right side of the graph shows that we have entered a global cooling phase that fits the historic pattern very well. The switch to the PDO cool mode to its cool mode virtually assures cooling global climate for several decades.

Three possible projections are shown in Fig. 24: (1) moderate cooling (similar to the 1945−1977 cooling); (2) deeper cooling (similar to the 1945−1977 cooling); or (3) severe cooling (similar to the 1790−1830 cooling). Only time will tell which of these will be the case, but at the moment, the sun is behaving very similar to the Dalton Minimum (sunspot cycles 4/5), which was a very cold time. This is based on the similarity of sun spot cycle 23 to cycle 4 (which immediately preceded the Dalton Minimum).

As the global climate and solar variation reveals themselves in a way not seen in the past 200 years, we will surely attain a much better understanding of what causes global warming and cooling. Time will tell. If the climate continues its ocean cycle cooling and the sun behaves in a manner not witnessed since 1800, we can be sure that climate changes are dominated by the sun and sea and that atmospheric CO_2 has a very small role in climate changes. If the same climatic patterns, cyclic warming and cooling, that occurred over the past 500 years continue, we can expect several decades of moderate to severe global cooling.

REFERENCES

Baldwin, M.P., Dunkerton, T.J., 2005. The solar cycle and stratospheric−tropospheric dynamical coupling. Journal of Atmospheric and Solar−Terrestrial Physics 67, 71−82.

Bjerknes, J., 1969. Atmospheric teleconnections from the equatorial Pacific. Monthly Weather Review 97, 153−172.

Boberg, F., Lundstedt, H., 2002. Solar wind variation related to fluctuations of the North Atlantic Oscillation. Geophysical Research Letters 29, 13.1–13.4.

Christy, J.R., Spencer, R.W., Braswell, W.D., 2000. MSU tropospheric temperatures: dataset construction and radiosonde comparisons. Journal of Atmospheric and Oceanic Technology 17, 1153–1170.

Clilverd, M.A., Clarke, E., Ulich, T., Rishbeth, H., Jarvis, M.J., 2006. Predicting Solar Cycle 24 and beyond. SpaceWeather 4.

Douglass, D.H., Christy, J.R., 2008. Limits on CO_2 climate forcing from recent temperature data of Earth. Energy and Environment.

Easterbrook, D.J., 2001. The next 25 years: global warming or global cooling? Geologic and oceanographic evidence for cyclical climatic oscillations. Geological Society of America, Abstracts with Program 33, 253.

Easterbrook, D.J., 2005. Causes and effects of abrupt, global, climate changes and global warming: Geological Society of America, Abstracts with Program 37, 41.

Easterbrook, D.J., 2006a. Causes of abrupt global climate changes and global warming predictions for the coming century: Geological Society of America, Abstracts with Program 38, 77.

Easterbrook, D.J., 2006b. The cause of global warming and predictions for the coming century: Geological Society of America, Abstracts with Program 38, 235–236.

Easterbrook, D.J., 2007. Geologic evidence of recurring climate cycles and their implications for the cause of global warming and climate changes in the coming century. Geological Society of America, Abstracts with Programs 39, 507.

Easterbrook, D.J., 2008a. Solar influence on recurring global, decadal, climate cycles recorded by glacial fluctuations, ice cores, sea surface temperatures, and historic measurements over the past millennium. Abstracts of American Geophysical Union Annual Meeting, San Francisco.

Easterbrook, D.J., 2008b. Implications of glacial fluctuations, PDO, NAO, and sun spot cycles for global climate in the coming decades: Geological Society of America, Abstracts with Programs 40, 428.

Easterbrook, D.J., 2008c. Correlation of climatic and solar variations over the past 500 years and predicting global climate changes from recurring climate cycles. Abstracts of 33rd International Geological Congress, Oslo, Norway.

Easterbrook, D.J., Kovanen, D.J., 2000. Cyclical oscillation of Mt. Baker glaciers in response to climatic changes and their correlation with periodic oceanographic changes in the northeast Pacific Ocean: Geological Society of America, Abstracts with Program 32, 17.

Fuller, S.R., 1980. Neoglaciation of Avalanche Gorge and the Middle Fork Nooksack River valley, Mt. Baker, Washington. M.S. thesis, Western Washington University, Bellingham, Washington.

Fuller, S.R., Easterbrook, D.J., Burke, R.M., 1983. Holocene glacial activity in five valleys on the flanks of Mt. Baker. Geological Society of America, Washington. Abstracts with Program 15, 43.

Gershunov, A., Barnett, T.P., 1998. Interdecadal modulation of ENSO teleconnections. Bulletin of the American Meteorological Society 79, 2715–2725.

Grootes, P.M., Stuiver, M., 1997. Oxygen 18/16 variability in Greenland snow and ice with 10^3 to 10^5-year time resolution. Journal of Geophysical Research 102, 26455–26470.

Grootes, P.M., Stuiver, M., White, J.W.C., Johnsen, S.J., Jouzel, J., 1993. Comparison of oxygen isotope records from the GISP2 and GRIP Greenland ice cores. Nature 366, 552–554.

Hanna, E., Cappelen, J., 2003. Recent cooling in coastal southern Greenland and relation with the North Atlantic Oscillation. Geophysical Research Letters 30.

Hanna, E., Jonsson, T., Olafsson, J., Valdimarsson, H., 2006. Icelandic coastal sea surface temperature records constructed: Putting the pulse on air–sea–climate interactions in the

Northern North Atlantic. Part I: Comparison with HadISST1 open-ocean surface temperatures and preliminary analysis of long-term patterns and anomalies of SSTs around Iceland. Journal of Climate 19, 5652–5666.

Hare, S.R., 1996. Low frequency climate variability and salmon production. PhD dissertation. School of Fisheries, University of Washington, Seattle, WA.

Harper, J.T., 1993. Glacier fluctuations on Mount Baker, Washington, U.S.A., 1940–1990, and climatic variations: Arctic and Alpine Research. Arctic and Alpine Research 4, 332–339.

Hathaway, D., 2006. Long range solar forecast, 2006 (Press release editor: Dr. Tony Phillips) http://science.nasa.gov/headlines/y2006/10may_longrange.htm.

Hurell, J.W., 1995. Decadal trends in the North Atlantic Oscillation, regional temperatures and precipitation. Science 269, 676–679.

Idso, C., Singer, S.F., 2009. Climate Change Reconsidered: 2009 Report of the Nongovernmental Panel on Climate Change (NIPCC). The Heartland Institute, Chicago, IL, p. 855.

IPCC-AR4, 2007. Climate Change: The Physical Science Basis. Contribution of Working Group I to the Fourth Assessment Report of the Intergovernmental Panel on Climate Change. Cambridge University Press.

Jones, P.D., Johnson, T., Wheeler, D., 1997. Extension in the North Atlantic Oscillation using early instrumental pressure observations from Gibralter and southwest Iceland. Journal of Climatology 17, 1433–1450.

Kerr, R.A., 2000. A North Atlantic climate pacemaker for the centuries,. Science 288, 1984–1986.

Labitzke, K., 2001. The global signal of the 11-year sunspot cycle in the stratosphere. Differences between solar maxima and minima: Meteorologische Zeitschift 10, 83–90.

Latif, M., Barnett, T.P., 1994. Causes of decadal climate variability over the North Pacific and North America. Science 266, 634–637.

Latif, M., Park, W., Ding, H., Keenlyside, N., 2009. Internal and external North Atlantic sector variability in the Kiel climate model. Meteorologische Zeitschrift 18, 433–443.

Mantua, N.J., Hare, S.R., Zhang, Y., Wallace, J.M., Francis, R.C., 1997. A Pacific interdecadal climate oscillation with impacts on salmon production: Bulletin of the American Meteorological Society 78, 1069–1079.

Marsh, N.D., Svensmark, H., 2000. Low cloud properties influenced by cosmic rays. Physical Review Letters 85, 5004–5007.

McLean, J.D., de Freitas, C.R., Carter, R.M., 2009. Influence of the Southern Oscillation on tropospheric temperature. Journal of Geophysical Research 114.

Miller, A.J., Cayan, D.R., Barnett, T.P., Graham, N.E., Oberhuber, J.M., 1994. The 1976–77 climate shift of the Pacific Ocean: Oceanography 7, 21–26.

Moore, G.W.K., Holdsworth, G., Alverson, K., 2002. Climate change in the North Pacific region over the past three centuries. Nature 420, 401–403.

Perlwitz, J., Hoerling, M., Eischeid, J., Xu, T., Kumar, A., 2009. A strong bout of natural cooling in 2008. Geophysical Research Letters 36.

Scafetta, N., West, B.J., 2007. Phenomenological reconstructions of the solar signature in the Northern Hemisphere. Journal of Geophysical Research 112, 5004–5007.

Shaviv, N.J., 2005. On climate response to changes in the cosmic ray flux and radiative budget. Journal of Geophysical Research–Space Physics 110, p. A08105.

Steffensen, J.P., Andersen, K.K., Bigler, M., Clausen, H.B., Dahl-Jensen, D., Goto-Azuma, K., Hansson, M.J., Sigfus, J., Jouzel, J., Masson-Delmotte, V., Popp, T., Rasmussen, S.O., Roethlisberger, R., Ruth, U., Stauffer, B., Siggaard-Andersen, M., Sveinbjornsdottir, A.E., Svensson, A., White, J.W.C., 2008. High-resolution Greenland ice core data show abrupt climate change happens in few years. Science 321, 680–684.

Stuiver, M., Brasiunas, T.F., 1991. Isotopic and solar records. In: Bradley, R.S. (Ed.), Global Changes of the Past. Boulder University, Corporation for Atmospheric Research, pp. 225−244.

Stuiver, M., Brasiunas, T.F., 1992. Evidence of solar variations. In: Bradley, R.S., Jones, P.D. (Eds.), Climate Since A.D. 1500. Routledge, London, pp. 593−605.

Stuiver, M., Grootes, P.M., 2000. GISP2 oxygen isotope ratios. Quaternary Research 54/3.

Stuiver, M., Quay, P.D., 1979. Changes in atmospheric carbon-14 attributed to a variable sun. Science 207, 11−27.

Stuiver, M., Grootes, P.M., Brasiunas, T.F., 1995. The GISP2 ^{18}O record of the past 16,500 years and the role of the sun, ocean, and volcanoes: Quaternary Research 44, 341−354.

Svensmark, H., Friis-Christensen, E., 1997. Variation of cosmic ray flux and global cloud cover—a missing link in solar−climate relationships. Journal of Atmospheric and Solar−Terrestrial Physics 59, 1125−1132.

Thompson, D.W.J., Wallace, J.M., 1998. The Arctic Oscillation signature in the wintertime geopotential height and temperature fields. Geophysical Research Letters 25, 1297−1300.

Thompson, D.W.J., Wallace, J.M., 2000. Annular modes in the extratropical circulation. Part 1: Month-to-Month variability. Journal of Climate 13, 1000−1016.

Tsonis, A.A., Hunt, G., Elsner, G.B., 2003. On the relation between ENSO and global climate change. Meteorology and Atmospheric Physics, 1−14.

Tsonis, A.A., Swanson, K.L., Kravtsov, S., 2007. A new dynamical mechanism for major climate shifts. Geophysical Research Letters 34.

Verdon, D.C., Franks, S.W., 2006. Long-term behaviour of ENSO: interactions with the PDO over the past 400 years inferred from paleoclimate records. Geophysical Research Letters 33.

Walker, G., Bliss, 1932. World Weather V. Memoirs Royal Meteorological Society 4, 53−84.

Wang, Y.M., Lean, J.L., Sheeley, N.R., 2005. Modeling the sun's magnetic field and irradiance since 1713. Astrophysical Journal 625, 522−538.

Wolter, K., 1987. The Southern Oscillation in surface circulation and climate over the tropical Atlantic, Eastern Pacific, and Indian Oceans as captured by cluster analysis: Journal of Climate and Applied Meteorology 26, 540−558.

Wolter, K., Timlin, M.S., 1993. Monitoring ENSO in COADS with a seasonally adjusted principal component index: Proceedings of the 17th Climate Diagnostics Workshop, Norman, OK, Oklahoma Climatological Survey, CIMMS and the School of Meteorology. University of Oklahoma, pp. 52−57.

Zhang, Y., Wallace, J.M., Battisti, D., 1997. ENSO-like interdecadal variability: 1990−1993. Journal of Climatology 10, 1004−1020.

Zhang, Y., Wallace, J.M., Battisti, D., 1997. ENSO-like interdecadal variability: 1990−1993. Journal of Climatology 10, 1004−1020.

Zhang, Y., Wallace, J.M., Iwasaka, N., 1996. Is climate variability over the North Pacific a linear response to ENSO? Journal of Climatology 9, 1468−1478.

Chapter 6

Setting the Frames of Expected Future Sea Level Changes by Exploring Past Geological Sea Level Records

Nils-Axel Mörner

Paleogeophysics & Geodynamics, Rösundavägen 17, 13336 Saltsjöbaden, Sweden

1. INTRODUCTION

Eustasy was once defined as "simultaneous changes in global sea level" (e.g., Fairbridge, 1961). With the concept of geoid changes (Mörner, 1976), global loading adjustment (e.g., Clark, 1980; Peltier 1998), and global redistribution of the ocean water masses (Mörner, 1984, 1988), this definition had to be changed. Hence, Mörner (1986) redefined eustasy simply as "changes in ocean level" (contrary to changes in land level) and regardless of causation factor.

Many variables control the changes in ocean level (Mörner, 1996a, 2000, 2005). The most significant parameters to drive a possible sea level rise today are the redistribution of water to the oceans by glacial melting (a process known

Evidence-Based Climate Science. DOI: 10.1016/B978-0-12-385956-3.10006-3

as glacial eustasy) and the expansion of the water column by heating up the water (a process known as steric expansion). Both these parameters can be quantified as to maximum rates and amplitudes (Mörner, 1989, 1996a,b). Those values may even be used to define the frames, inside which we have values of possible changes and outside which we have values that are not anchored in physics of sea level variability. This is the topic of the present paper, initiated by the fairly careless dropping of values of assumed sea level rise by year 2100 that often seem, by far, to exceed the frames of possible changes.

2. SEA LEVEL CHANGES BY YEAR 2100

The first predictions by the IPCC (Intergovernmental Panel on Climate Change) did not hesitate to claim a sea level rise as high as 2−3 m by the year 2100 (Hoffman et al., 1983; Kaplin, 1989; IGBP, 1992). The values presented in the IPCC reports have diminished successively to +47 cm ± 39 cm in the 2001 report and to +37 cm ± 19 cm in the 2007 report (IGCP, 2001, 2007). The INQUA Commission on Sea Level Changes and Coastal Evolution, on the other hand, objected to values exceeding +20 cm with the expected value set at +10 cm ± 10 cm (INQUA, 2000), a value later updated at +5 cm ± 15 cm (Mörner, 2004). Recently, a sea level rise of 2 m, or even more by 2100, has again been claimed due to exceptional melting of the ice cap in Greenland (ACID, 2004) or Antarctica (e.g., Rapley, 2007). It is significant, however, that the Antarctic ice cap has expanded, not decreased, in the last 20 years (D'Aleo, 2007) and that the Greenland ice cap was roughly the same dimension during the Holocene Climatic Optimum and probably even during the Last Interglacial when climate was about 4−5° warmer than today (Willerslev et al., 2007). Despite all this, Rahmstorf (2007) proposed a sea level rise of 50−120 cm by 2100.

3. FRAMES OF GLACIAL EUSTATIC RISE IN SEA LEVEL

For more than 25 years, I have been engaged in the quantification of the various eustatic variables. The maximum glacial eustatic rate has been established at 10 mm year^{-1} or 1.0 m in a century (Mörner, 1983). Subsequent studies confirmed this value (Mörner, 1987, 1989, 1996a,b). The relations among rates and amplitudes of different eustatic variables are given in Fig. 1. This value is primarily obtained from the rates recorded at the glacial eustatic rise after the last glaciation maximum (LGM) of the Last Ice Age at around 20 ka with a sea level low-stand in the order of −120 m.

The mean rate of glacial eustatic rise was 120 m in about 13,000 years from 18 to 5 ka or 9.2 mm year^{-1}. The rise from 18 to 12 ka was about 60 m or 10.0 mm year^{-1}. The Holocene rise from 10 to 5 ka was 50 ± 10 m or 10.0 mm year^{-1} ± 2.0 mm year^{-1}. Therefore, a round figure of 10.0 mm year^{-1} can be held as a quite good estimate of the maximum rate of sea level rise after the last glaciation maximum.

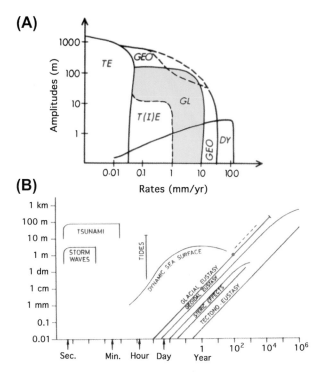

FIGURE 1 Rates and amplitudes of different sea level variables. (A) Estimates of 1983 and 1989 (Mörner, 1983, 1989) with GL (yellow field) denoting glacial eustasy peaking at a rate of $10\,\mathrm{mm\,year}^{-1}$. (B) Improved estimates of 1996 (Mörner, 1996b) with variables from seconds up to a million year. Glacial eustasy (red line) is still peaking at a rate of $10\,\mathrm{mm\,year}^{-1}$ or $1.0\,\mathrm{m}$ in a century (red dots).

The predominant contribution to the post-LGM rise in sea level came from the huge continental ice caps over North America and Northwest Europe. In America, the ice cap reached down to about Latitude 40° N and in Europe down to about Latitude 53° N; i.e., mid-latitudes, which would be impossible under present-day climate conditions. The post-LGM rise in temperature was significant and the ice melted as fast as possible — still, sea level did not rise faster than $10\,\mathrm{mm\,year}^{-1}$ ($1.0\,\mathrm{m}$ in a century). Therefore, this value seems to set a natural frame for possible maximum rates of sea level rise.

The volume of the post-LGM ice melting was in the order of 50.7 million km^3 (Flint, 1971), and most of these masses were melting under strong climate forcing in mid-latitude positions. Still, it took about 12,000−13,000 years to melt the main continental ice caps of the Last Ice Age.

Today, there are two major continental ice caps left—Antarctica (~21.5 million km^3) and Greenland (~2.4 million km^3), both in high-latitude positions. In case of a hypothetical melting, the rate of melting must be significantly lower than that of the big mid-latitude ice caps at LGM.

The ice cap in Antarctica was assumed even by the IPCC to increase rather than decrease in the near future (IGCP, 2001). The recent claim of a rapid

FIGURE 2　The three factors controlling the rate of melting of continental ice
caps. A fourth factor is time: it took about
12,000—13,000 years to melt the huge
ice caps of the Last Ice Age.

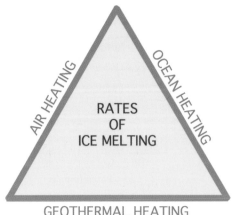

ongoing melting (e.g., Rapley, 2007) is strongly contradicted by the observed
increase in ice cover during the last 20 years (D'Aleo, 2007).

The Greenland ice cap was recently claimed to be under rapid melting
(ACID, 2004). However, the Greenland ice cap seems to have been roughly the
same size in mid-Holocene time when temperature was about 2.5 °C warmer
than today in the northern hemisphere. Recent studies (Willerslev et al., 2007)
also record a persisting ice cap during the Last Interglacial when temperature
was some 4—5 °C warmer than today. Therefore, a rapid down-melting of the
whole ice cap seems highly unlikely.

Physically, the melting of ice is a function of the calories available in the
process. The heat (calorie) input comes from the air, the sea, and the under-
ground (Fig. 2). All of these three sources of energy are significantly larger in
the mid-latitude positions than in the polar to high-latitude areas. Therefore,
any melting of the ice caps in Antarctica and Greenland (totally or partly) must
be significantly slower than that of the huge mid-latitude ice caps at LGM.

Even in the earliest part of the Holocene, when temperature rose very rapidly
and the rate of ice retreat was fast, the glacial eustatic rise did not exceed
10 mm year^{-1}. So, for example, the ice recession in the Stockholm area (in the
early Holocene some 400 years after the end of the Younger Dryas period) was
about 300 m year^{-1}, which is a very rapid recession, especially if one considers
that the outward ice-flow at the same time was in the order of 500 m year^{-1},
implying a total ice-front recession in the order of 800 m year^{-1} (Fig. 3).

In conclusion, the observed rate of post-LGM glacial eustatic rise in sea
level—10 mm year^{-1} or 1.0 m in a century—appears to be a very good value of
the maximum possible rate of sea level rise from glacial melting.

It might be appropriate to stress that any melting of the Arctic summer ice
has insignificant effects on global sea level, because the ice cover is thin and is
already floating in the sea. Similarly, in view of present melting of some alpine

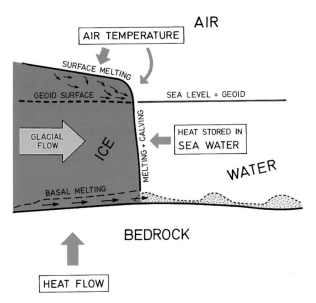

FIGURE 3   Mode of ice melting, controlling the rate of ice retreat with an actual example from the Stockholm area in Sweden, deglaciated some 11,000 cal. years B.P.: the ice margin retreated at a rate of about 300 m year^{-1}, despite the fact that the ice-flow outwards was in the order of 500 m year^{-1} giving a total ice melting of about 800 m year^{-1}, which is very much. This period represents the earliest Holocene time when temperature rose significantly. Still, global sea level did not rise faster than 10 cm year^{-1}.

glaciers, one should also remember that the extensive expansion of alpine glaciers during the so-called Little Ice Ages of the last 600 years had seemingly little to insignificant effects on global sea level.

4. FRAMES OF STERIC EXPANSION OF THE WATER COLUMN

The amount of expansion of a water column that is heated can fairly easily be calculated (Fig. 4.). In the oceans, only the upper couple of hundred meters are likely to become heated. Such a heating is hardly likely to amount to more than a degree or two. Figure 4 gives the expansion effects (i.e., sea level rise) with respect to different amount of heating and height of the water column affected. An expansion (sea level rise) of more than some 5–10 cm seems hardly likely (Mörner, 1996a; Nakibogul and Lambeck, 1991).

Another factor of fundamental importance is the available water depth. In the littoral zone, the water depth is so small that any heating expansion will be more or less negligible (Fig. 4). At the shore (the land/sea interface), the effect will always remain zero. Therefore, thermal expansion will not affect coastal sea level. There seems to exist a misunderstanding that a water expansion at sea will flood landwards. What is deformed, is the dynamic sea

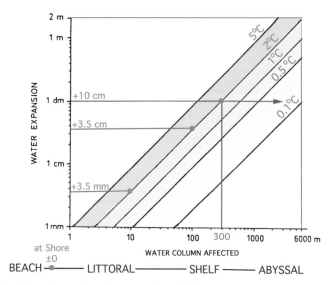

FIGURE 4 Thermal expansion of a water column of different size (depth) with respect to various degrees' of heating (Mörner, 1996a). In the oceans, the water column is never heated uniformly with depth due to its stratification and slow vertical mixing. At the base, a schematic depth zonation is added, illustrating the decreasing effect towards the coast and the absence of any effect at the very shoreline.

level, which is a highly irregular surface due to all interacting dynamic variables.

The warming at the Younger Dryas to early Holocene transition was both rapid and drastic (e.g., Alley, 2000). Still we are unable to identify any sea level effect driven by steric expansion of the water masses.

The Holocene sea level oscillations observed have sometimes been proposed to be the steric effects of changes in ocean surface temperature and/or salinity (Schofield, 1980). Nowadays, it seems quite clear, however, that such sea level oscillations are the function of dynamic redistribution of the water masses (Mörner, 1996a, 2005, 2007).

During the ENSO-year 1998, the Maldives were affected by a strong coral bleaching due to over-heating of the water. The sea level effect was in the order of 3 cm (Mörner, 2011).

5. SETTING THE FRAMES OF FUTURE SEA LEVEL CHANGES

In view of the data presented, it seems justifiable to apply physical frames of possible sea level changes in the near future. In Fig. 5, the value 10 mm year^{-1} is used as the frame value. Possible future changes in sea level should lie inside (probably well inside) this box (blue). Values falling outside the frame should be discarded as unrealistic.

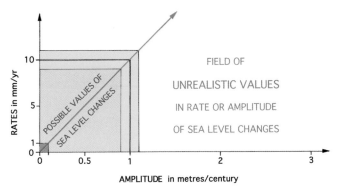

FIGURE 5 Rates and amplitudes of expected future changes in sea level with the frame of physically possible changes set at 10 ± 1 mm year^{-1} or 1.0 ± 0.1 m in a century, and with the observed rate of changes in the last 300 years of 1 mm year^{-1} or 10 cm in a century (Mörner, 1996a, 2004) marked in lower right corner. For reasons discussed in the text, reliable estimates of sea level changes by year 2100 must lie well within the frame given. Values falling outside should simply be considered unreliable or exaggerated. At the same time, however, even a possible rise in the order of half a meter or so by year 2100 would pose serious threats to some low-lying coasts and islands around the globe.

The sea level changes observed over the last 300 years have oscillated within a range of about 1.1 mm year^{-1} lacking sign of any long-term trend (Mörner, 2004, 2005, 2008). For the last half a century, the situation has remained highly controversial, however (Mörner, 2010a). The excellent northwest European tide gauge records give little or no rise (Woodworth, 1990; Mörner, 1996a, 2004, 2010a). Selected Pacific and Indian Ocean tide gauges (Church et al., 2007) give 1.4 mm year^{-1}. Proposed global mean tide gauges (Holgate, 2007) give 1.45 mm year^{-1} with the actual rate for the last 40 years being only 1.2 mm year^{-1}. All values lie well within the lower half of the blue square set by the frames in Fig. 5.

Satellite altimetry is an important new tool because it does no longer limit our records to the shores of the world but covers the whole ocean surface below Latitude 60° N and S. Whilst the satellite altimetry groups themselves (Cazenave and Llovel, 2010; Nicholls and Casenave, 2010; NOAA, 2008) give rates of +3.1 to 3.4 mm year^{-1}, Mörner (2004) gives a value of ± 0 mm year^{-1}. The difference in interpretation is illustrated in Fig. 6 (Mörner, 2008, 2010a). The satellite readings need some technical adjustment. This gives an ongoing rate of sea level change of ± 0 mm year^{-1} (Mörner, 2008, 2010a; Aviso, 2000) or a slight rise of <0.5 mm year^{-1} (Mörner, 2005). I term this as "the instrumental record" (Mörner, 2008). Other persons seeking a rise in sea level add "personal calibrations" (primarily a tide gauge factor), by that arriving at an "interpretational record" of about +3.0 mm year^{-1} (Cazenave and Llovel, 2010; Nicholls and Casenave, 2010; Aviso, 2003, 2008). Even the "calibrated records" lie well within the lower half of the square set by the frames in Fig. 5.

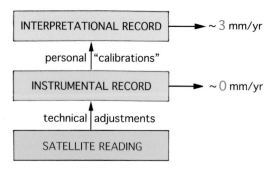

FIGURE 6   Satellite altimetry instrumental readings are in need of technical adjustments in order
to provide an "instrumental record" (Mörner, 2008, 2010a). This record gives no rise in sea level
(Mörner, 2004, Fig. 2; Aviso, 2000), or at the most a very slight rise of <0.5 mm year^{-1} (Mörner,
2005, Fig. 5). By applying additional "calibrations" (of subjective character), an "interpretational
record" is established (e.g., Cazenave and Llovel, 2010; Nicholls and Casenave, 2010; NOAA,
2008), which shows a rise; today in the order of +3.0 mm year^{-1} (Aviso, 2008).

6. SOLAR CYCLES IN THE NEAR FUTURE

Finally, discussing the next centennial changes in sea level, one should, at least,
consider the long-term changes in solar activity. From the cyclic repetition
between solar maxima and minima, one can infer that there will be a future
solar minimum in the middle of this century (Mörner, 2006, 2010b; Easterbrook, 2007).

During the previous solar minima—the Dalton, Maunder, and Spörer
Minima—climate was cold and known as "Little Ice Age". In analogy, by about
2040–2050 similar climatic conditions might reappear. This might even lead to
a minor, decadal, lowering in sea level. Therefore, my estimate of the possible
changes in sea level by years 2100 is +5 cm ± 15 cm (Mörner, 2004).

7. THE STATE OF FEAR

A rise in sea level will always pose a threat to low-lying coastal areas. This is
also something that coastal dwellers through time seem to have learned to cope
with (e.g., the changes in sea level affecting the Maldives in the last 4000 years;
Mörner, 2007, 2011). There are, of course, limits of the capacity of "adapta-
tion". The more reasons there are to be realistic in our predictions and estimates. Exaggerations just feed a "state of fear", to borrow the title of Crichton's
book (Crichton, 2004).

The main purpose of this paper is to provide a tool of discrimination
between reasonable and unreasonable sea level predictions for this century (up
to 2100). Using the frames set in Fig. 5, one should exclude all claims
exceeding 1 m rise in a century or 10 mm year^{-1} (for example, the recent claim
by Black (2008) of a rise by 2100 of 1.5 m). Even values in the order of

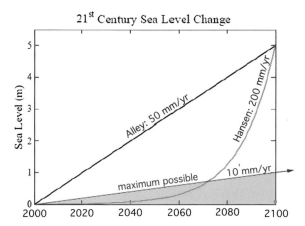

FIGURE 7 The sea level changes within this century according to Hansen and Sato (2011). A rise to +5 m by year 2100 is totally out of the question. The maximum possible rise allowed for from geology and physics is 10 mm year^{-1} (or 1 m per century) as given in Fig. 5. The probable change by 2100 is marked by a red dot. The claim by Hansen and Sato is sheer disinformation without any relation to present-day knowledge in the science of sea level change.

0.5–1.0 m (5–10 mm year^{-1}) seem less probable (in view of what is said above about the rate of melting).

Hansen and Sato (2011) have recently claimed that global sea level will rise to +5 m by year 2100. They even proposed an exponential increase in the glacier melting giving a 4 m rise in sea level in 20 years from 2080 to 2100, which is a rate of 200 mm year^{-1} (Fig. 7). This rate far exceeds the "ultimate frame of possible sea level rise" as given in Fig. 5. Their claim violates geology (it does not concur with our experience and observations), physics (ice can simply not melt that fast), and scientific ethics (one should know what one is talking about and not drop idle talk just for the sake of promoting a view). Hence, this contribution has only one message to offer; it provides a shocking example of disinformation and should be immediately dismissed as such.

Values ranging between 0 m and 0.5 m (0–5 mm year^{-1}) must be classified as possible, though our best estimate is +5 cm ± 15 cm by year 2100 (Mörner, 2004).

Figure 8 shows coastal problems plotted against the possible future sea level rise. The scale is, of course, arbitrary. A rise from 0 cm to +25 cm by the year 2100 seems "OK", from +25 cm to +75 cm "bad", and above +75 cm "disastrous". A rise of up to +1 m may be theoretically possible (Fig. 4), but is quite unlikely to exceed about 0.5 m.

By the frames set in Fig. 5, there seems little reason for the expectation of a sea level rise causing disastrous effects (Fig. 8). Still, even a minor rise may cause problems in low areas where the balance between sea and land remains delicate. In general, however, we seem to be exposed to extensive exaggerations (Mörner, 2010a).

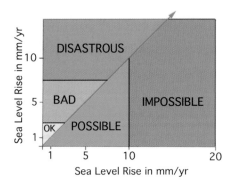

FIGURE 8   The ultimate frame of possible sea level changes in this century is set at 1.0 m by year 2100 or 10 mm year^{-1} (Fig. 5). Even rates above 0.5 mm year^{-1} are classified as "not likely". This, in its turn, may be converted into coastal effects (vertical zonation); <25 cm = ok, 25–75 cm = bad, and >75 cm = disastrous. In conclusion, this may pose future problems but in no way as serious as is often claimed in "state of fear scenarios". Besides, our own best estimate of a possible sea level rise by year 2100 is only +5 cm ± 15 cm (INQUA, 2000; Mörner, 2004).

8. CONCLUSIONS

The rate of post-glacial melting of the big continental ice caps in mid-latitude position provides an excellent maximum value for all talk of what will happen in the next 100 years. This value provides us with a tool of discriminating between realistic proposals and unrealistic claims that should be discarded or, at least, be taken with great care. The following conclusions are drawn:

1. The prediction of sea level changes within this century (i.e., up to year 2100) must be well within the frames set by the post-LGM rates of sea level rise; that is ≤1.0 m (or <10 mm year^{-1}).
2. Any rate of melting of the Greenland and/or Antarctic ice caps must be well below that of the melting of the major LGM ice caps.
3. The effect of thermal expansion must be well below that of the major post-LGM warming pulse at the "terminations" and especially at the rapid warming of the early Holocene. At the coast, the effect is negligible to zero.

ACKNOWLEDGMENTS

I acknowledge very constructive collaborative work within the INQUA Commission on Sea Level Changes and Coastal Evolution, which I chaired 1999–2003, including our discussions and conclusion at five international meetings with respect to present-to-future sea level changes (INQUA, 2000). I also acknowledge the fine teamwork within our Maldives International Sea Level Research Project, which brought about new and fundamental observational facts (Mörner, 2007, 2011; Mörner et al., 2004).

REFERENCES

ACID, 2004, Impact of a warm arctic: arctic climate impact assessment, *ACIA Overview Report*, Cambridge Univ. Press (2004) http://amap.no/acia/.

Alley, R.B., 2000. The Younger Dryas cold interval as viewed from central Greenland. Quaternary Science Reviews 19, 213−216.

Aviso, 2000. Observing the oceans by altimetry. www.aviso.cis.cnes.fr.

Aviso, 2003. Observing the oceans by altimetry. www.aviso.cis.cnes.fr.

Aviso, 2008. Mean sea level as seen by altimeters. www.avsio.oceanobs.com.

Black, R., 2008. Forecast for Big Sea Level Rise. BBC News.

Cazenave, A., Llovel, W., 2010. Contemporary sea level rise Annual Revue. Marine Science 2, 145−173.

Church, J.A., White, N.J., Hunter, J.R., 2007. Sea-level rise at tropical Pacific and Indian Ocean islands. Global Planetary Change 53, 155−168.

Clark, J., 1980. A numerical model for worldwide sea level changes on a viscoelastic Earth. In: Mörner, N.-A. (Ed.), Earth Rheology, Isostasy and Eustasy. John Wiley and Sons, pp. 525−534.

Crichton, M., 2004. State of Fear. HarperCollins Publishers, New York.

D'Aleo, J., 2007. A new record of Antarctic total ice thickness? http://icecap.us/index.php/go/joes-blog/.

Easterbrook, D.J., April 12−18, 2007. Global warming and CO_2: inconvenient truth or incongruent reality? Whatcom Independent, Bellingham.

Fairbridge, R.W., 1961. Eustatic changes in sea level: Physics and Chemistry of the Earth 4, 99−185.

Flint, R.F., 1971. Glacial and Quaternary Geology. Wiley & Sons.

Hansen, J.E., Sato, M., 2011. Paleoclimate implications for human-made climate change. www. columbia.edu/~jeh1/mailings/2011/20110118_MilankovicPaper.pdf.

Hoffman, J.S., Keyes, D., Titus, J.G., 1983. Projecting Future Sea Level Rise: US Environmental Protection Agency. Government Printing Office, Washington, DC, p. 266.

Holgate, S.J., 2007. On the decadal rates of sea level change during the twentieth century. Geophysical Research Letters 34.

IGBP, 1992. Global Change. Reducing Uncertainties. IGBP report.

IGCP, 2001. Climate Change. Cambridge Univ. Press, Oxford.

IGCP, 2007. Climate Change. Cambridge Univ. Press, Oxford.

INQUA, 2000. The Commission on "Sea level changes and coastal evolution" www.pog.su.se/sea(2000)www.pog.nu(2005)

Kaplin, P.A., 1989. Shoreline Evolution During the Twentieth Century: Oceanography 1988. UNAM Press, Mexico, p. 208.

Mörner, N.-A., 1976. Eustasy and geoid changes. Journal of Geology 84, 123−151.

Mörner, N.-A., 1983. Sea level. In: Gardner, R.A.M., Scoging, H. (Eds.), Mega-Geomorphology. Oxford Univ. Press, Oxford, pp. 79−92.

Mörner, N.-A., 1984. Planetary, solar, atmospheric, hydrospheric and endogene processes as origin of climatic changes on the Earth. In: Mörner, N.-A., Karlén, W. (Eds.), Climatic Changes on a Yearly to Millennial Basis. Reidel, Dordrecht, pp. 637−651.

Mörner, N.-A., 1986. The concept of eustasy. A redefinition. Journal of Coastal Research SI-1, 49−51.

Mörner, N.-A., 1987. Eustasy, geoid changes and dynamic sea surface changes due to the interchange of angular momentum. In: Qin, Y., Zhao, S. (Eds.), Late Quaternary Sea Level Changes. China Ocean Press, Beijing, pp. 26−39.

Mörner, N.-A., 1988. Terrestrial variations within given energy, mass and momentum budgets; paleoclimate, sea level, paleomagnetism, differential rotation and geodynamic. In:

Stephenson, F.R., Wolfendale, A.W. (Eds.), Secular Solar and Geomagnetic Variations in the Last 10,000 Years. Kluwer, Dordrecht, pp. 455−478.

Mörner, N.-A., 1989. Global changes; the lithosphere; internal processes and Earth's dynamicity in view of Quaternary observational data. Quaternary International 2, 55−61.

Mörner, N.-A., 1996a. Sea level variability. Zeitschrift fur Geomorphogie N.F. 102 (Suppl.), 223−232.

Mörner, N.-A., 1996b. Rapid changes in coastal sea level. Journal of Coastal Research 12, 797−800.

Mörner, N.-A., 2000. Sea level changes and coastal dynamics in the Indian Ocean. Integrated Coastal Zone Management 1, 17−20.

Mörner, N.-A., 2004. Estimating future sea level changes. Global Planetary Change 40, 49−54.

Mörner, N.-A., 2005. Sea level changes and crustal movements with special aspects on the eastern Mediterranean. Zeitschrift fur Geomorphologie 137 (Suppl.), 91−102.

Mörner, N.-A., 2006. 2500 years of observations, deductions, models and geoethics. Boletin de la Sociedad Geologica Italiana 125, 259−264.

Mörner, N.-A., 2007. Sea level changes and tsunamis, environmental stress and migration overseas. The case of the Maldives and Sri Lanka. Internationales Asienforum 38, 353−374.

Mörner, N.-A., 2008. Reply to Coment by Nerem et al. (2007) on "Estimating future sea level changes from past records" by Nils-Axel Mörner (2004). Global Planetary Change 62, 219−220.

Mörner, N.-A., 2010a. No alarming sea level rise. A great sea level humbug revealed. 21st Century Science and Technology, 7−17. Winter 2010/11.

Mörner, N.-A., 2010b. Solar minima, Earth's rotation and Little Ice Ages in the past and in the future. The North Atlantic−European case. Global Planetary Change 72, 282−293.

Mörner, N.-A., 2011. The Maldives: a measure of sea level changes and sea level ethics, Evidence-Based Climate Science, pp. 197−210.

Mörner, N.-A., Tooley, M.J., Possnert, G., 2004. New perspectives for the future of the Maldives. Global Planetary Change 40, 177−182.

Nakibogul, S.M., Lambeck, K., 1991. Secular sea-level changes. In: Sabadini, R., Lambeck, K., Boschi, E. (Eds.), Glacial Isostasy, Sea-Level and Mantle Rheology. Kluwer Academic Publ. Press, pp. 237−258. NATO C-334.

Nicholls, R.J., Casenave, A., 2010. Sea-level rise and its impact on coastal zones. Science 328, 1517−1520.

NOAA, 2008. The NOAA satellite altimetry program: closing the sea level rise budget with altimetry. Argos and Grace. www.oco.noaa.gov.

Peltier, W.R., 1998. Postglacial variations in the level of the sea: implications for climate dynamics and solid-earth geophysics. Reviews of Geophysics 38, 603−689.

Rahmstorf, S., 2007. A semi-empirical approach to projecting future sea level rise. Science 315, 368−370.

Rapley, C., 2007. Sea rise seen outpacing forecasts due to Antarctica. Reuter, 22 August, from meeting in Ny Alesund, Norway.

Schofield, J., 1980. Postglacial transgressive maxima and second-order transgression of the southwest Pacific Ocean. In: Mörner, N.-A. (Ed.), Earth Rheology, Isostasy and Eustasy. John Wiley and Sons, pp. 517−521.

Willerslev, E., 29 others, 2007. Ancient biomolecules from deep ice cores reveal a forested southern Greenland. Science 317, 111−114.

Woodworth, P., 1990. A search for accelerations in records of European mean sea level. International Journal of Climatology 10, 129−143.

Chapter 7

The Maldives: A Measure of Sea Level Changes and Sea Level Ethics

Nils-Axel Mörner

Paleogeophysics & Geodynamics, Rösundavägen 17, 13336 Saltsjöbaden, Sweden

Chapter Outline

1. INTRODUCTION

The Maldives is an island nation in the Indian Ocean. It consists of some 1,200 low islands arranged in some 20 larger atolls. The islands extend from Latitude 7° N to Latitude 1° S (Fig. 1). Like any coastal nation, it has always been threatened by great waves at extreme storms or, even worse, at tsunami events

Evidence-Based Climate Science. DOI: 10.1016/B978-0-12-385956-3.10007-5
197

FIGURE 1 The Maldives and its main atolls. Black dots mark sites studied by our group. The Island of Minicoy belongs to the Laccadives.

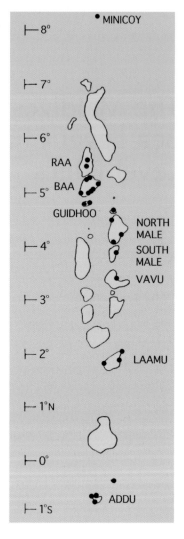

(Mörner et al., 2008). In the past 4,000 years, it has experienced several short-term sea level highs in the order of +0.6−1.2 m (Fig. 2; Mörner, 2007a). We do not blame glacial eustasy for those oscillations, rather ocean dynamic factors like drastic changes in evaporation/precipitation or redistributions of the water masses. The islands have been inhabited for, at least, 1,500−2,000 years.

In the media, we read about the Maldives; either as "the paradise" for tourists, or as the nation soon to become flooded by a rapidly rising sea level as a function of global warming (e.g., IPCC, 2007).

When I, in 1999, became president of the INQUA commission Sea Level Changes and Coastal Evolution I launched an international sea level project in

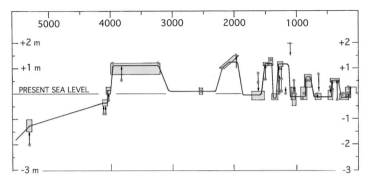

FIGURE 2 The new sea level curve of the Maldives (Mörner, 2007a). The recorded oscillations were driven by regional oceanographic-climatic factors. The islands have been inhabited since about A.D. 400−500 (the last 500 years are enlarged and simplified in Fig. 7A).

the Maldives partly because this is an area where multiple sea level parameters interact (Mörner, 2000) and partly because this was a key area for the proposed sea level rise as a function of global warming (IPCC, 2001). We have made three major expeditions to the Maldives and three shorter visits. Quite rapidly, we were able to conclude that sea is not at all in a rapidly rising mode over the Maldives.

When I launched the project, I could read on the web that the Maldives was on its way to become flooded, and that this was a general consensus and that "2500 scientists cannot be wrong". This was most surprising, as very few if any proper sea level investigation had been performed in the area. This meant that none of the 2,500 scientists had actually been there, and none of them was a true sea level specialist. A true sea level study at the spot itself was obviously urgently needed; hence our multi-parameter sea level project.

When I, as expert reviewer, reviewed the sea level chapter of the IPCC assessment (2001), I was struck that none of the 33 authors was a sea level specialist (INQUA, 2000). Therefore, this chapter was not a product of the international sea level community but rather of selected persons who had "the correct believe" and hence could be expected to provide the answer wanted (today we may call it a "*sea-level-gate*").

Whilst most changes have pros and cons, there is nothing good that can come out of a sea level rise. Therefore, this poses the only real threat. This is probably why this question attracts so much emotion and remarkable behavior.

2. THE MALDIVES SEA LEVEL PROJECT

We had major research expeditions in 2000, 2001, and 2005, and shorter visits in 2000, 2001, and 2003. Our group consisted of a number of international sea level experts. We had very good collaboration with the local people who helped us very much. Our results have been published in solid peer-reviewed papers

(Mörner et al., 2004; Mörner, 2007a) besides several reports elsewhere. Our findings with respect to paleo-tsunamis are presented separately (Mörner et al., 2008; Mörner and Dawson, 2011). Our new sea level curve of the Maldives (Fig. 2) is closely discussed elsewhere (Mörner, 2007a). In this paper, I will deal with some of the findings with respect to the present, on-going sea level situation.

The composition of our research team is given in the Acknowledgments.

3. PROBLEMS

Contrary to our expectations, we came to face several problems in our research project in the Maldives. Some of those will be exposed here because they have a message to tell of how complicated it is to present data that do not concur with the scenario of IPCC.

3.1. A Tree Cannot Lie

The island of Male is the capital of the Maldives. Close-by lies the island of Viligili, former an island for prisoners. At the shore, there was a tree well off the coast in a very delicate growing position (Fig. 3). Any rise in sea level would have destroyed it. We were struck by its position when we saw it for the first time in 2000. It gave evidence against any rapid rise in sea level.

FIGURE 3   The tree on the island of Viligili in 2000 (cf. Mörner, 2007b). It has a very delicate position. This position has remained since the early 1950s (maybe late 1940s). In 2003, the tree (held by us as a strong evidence of 50 years of sea level stability) was vandalized "*by a group of Australian scientists*" (as observed by local people from the house in the background).

Then local people told us that this tree has been in the same position since late 40s, and that it had been a marker for prisoners returning to freedom. This means that the tree had remained in its delicate position for, at least, 50 years. Consequently, sea level cannot have risen in any significant amount for the last 50 years.

This was a hard fact against the IPCC sea level scenario.

We discussed it with local people in the government (very much pro-IGCP), included it in our research reposts and on the web (INQUA, 2000).

3.2. Vandalism in 2003

In 2003, we returned to produce a TV-documentary of our observational field evidence of a non-rising sea level in the Maldives (Mortensen, 2004). To my disappointment, I found the tree fallen down at the shore, still green. Finally the sea had taken it, I assumed. Soon, I learned the truth, however. A nearby restaurant had a nice view of the tree, and we used to rest there and watch the tree. The people running the restaurant told what actually had happened with the tree: *"it was pulled down by a group of Australian scientists"* (Mörner, 2007b; Murphy, 2007).

When an issue has gone so far away from reality that you allow yourself to destroy a piece of evidence just because it contradicts your own believe, we are indeed far away from the normal ethics and theory formulations of science (Mörner, 2006, 2008).

3.3. The Governmental Attitude in Male

Our sea level research program in the Maldives started as a collaboration program between INQUA and the Government in Male. When our observational facts accumulated and started, with increasing strength, to indicate that sea was by no means in a drastically rising mode in the Maldives, we started to get into trouble with the governmental agencies.

The president of the Maldives was putting much effort into the international claim that the west had polluted the air so that the globe was heating up, the glaciers were melting, and the sea was rising, soon to flood the islands of the Maldives. Therefore, the west had to compensate with money and investments. Our findings were interpreted as violating their claims and hence negative, not to say anti-governmental.

When we in 2000 had become convinced by our own observational facts that sea was not at rapidly rising, but virtually stable for the last 30 years, we wanted to share those very good news with the people of the Maldives. A reporter at the Male-TV shot an interview but it was censured by the government and not to be shown in TV (Mörner, 2007b).

Our subsequent studies were performed with an uneasy relation to the governmental agencies. The people of the Maldives, however, helped us in all kind of ways, and we were able to conduct our programs, make the observations wanted, and collect the samples needed (Mörner et al., 2004; Mörner, 2007a).

3.4. "Confirmed in Private"

SASNET (South Asian Studies Network) is a part of the Swedish governmental agencies for international assistance SIDA/SAREC. Our Maldives Research Project has obtained founding from SASNET for the 2001 expedition and I have published some reports on our findings in their journal "South Asia". In 2009, Lars Eklund of SASNET made an official visit to the Maldives. In his report from the visit (Eklund, 2009), he has a passage on "Flooding or not?" ending with the following both interesting and revealing paragraph:

"In June 2004, Prof. Mörner published his research results in an article titled *"The Maldives Project: a future free from sea-level flooding"* in the Contemporary South Asia magazine. However, the Maldivian government did not react positively to these findings since they went against the official policy, even though the facts presented seem to be beyond dispute and are confirmed in private by individual Maldivian researchers".

3.5. The Moscow Meeting in 2004

An English delegation under the leadership of Sir David King set out to "reform" the Russians in the issue of global warming and related problems. A meeting at the Russian Academy of Science was arranged. The chairman of the session, Andrei Illarionov, had some external experts to join the meeting. I was one of the invited. When King, in his talk, slipped into a quite unfounded picture of global sea level change, I, of course, had to object. The sea level picture was later highlighted in my own talk entitled: "Flooding concept called off — New facts from the Maldives". Whilst Illarionov and most others gave vivid appreciation of my talk, the English delegation was silent and obviously disturbed.

Quite some time later, my attention was called to a letter available on the Web from John Clague, president of INQUA, to Yuri Osipov, president of the Russian Academy of Science. In this letter, it was said that I had been "claiming that I was president of the Commission on Sea Level Changes of INQUA" by this "misrepresented his position". Furthermore, Clague stated that "nearly all of the researcher" in INQUA "agree that humans are modifying Earth's climate, a position diametrically opposed to Dr. Mörner's point of view".

Clague had not contacted me on the issue, just sent his mail with copy to David King. The fact, however, is that I very clearly on the first power-point picture (which I still have) has stated "President (1999—2003) of the INQUA Commission on *Sea Level Changes and Coastal Evolution*". When I pointed this out, he promised to withdraw the letter from the web (but gave no apology).

What concerns the view "of nearly all researchers of INQUA", our commission had a network of some 300–400 sea level specialists. The issue of present sea level movements and expected changes by year 2100 had been up at five of our international commission meetings, and being up on our web for discussions for several years (INQUA, 2000). Still, our commission agreed on a "most likely" estimate of sea level change by year 2100 of $+10$ cm $\pm$ 10 cm (i.e., quite different to the values given by IPCC). Another fact with respect to Clague's statement "nearly all" comes from a meeting by the European Science Foundation on Glacial–Interglacial Sea level Changes in Four Dimensions, held in St. Andrews in 2001. At this meeting there were some 100 sea level specialists, when Titus (a proponent of large-scale flooding) asked for a vote on how many were for vs. against the IPCC sea level scenario, only one was for and all the rest (some 99) against. So, when Clague talked about "nearly all", he was by no means anchored in facts.

Long later (in 2008), was I to learn that Clague is a very strong proponent of IPCC. This may explain his action, but can hardly excuse it. It seems appropriate to pose the question: "who misused his position?"

I considered all this to be nonsense up to recently, when I understood that it provides an interesting piece of information of how the debate was twisted in the most questionable spirit that "the end justifies the means". The same is the case with the tree (Fig. 3), vandalized in 2003 "by a group of Australian scientists". With respect to recent talks about a "climate-gate", we may speak about a "clague-gate" and "tree-gate", too.

3.6. The President Enters the Scene

In 2009, the new President of the Maldives, Mohamed Nasheed, entered the scene with very firm statements on his nation soon to be drowned, and with noticeable (not to say "exhibitionistic") official actions like standing in the sea saying "we are drowning" or having a submarine cabinet-meeting supposed to illustrate the flooding to come.

I tried to communicate the observational fact that sea is not at all in an alarming rising mode in the Maldives, and I even sent an open letter to the President (Mörner, 2009) — but no reply.

The reason behind the president's fixation on the rising sea level concept is economical, and certainly not scientific. From the previous president he inherited the idea that "the west" would provide extensive economical support to their "drowning nation" because it was the fault of their overuse of CO_2 producing industrial activities, and, indeed, a lot of money have "flooded" the nation.

For the people of the Maldives, the situation is another. They have to live, work, and raise kids under the false conception that there is no future for them on their own islands. This is an immense psychological burden. In the name of decency, this burden must be lifted from their shoulders now when we know that the rising sea idea does not concur with observational facts.

4. FIELD EVIDENCE IN THE MALDIVES

When we, in 2000, saw the tree (Fig. 3) standing nearly "with its feet in water", we started to realize that there was, indeed, no on-going alarming sea level rise in the Maldives. When we began to examine the coasts of the individual islands, this picture was strengthened.

4.1. The Sea Level Fall in the 1970s

We soon found clear evidence of a significant and rapid drop in sea level. On island after island, we observed a recent redeposition of sand graded to a lower level than previously. In Addu, we studied (cored and leveled), a lake named "Queen's bath", and were able to distinguish the present high- and mean-tide levels and an elevated "sub-recent" high-tide level. On the island of Guidhoo, we drilled two fens and established rise in 1790 to a level 20−30 cm above the present zero. This level was kept up to the 1970s, when sea fell causing some lakes to dry out. On the island of Lamuu, we established a very clear picture with respect to morphology; an old rock-cut platform, now at +20 cm is today abundant and a new and lower platform is in the process of being cut. Maps drawn in 1922 give the higher, now abundant, sea position. From local fishermen, we were informed that sea fell in the 1970s, because a previous sailing route became too shallow in the 1970s. We have presented our observational data in two papers (Mörner et al., 2004; Mörner, 2007a) and one booklet in color (Mörner, 2007b, pp. 9−11).

4.2. The Stability of the Last 30 Years

The fact that sea level fell by 20 cm in the 1970s implies that we got a fresh zero-level to explore for possible traces of movements. Nowhere, do we see any trace of a tendency of a change in sea level over the past 30 years. Wherever we look (i.e., lagoonal environment, rock-cut platform shores, sandy beaches, singly beaches), we find nothing but clear indications of a post-1970 stability in shoreline position and sea level (e.g., Mörner, 2007b, pp. 10−11). One example of this stability is shown in Fig. 4 from an erosive sandy beach environment.

4.3. Tide-Gauge Records

Tide-gauges do not offer straightforward information on sea level trends, on the contrary they have to be treated with care. One set of problems comes from local compaction and ground motions. Another comes from cyclic changes (not least the 18.6-year cycle) and inter-annual signals like ENSO events.

The tide-gauge in Male is useless because of repeated ground deformations due to the loading by enormous buildings in recent years. In year 2001, there were three other tide-gauges in operation (Fig. 5). None of those

FIGURE 4   A general characteristic of the islands of the Maldives is that there is a sub-recent (pre-1970) abandoned beach-ridge and shore plane, now starting to become overgrown (Mörner et al., 2004; Mörner, 2007a,b). At this site, there is active coastal erosion. The redeposition of sand is downward—outward to a lower sea level position. This shore (here marked by sea-weed at MHTL) has remained quite stable, at least, for the last 30 years (Mörner et al., 2004).

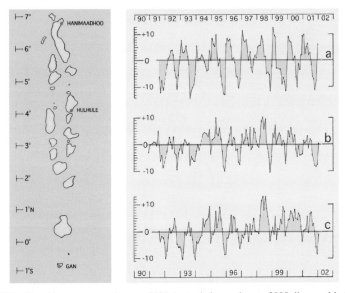

FIGURE 5   The tide-gauge records up to 2001 (extended record up to 2008 discussed in Mörner, 2010a) recording an absence of any rising trend.

recorded any long-term trend; rather they all record incomplete cyclic motions with a few ENSO events superimposed. There is no rapid rise in sea level to be seen.

Today, when eight more years has been added, I have revisited the records with more or less the same result (Mörner, 2010a).

4.4. Minicoy, Just North of the Maldives

The island of Minicoy (Maliku) is the southernmost island in the Laccadives, located some 120 km to the north of the Maldives. It is an atoll island very similar to the islands in the Maldives. The local people testify (1) that sea level is not at all rising (on the contrary they have gained land), (2) that they are amused to hear what the President in the Maldives keep on claiming, and (3) that they understand that *"it is all a matter of money"* (all according to personal information from the Danish social-anthropologist Nils Finn Munch-Petersen after a visit in December, 2010).

In 1992, Mr. Ali Manikfan took Dr. Munch-Petersen down to the shore and showed him that the island, in fact, had gained — not lost — land (Fig. 6). Just as in many island in the Maldives (Fig. 4), sea had fallen and left the old shore to become overgrown and invaded by land-snails (*Ipomea biloba*).

This is a good measure of value of what is politics and what are actual facts. It is, of course, also a matter of ethics.

FIGURE 6   One of the locals of Minicoy, Ali Manikfan, showing how much the island has grown. What was once the beach is now located well above the wave-washing zone, is becoming overgrown (just as many beaches in the Maldives; Fig. 4; Mörner et al., 2004; Mörner, 2007a,b) and invaded by land-snails (photo: N.F. Munch-Petersen, 1992).

5. COMPARISONS

In 2009, a complimentary study was performed in Bangladesh (Mörner, 2010b). This region is notorious for its flooding disasters; fluvial flooding due to heavy precipitation as well as coastal flooding in association with cyclones. Hard observational facts indicate that there is no on-going factor of global sea level rise.

The sea level curve of the last 400 years is strikingly similar to the one of the Maldives (Fig. 7). Even a recent sea level lowering is recorded occurring some 50 years ago.

Having established this new sea level record, I investigated the tide-gauge records in India (Mörner, 2010b). There are records on opposite side of the Decca Plateau; a record from 1878 in Bombay on the west coast and a record from 1937 in Viskhapatnam on the east coast. Both records give a similar picture: a rise 1900—1955, a 12-cm drop 1955—1962, and stability thereafter.

The Indian records fit the new Bangladesh curve perfectly well. They also fit the Maldives record well with the difference that the fall in sea level is dated at 1955—1962 in India and in the 1970s in the Maldives. Either this is an effect of a laterally moving change or a minor dating error in the Maldives. The evidence

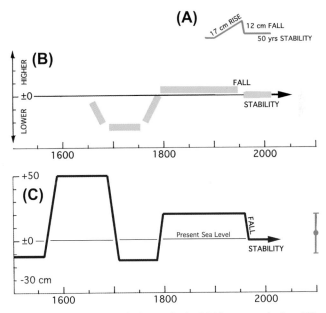

FIGURE 7 *At the base (C):* sea level changes in the Maldives over the last 500 years (from Mörner, 2009). This curve is based on multiple sea level criteria (Mörner, 2007a). *In the middle (B):* the new sea level curve from Bangladesh (Mörner, 2010b). *At the top (A):* the tide-gauge record from Bombay (Mörner, 2010b). In all three regions, there is a sea level fall followed by 40—50 years of stability.

of a sea level fall in the Maldives is strong. The date, however, comes from observations made by local fishermen (Mörner et al., 2004) and may well be pushed back by 10–15 years.

The important thing is that a lack of a sea level rise component is now recorded in the Maldives, in Bangladesh, and in India for the last 40–50 years, and that this period of stability was preceded by a sharp sea level drop in the order of 10–20 cm.

Even on a global scale, there seems to be a striking absence of a present sea level rise component; i.e., in other key-sites like Tuvalu and Vanuatu, in "test-sites" as Venice and North-west Europe, and even in basic satellite altimetry records (Mörner, 2007b, 2010a, 2010c, 2011).

6. CONCLUSIONS

Observational facts do not verify the story of a rapidly rising sea level in the Maldives. On the contrary, stability in sea level is well documented for the last 30–40 years. This is our firm conclusion: sea level is not in a rising mode over the Maldives today. This is a well-known fact for the locals in the Laccadives, to the north of the Maldives. The same picture is recorded in Bangladesh and India, indicating a regional dimension of an absence of a presently on-going sea level rise.

This conclusion is opposite to the scenario proposed by IPCC. As their idea is not based on actual field studies only modeling, our observational facts should be held superior.

During our research in the Maldives, we were confronted with several "remarkable" events, and we have to draw the conclusion that some proponents of IPCC take the liberty to act in a dark Medieval way where "the goal justifies the means".

ACKNOWLEDGMENTS

The Maldives Sea Level research project was initiated as a part of the INQUA Commission on Sea Level Changes and Coastal Evolution. The project includes six visits to the Maldives: (1) in 2000; setting up the project, (2) in 2000; main expedition 1, (3) in 2001; the "Reef Woman" investigation, (4) in 2001; main expedition 2, (5) in 2003; recording a TV-documentary, and (6) in 2005; main expedition 3. Our research group included: N.-A. Mörner (main project leader), J. Laborel (diving leader, biozonation), M.J. Tooley (fen coring), S. Dawson (tsunami research), W. Allison, J. Collina, F. Laborel, C. Rufin, S. Islam, B. Lembke, D. Dominey-Howes, A. Dupuch, M. Banfield and our local friends from the National Centre of Linguistic and Historical Research (N. Mohamed et al.), from Ecocare (M. Zhair, M. Manik, H. Maniku), from Sea Explorer and Whale Submarine. I am indebted to Nils Finn Munch-Petersen for providing valuable information on the Island of Minicoy. For the Bangladesh study, I primarily want to thank the local guide Nazrul Islam Bachchu for very good assistance in the field.

REFERENCES

Eklund, L., 2009. SASNET visit to Maldives 6—8 February 2009 (Report by Lars Eklund). http://www.sasnet.lu.se/maldives09.html.

IPCC, 2001. Climate Change. Cambridge Univ. Press, Oxford.

IPCC, 2007. Climate Change. Cambridge Univ. Press, Oxford.

INQUA, 2000. The Commission on "Sea level changes and coastal evolution". www.pog.su.se/sea (2000), www.pog.nu (2005).

Mortensen, L., 2004. Doomsday Called Off. TV-documentary. See also: "Maldives will avoid extinction". Danish TV, Copenhagen. http://climateclips.com/archives/117.

Murphy, G., 2007. Sea-level expert: it's not rising! Interview: Dr. Nils-Axel Mörner. 21st Century Science & Technology Fall 2005, 25—29.

Mörner, N.-A., 2000. Sea level changes and coastal dynamics in the Indian Ocean: Integrated Coastal Zone Management 1, 17—20.

Mörner, N.-A., 2006. 2500 years of observations, deductions, models and geoethics: Boletin de la Sociedad Geologica Italiana 125, 259—264.

Mörner, N.-A., 2007a. Sea level changes and tsunamis, environmental stress and migration overseas. The case of the Maldives and Sri Lanka: Internationales Asienforum 38, 353—374.

Mörner, N.-A., 2007b. The Greatest Lie Ever Told. P&G-print (Mörner), 2007 first ed., 2009 second ed., 2010 third ed., pp. 1—20.

Mörner, N.-A., 2008. The Sun in the centre and no danger of any on-going global flooding. Quaternary studies (Maria Assuncao Araujo, Ed.), Journal of the Portuguese Association for Quaternary Research (APEQ) 5, 3—9.

Mörner, N.-A., 2009. Open letter to President Mohamed Nasheed of the Maldives. New Concepts in Global Tectonics Newsletter (No. 53), 80—83.

Mörner, N.-A., 2010a. Some problems in the reconstruction of mean sea level and its changes with time. Quaternary International 221, 3—8.

Mörner, N.-A., 2010b. Sea level changes in Bangladesh. New observational facts. Energy & Environment 21 (3), 249—263.

Mörner, N.-A., 2010c. No alarming sea level rise. Nature against IPCC. Observations vs Models. In: The Fourth Heartland Conference on Climate, ppt-presentation. http:/www.heartland.org/events/2010Chicago/program.html.

Mörner, N.-A., 2011. The great sea level humbug. No alarming sea level rise. 21st Century Science & Technology Winter 2010/11, 7—17.

Mörner, N.-A., Tooley, M.J., Possnert, G., 2004. New perspectives for the future of the Maldives. Global and Planetary Change 40, 177—182.

Mörner, N.-A., Laborel, J., Dawson, S., 2008. Submarine sandstorms and tsunami events in the Indian Ocean. Journal of Coastal Research 24, 1608—1611.

Mörner, N.-A., Dawson, S., 2011. Traces of tsunami events in off shore and on shore environments. Case studies in the Maldives, Scotland and Sweden. In: Mörner, N.-A. (Ed.), The Tsunami Threat: Research & Technology. InTech. Publ., pp. 371—388.

Arctic Sea Ice

Steve Goddard

Real Science Blog

1. INTRODUCTION

The National Snow and Ice Data Center (NSIDC) often displays graphs like Fig. 1, which appear to indicate a linear trend of ice loss in the Arctic—going back to the start of the satellite record in 1979.

This decline has been widely attributed to melt caused by higher temperatures—due to CO_2 buildup in the atmosphere. There is no question that there is less ice in the Arctic now than there was 30 years ago, but the evidence tying it to CO_2 is thin—as shown below.

During the 1970s, an unusually large amount of sea ice existed in the Arctic. So much so that some climatologists were actually considering spreading soot

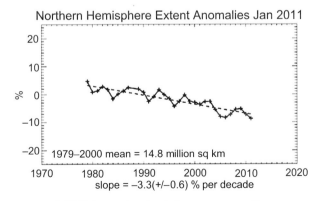

FIGURE 1 Northern hemisphere extent anomalies.

Evidence-Based Climate Science. DOI: 10.1016/B978-0-12-385956-3.10008-7

FIGURE 2   Newsweek article
on global cooling from 1975.

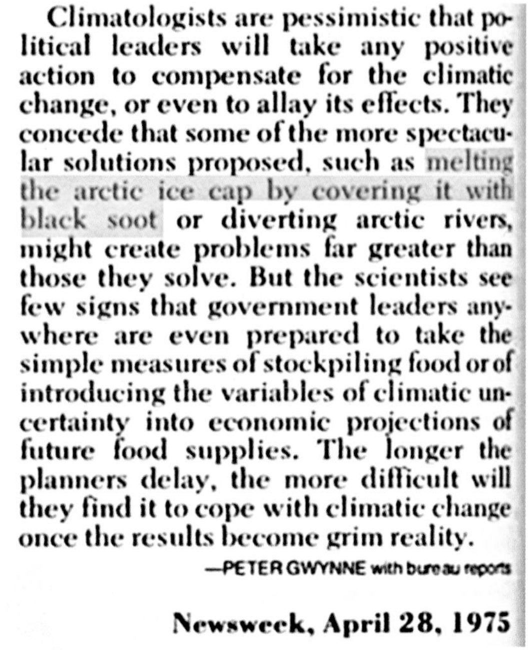

Climatologists are pessimistic that political leaders will take any positive action to compensate for the climatic change, or even to allay its effects. They concede that some of the more spectacular solutions proposed, such as melting the arctic ice cap by covering it with black soot or diverting arctic rivers, might create problems far greater than those they solve. But the scientists see few signs that government leaders anywhere are even prepared to take the simple measures of stockpiling food or of introducing the variables of climatic uncertainty into economic projections of future food supplies. The longer the planners delay, the more difficult will they find it to cope with climatic change once the results become grim reality.

—PETER GWYNNE with bureau reports

Newsweek, April 28, 1975

over the Arctic to melt the ice in order to prevent additional global cooling. The article in Fig. 2 appeared in Newsweek on April 28, 1975.

The Deseret News reported, in 1970, that sea ice had the highest coverage of the last century (Fig. 3).

Given that the NSIDC satellite record starts at a local maximum, it is not surprising that coverage has since declined. We also know that Arctic warming/cooling is cyclical. Periods of strong warming and melt occurred earlier in the century. During the 1920s, "extraordinary warmth" was reported in the Arctic, and it was hypothesized that the Arctic ice cap might disappear. The Pittsburgh Press reported on this in December 1922 (Fig. 4).

90 year old observations of disappearing ice, extraordinary warmth, changes in flora and fauna—the same worries which we hear people verbalizing about the Arctic now. The GISS (Goddard Institute for Space Studies)

DESERET NEWS

SALT LAKE CITY, UTAH

Our Phone Numbers
News Tips ————521-4400
Home Delivery ————521-2840
Information ————524-4445
Sports Scores ————524-4448
Classified Ads Only ——521-3355
Editorial Offices—31 E. 1st South

GES   10c   THE MOUNTAIN WEST'S FIRST NEWSPAPER   MONDAY, JANUARY 26, 1970

FIGURE 3   Article in the salt lake city Deseret News in 1970.

temperature graph for Reykjavik, Iceland shows that warming in the 1920s and 1930s was essentially identical to the warming seen over the last two decades (Fig. 5).

This same pattern is seen in most of GISS eastern Arctic temperature records. In the western Arctic (Pacific side) the pattern is different. Temperatures rose sharply in 1977 with the shift in the PDO (Pacific Decadal Oscillation) (Fig. 6) and have been flat since.

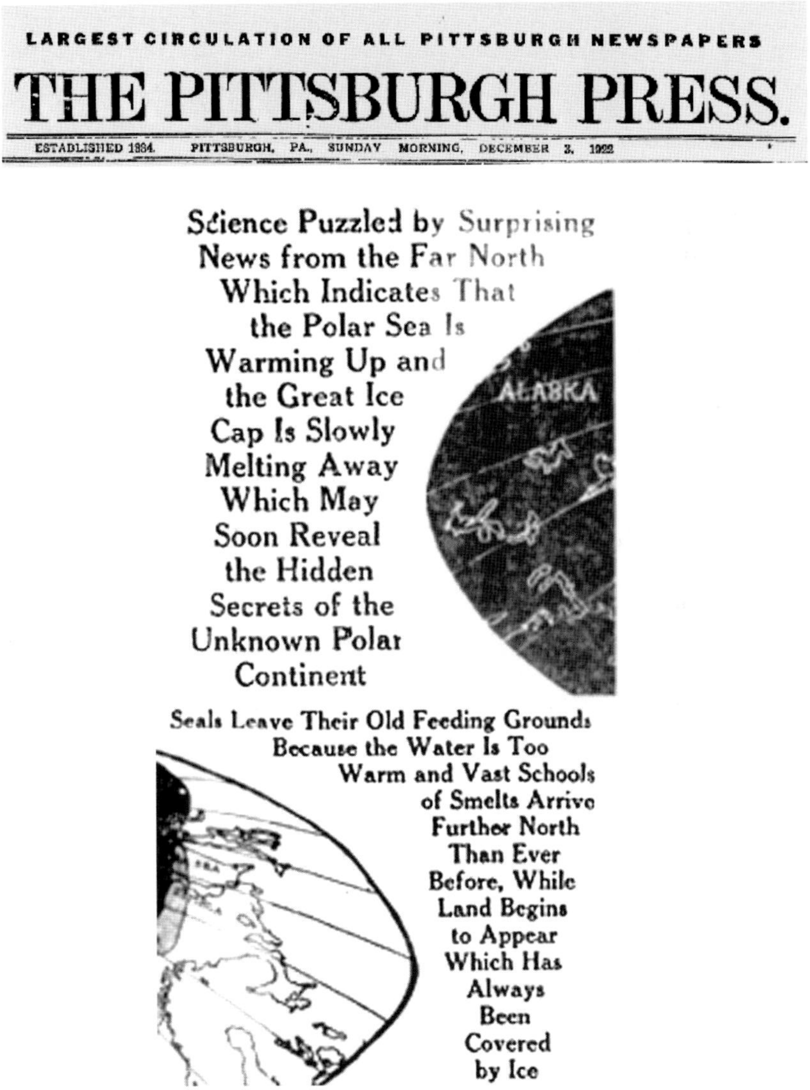

LARGEST CIRCULATION OF ALL PITTSBURGH NEWSPAPERS

THE PITTSBURGH PRESS.

ESTABLISHED 1884.   PITTSBURGH, PA., SUNDAY MORNING, DECEMBER 3, 1922

Science Puzzled by Surprising News from the Far North Which Indicates That the Polar Sea Is Warming Up and the Great Ice Cap Is Slowly Melting Away Which May Soon Reveal the Hidden Secrets of the Unknown Polar Continent

Seals Leave Their Old Feeding Grounds Because the Water Is Too Warm and Vast Schools of Smelts Arrive Further North Than Ever Before, While Land Begins to Appear Which Has Always Been Covered by Ice

FIGURE 4   Article in the Pittsburgh Press, December, 1922.

Photograph of Land Which Has Appeared for the First Time in Greenland, and Which Has Always Heretofore Been Buried Under the Glaciers Which Are Now Melting Away.

FIGURE 4 *(Continued)*

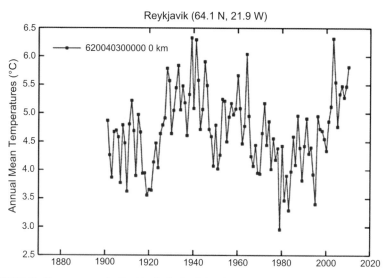

FIGURE 5 Temperatures for Reykjavik, Iceland showing that warming in the 1920s and 1930s was essentially identical to the warming seen over the last two decades (GISS).

Considering just a linear trend can mask some important variability characteristics in the time series. Figure 6 shows clearly that this trend is non-linear: **a linear trend might have been expected from the fairly steady observed increase of CO_2**

Neither of the graphs above (from both sides of the Arctic) lends any support to the idea that CO_2 is the cause of the recent warming. Quite the opposite—Arctic warming appears to be cyclical, rather than linear.

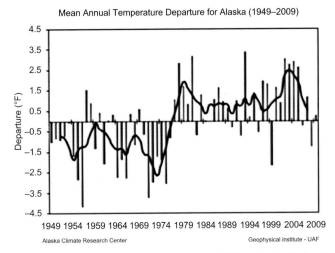

Mean Annual Temperature Departure for Alaska (1949–2009)

Alaska Climate Research Center   Geophysical institute - UAF

FIGURE 6   Alaska temperatures since 1949. (From The University of Alaska, Geophysical Institute).

Looking closer at recent ice loss, we see some interesting patterns. The thickness of the ice is largely dependent on its age. Every winter, surviving ice gets thicker, but much of the sea ice gets blown out into the North Atlantic and melts in the warmer water. The amount of ice lost is largely dependent on wind patterns.

The NSIDC map (Fig. 7) shows that in 1988, there was a lot of 5-year-old ice (marked as red) in the Arctic.

However, by 1996 the vast majority of the older ice was lost. It didn't melt in place but rather was moved out the Arctic by the wind. NSIDC animation shows this movement very clearly (Fig. 8).

Figure 9 overlays NSIDC Arctic sea ice extent, on the 10-year running mean of ENSO (El Nino Southern Oscillation). Note that the period from 1988 to 1996 (when most of the old ice was lost) was dominated by strong El Nino events. Coincidentally, the NSIDC satellite data began measurement right after the PDO shift of 1977. A positive PDO is dominated by warm El Nino events, and a negative PDO (pre-1977) is dominated by cold La Nina events.

We see a correlation between ocean cycles and loss of Arctic sea ice, which is often incorrectly attributed to rises in atmospheric CO_2.

Now, let's look at more recent developments with the ice. The NSIDC map (Fig. 10) shows ice extent on February 25, 2011 with the 1979−2000 median indicated by a red line. Note that most of the missing ice is located near Quebec, Newfoundland, and the Sea of Okhotsk. These are not part of the Arctic, which makes claims of "missing Arctic ice" somewhat dubious. There is also a deficiency in the Bering Sea caused by southerly winds which pushed the ice edge to the north during the prior week.

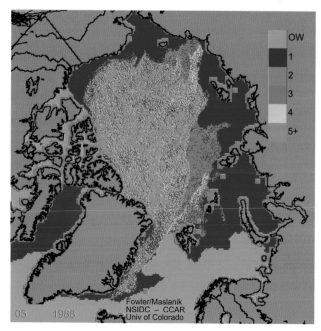

FIGURE 7   Age of ice in years—February 1988. Red indicates 5-year-old ice (NSIDC).

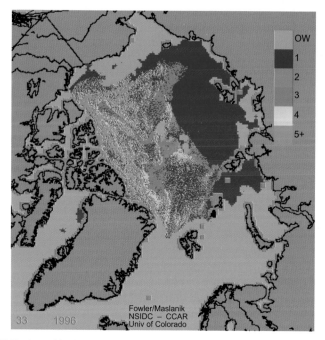

FIGURE 8   Age of ice in years—September, 1996. Red indicates 5-year-old ice (NSIDC).

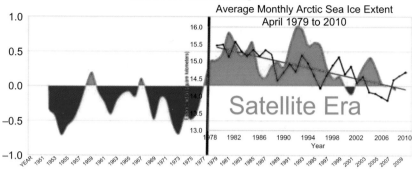

FIGURE 9   Average monthly Arctic sea ice extent, 1979–2010 (NSIDC).

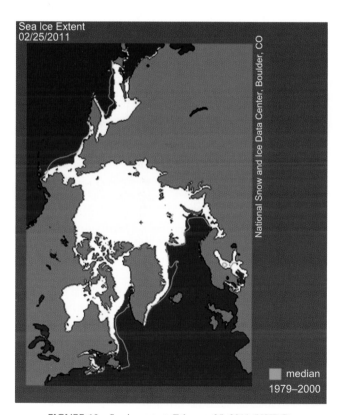

FIGURE 10   Sea ice extent, February 25, 2011 (NSIDC).

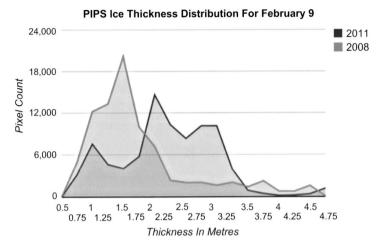

FIGURE 11   Comparison of February 9 ice thickness between 2008 and 2011.

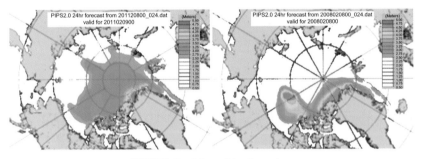

FIGURE 12   Maps of Arctic sea ice.

The areas of ice deficiency never have ice in the summer, so it is impossible to surmise anything about the "health" of *Arctic* ice from this map—or what the likely behavior is for the summer of 2011. The Arctic Basin is completely full of ice, as it always is this time of year. In order to make forecasts for the summer, we need to look at the thickness and age of the ice. The US Navy operational ice model is called PIPS and provides estimates of ice thickness which are updated daily. Figure 11 compares February 9 ice thickness between 2008 and 2011.

The ice has thickened considerably over the last 3 years. Most of the ice is now 2–3 m thick. In 2008, most of the ice was less than 2 m thick. The location of the thick ice can be seen in the modified PIPS maps below (Fig. 12), where ice less than 2.5 m thickness has been removed.

February 9, 2011 February 9, 2008

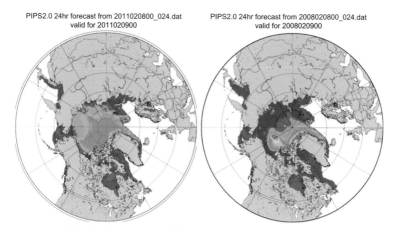

PIPS2.0 24hr forecast from 2011020800_024.dat
valid for 2011020900

PIPS2.0 24hr forecast from 2008020800_024.dat
valid for 2008020900

FIGURE 13   Original PIPS maps without the thin ice removed.

Note that in 2008, there was no thick ice directly over the North Pole, which lead Mark Serreze at NSIDC to forecast an ice-free summer at the pole.

The original PIPS maps without the thin ice removed, are shown in Fig. 13.

Since 2008, the amount of thick ice has greatly increased. Looking again at the PDO graph in Fig. 9, we can see that the increase in ice thickness is coincident with ENSO going dominantly negative in 2008.

The NSIDC graph below shows ice loss due to melt and transport out of the Arctic since 1993. The vast majority of ice loss prior to 2004 was due to ice being blown out of the Arctic near Greenland. From 2004 to 2007 there was also a lot of melt, which led to the widely publicized 2007 "record low" (Fig. 14).

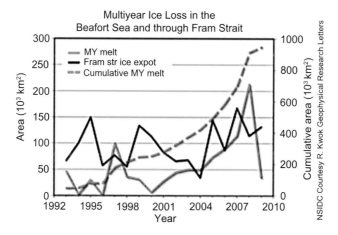

FIGURE 14   Ice loss due to melt and transport out of the Arctic since 1993.

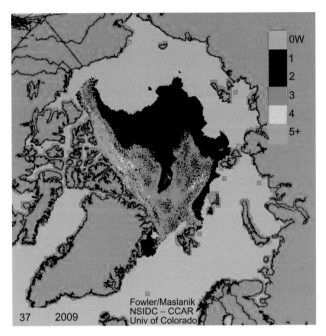

FIGURE 15   Age of ice in years, September 2009 (NSIDC).

The 30-year period from 1977 to 2007, which was dominated by El Nino and ice loss, appears to have come to an end. Now let's look at the prospects for summer 2011, compared to summer 2010. Figure 15 shows that at the end of the 2009 summer, most of the 3-year-old ice was located in vulnerable locations in the eastern Arctic where it was likely to get blown out into the North Atlantic during the winter.

Figure 16 is the equivalent map for the following spring of 2010. Much of the ice was indeed lost to the Atlantic Ocean east of Greenland.

By contrast, ice in early autumn 2010 was much better situated to survive the winter. Figure 17 shows the condition of the ice at the end of September, 2010 after the ice had another birthday. Note that most of the 3+ year old ice was located on the Pacific side, where there is no chance of it being lost over the winter.

We can infer from this that the 2011 ice is thicker and going to be much harder to melt through than ice during the last three summers. Also interesting to note in Fig. 18 (measured from NSIDC maps) is that the amount of 3-year-old ice has nearly doubled since 1996. From 1988 to 1996, ice was being blown out of the Arctic faster than it was accumulating. Since 2008, we have been seeing a net buildup of multi-year ice. The 30-year trend has been reversed in the last few years.

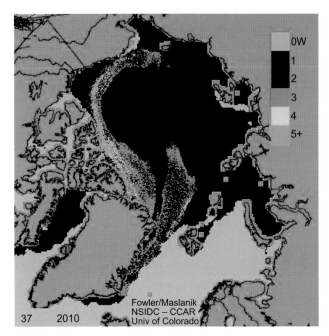

FIGURE 16   Age of ice in years, April 2010 (NSIDC).

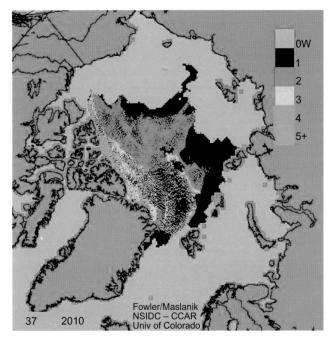

FIGURE 17   Age of ice in years (after aging one more year), September 2010 (NSIDC).

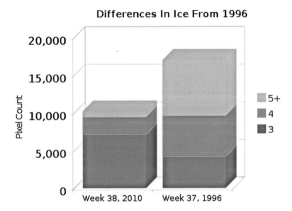

FIGURE 18   Age of ice at the end of summer 1996 and 2010 (NSIDC).

2. SUMMARY

Temperatures and ice loss/gain in the Arctic appear to be cyclical rather than linear. In the eastern Arctic, there was a very similar period of warming and ice loss from 1920 to 1940. This was followed by a period of ice gain until the 1980s. In the western Arctic, temperatures are controlled by the PDO. Western Arctic temperatures rose in 1977 and appear to be on the way back down again. There is no correlation with CO_2. It is an unfortunate coincidence that satellites came on line right at the peak extent for the century, which has confused many people into believing we are seeing a long-term linear trend of ice loss.

Since the minimum in 2008, we have seen a significant gain in the amount of multi-year ice. Navy maps show that Arctic sea ice has been increasing in volume for 3 years. At some point this is likely to be reflected as an increase in extent, which will break the downwards trends shown in the NSIDC maps.

Chapter 9

Have Increases in CO_2 Contributed to the Recent Large Upswing in Atlantic Basin Major Hurricanes Since 1995?

William M. Gray and Philip J. Klotzbach

Department of Atmospheric Science, Colorado State University, Fort Collins, CO 80523, USA

Chapter Outline

Evidence-Based Climate Science. DOI: 10.1016/B978-0-12-385956-3.10009-9

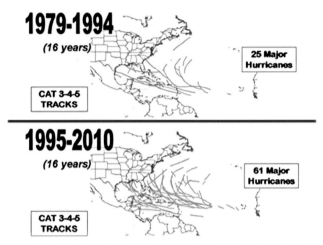

FIGURE 1   The tracks of major (Category 3-4-5) hurricanes during the 16-year period of 1995—2010 when the Atlantic thermohaline circulation (THC) was strong vs. the prior 16-year period of 1979—1994 when the THC was weak. Note that there were approximately 2.5 times as many major hurricanes when the THC was strong as when it was weak.

1. INTRODUCTION

The U.S. landfall of major hurricanes Dennis, Katrina, Rita, and Wilma in 2005 and the four Southeast landfalling hurricanes of 2004 — Charley, Frances, Ivan, and Jeanne, raised questions about the possible role that global warming played in those two unusually destructive seasons. In addition, three category 2 hurricanes (Dolly, Gustav, and Ike) pummeled the Gulf Coast in 2008 causing considerable devastation. Some researchers have tried to link the rising CO_2 levels with SST increases during the late 20th century and say that this has brought on higher levels of hurricane intensity.

These speculations that hurricane intensity has increased in partial response to human activity have been given much media attention and was supported by the recent IPCC-AR4 report. Observational data do not support such an assessment.

Because there has been such a large increase in Atlantic basin major hurricane activity since 1995 in comparison with the very inactive prior 16-year period of 1979—1994 (Fig. 1) as well as the prior quarter-century period of 1970—1994 — it has been tempting for many who do not have a strong background of hurricane information to jump on this recent increase in major hurricane activity as strong evidence of a human influence on these systems. It should be noted, however, that the last 16-year active major hurricane period of 1995—2010 has not been more active than the earlier 16-year period of 1949—1964 when the Atlantic Ocean circulation conditions were similar to what has been observed over the last 16 years. These earlier

TABLE 1   Comparison of annual Atlantic basin hurricane activity in two 16-year periods when the Atlantic Ocean THC was strong vs. an intermediate period (1970–1994) when the THC was weak

	THC (or AMO)	SST (10–15°N; 70–40°W)	Avg. CO_2, ppm	NS	NSD	H	HD	MH	MHD	ACE	NTC
1949–1964 (16 years)	Strong	27.93	319	10.1	54.1	6.5	29.9	3.8	9.5	121	133
1970–1994 (25 years)	Weak	27.60	345	9.3	41.9	5.0	16.0	1.5	2.5	68	75
1995–2010 (16 years)	Strong	28.02	373	14.6	74.1	7.8	32.0	3.8	9.4	140	153
Per year ratio Strong/weak THC		0.35 °C	~0	1.3	1.5	1.4	1.9	2.5	3.7	1.9	1.9

active conditions occurred even though atmospheric CO_2 amounts were lower than now.

Table 1 shows how large were the Atlantic basin hurricane variations between strong and weak THC periods. Note especially how large the ratio is for major hurricane days (3.7) during strong vs. weak THC periods. Normalized U.S. hurricane damage studies by Pielke and Landsea (1998) and Pielke et al. (2008) show that landfalling major hurricanes account on average for about 80–85% of all hurricane-related destruction even though these major hurricanes make up only 20–25% of named storms.

Although global surface temperatures increased during the late 20th century, there are no reliable data to indicate increased hurricane frequency or intensity in any of the globe's other tropical cyclone basins since 1979. Global Accumulated Cyclone Energy (ACE) shows significant year-to-year and decadal variability but a distinct decreasing trend during the last 20 years (Fig. 2). Similarly, Klotzbach (2006) found no significant change in global TC activity during the period from 1986 to 2005.

2. CAUSES OF THE UPSWING IN ATLANTIC BASIN MAJOR HURRICANE ACTIVITY SINCE 1995

The Atlantic Ocean has a strong multi-decadal signal in its hurricane activity which is due to multi-decadal variations in the strength of the THC (Fig. 3). The oceanic and atmospheric response to the THC is often referred to as the Atlantic Multi-decadal Oscillation (AMO). We use the THC and AMO interchangeably

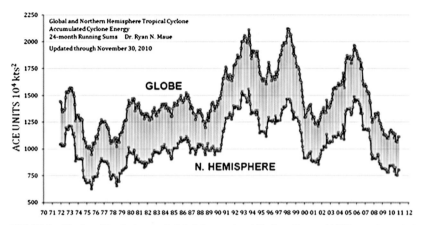

FIGURE 2   Northern Hemisphere and global Accumulated Cyclone Energy (ACE) over the period from 1971 to November 2010. Figure has been adapted from Ryan Maue, Center for Ocean-Atmospheric Prediction Studies, Florida State University.

throughout the remainder of this discussion. The strength of the THC can never be directly measured, but it can be diagnosed, as we have done, from the magnitude of the sea surface temperature anomaly (SSTA) in the North Atlantic (Fig. 4) combined with the sea level pressure anomaly (SLPA) in the Atlantic between the latitudes of the equator and 50 °N (Klotzbach and Gray, 2008). Background information on the nature of the THC and its variations can be found in papers by Broecker (1991), Bjerknes (1964), Curry and McCartney (2001), Hurrell (1995), and van Loon and Rogers (1978).

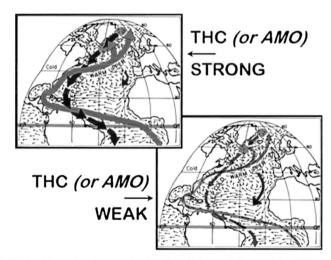

FIGURE 3   Illustration of strong (top) and weak (bottom) phases of the THC or AMO.

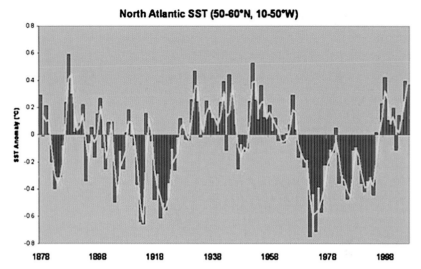

FIGURE 4 Long-period portrayal (1878–2006) of North Atlantic sea surface temperature anomalies (SSTA). The red (warm) periods are when the THC (or AMO) is stronger than average and the blue periods are when the THC (or AMO) is weaker than average.

The THC (or AMO) is strong when there is an above-average poleward advection of warm low-latitude waters to the high latitudes of the North Atlantic. This water can then sink to deep levels when it reaches the far North Atlantic in a process known as North Atlantic Deep Water Formation (NADWF). The water then moves southward at deep levels in the ocean. The amount of North Atlantic water that sinks is proportional to the water's density which is determined by its salinity content as well as its temperature. Salty water is denser than freshwater especially at water temperatures near freezing. There is a strong association between North Atlantic SSTA and North Atlantic salinity as calculated from the Simple Ocean Data Assimilation (SODA) reanalysis (Fig. 5). High salinity implies higher rates of NADWF (or subsidence) and thus a stronger flow of upper-level warm water from lower latitudes as replacement. See the papers by Gray et al. (1996), Goldenberg et al. (2001), and Grossmann and Klotzbach (2009) for more discussion.

3. WHY CO_2 INCREASES ARE NOT RESPONSIBLE FOR ATLANTIC SST AND HURRICANE ACTIVITY INCREASES

Theoretical considerations do not support a close climatological relationship between SSTs and hurricane intensity. In a global warming world, the atmosphere's upper air temperatures will warm or cool in unison with longer-period SST changes. Vertical lapse rates will thus not be significantly altered in a somewhat warmer tropical oceanic environment. We have no plausible

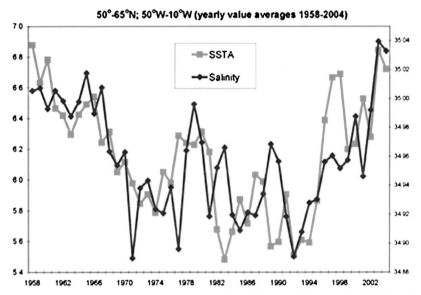

FIGURE 5   Illustration of the strong association of annually-averaged North Atlantic SSTA and North Atlantic salinity content between 1958 and 2004. Salinity data are from SODA as discussed in the text.

physical reasons for believing that Atlantic hurricane frequency or intensity will significantly change if global or Atlantic Ocean temperatures were to rise by $1-2\,^{\circ}C$. Without corresponding changes in many other basic features, such as vertical wind shear or mid-level moisture, little or no additional TC activity should occur with SST increases.

3.1. Confusing Time Scales of SST Influences

A hurricane passing over a warmer body of water, such as the Gulf Stream, will often undergo intensification. This is due to the sudden lapse-rate increase which the hurricane's inner-core experiences when it passes over warmer water. The warmer SSTs cause the hurricane's lower boundary layer temperature and moisture content to rise. While these low-level changes are occurring, upper-tropospheric conditions are often not altered significantly. These rapidly occurring lower- and upper-level temperature differences cause the inner-core hurricane lapse rates to increase and produce more intense inner-core deep cumulus convection. This typically causes a rapid increase in hurricane intensity. Such observations have led many observers to directly associate SST increases with greater hurricane potential intensity. This is valid reasoning for day-to-day hurricane intensity change associated with hurricanes moving over warmer or colder patches of SST. But such direct reasoning does not hold for conditions occurring in an overall climatologically warmer (or cooler) tropical

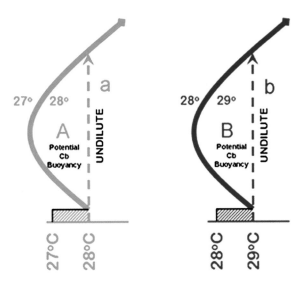

FIGURE 6   Illustration of how SST increases of 1 °C will bring about higher planetary boundary layer (PBL) temperature and moisture increases that will also occur in small amounts throughout the troposphere. The combination of these changes is such that potential buoyancy for cumulonimbus (Cb) development is not significantly altered by increases in SST alone — area 'B' is no larger than area 'A' even though area 'B' has a higher SST.

oceanic environment where broad-scale global and tropical rainfall conditions are not expected to significantly vary. During long-period climate change, temperature and moisture conditions rise at both lower and upper levels. Lapse rates are little affected (Fig. 6).

Any warming-induced increase in boundary layer temperature and moisture will be (to prevent significant global rainfall alteration) largely offset by a similar but weaker change through the deep troposphere up to about 10 km height. Upper-tropospheric changes are weaker than boundary layer changes, but they occur through a much deeper layer. These weaker and deeper compensating increases in upper-level temperature and moisture are necessary to balance out the larger increases in temperature and moisture which occur in the boundary layer. Global and tropical rainfall would be altered significantly if broad-scale lapse rates were ever to be altered to an appreciable degree.

Thus, we cannot automatically assume that with warmer global SSTs that we will necessarily have more intense hurricanes due to lapse-rate alterations. We should not expect that the frequency and/or intensity of Category 4−5 hurricanes will necessarily change as a result of changes in global or individual storm basin SSTs. Historical evidence does not support hurricanes being less intense during the late 19th century and the early part of the 20th century when SSTs were slightly lower.

3.2. CO_2 Influence on Hurricane Activity

We have been performing research with the International Satellite Cloud Climatology Project (ISCCP) and the NOAA National Centers for Environmental Prediction/National Center for Atmospheric Research (NCEP/NCAR) Reanalysis data sets. We have used this data to make an annual average of the global tropical ($30°N-30°S$; $0-360°$) energy budget (Fig. 7) for the years from 1984 to 2004. Note that the various surface and top of the atmosphere energy fluxes are very large. For the tropical surface, for instance, there are 637 W m^{-2} units of downward incoming solar and infrared (IR) energy. This downward energy flux is largely balanced by an upward surface energy flux of 615 W m^{-2} which is due to upward fluxes from IR radiation, evaporated liquid water, and sensible heat. Similar large energy fluxes are present at the top of the atmosphere and within the troposphere.

It has been estimated that a doubling of CO_2 (from the pre-industrial period) without any feedback influences would result in a blockage of OLR to space of about 3.7 W m^{-2}. The currently-measured value of CO_2 in the atmosphere is 385 parts per million by volume (ppmv). If we take the background pre-industrial value of CO_2 to be 285 ppmv, then by theory we should currently be having (from CO_2 increases alone) about $(100/285) \times 3.7 = 1.3$ W m^{-2} less OLR energy flux to space than was occurring in the mid-19th century.

This reduced OLR of 1.3 W m^{-2} is very small in comparison with most of the other tropical energy budget exchanges. Slight changes in any of these other

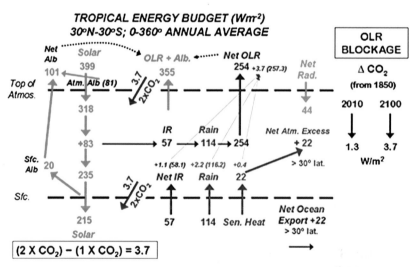

FIGURE 7   Vertical cross-section of the annual tropical energy budget as determined from a combination of ISCCP and NCEP/NCAR reanalysis data over the period from 1984 to 2004. Abbreviations are **IR** for longwave infrared radiation, **Alb** for albedo, and **OLR** for outgoing longwave radiation. The tropics receives an excess of about 44 W m^{-2} radiation energy which is convected and exported as sensible heat to latitudes poleward of 30°. Estimates are about half (22 W m^{-2}) of this excess is transported by the atmosphere and the other half is transported by the oceans. Note, on the right, how small an OLR blockage has occurred up to now due to CO_2 increases ($\sim$1.3 W m^{-2}) and the blockage of 3.7 W m^{-2} that will occur from a doubling of CO_2 by the end of this century.

larger tropical energy budget components could easily negate or reverse this small CO_2-induced OLR blockage. For instance, an upper-tropospheric warming of about 1 °C with no change in moisture would enhance OLR sufficient that it would balance the reduced OLR influence from a doubling of CO_2. Similarly, if there were a reduction of upper-level water vapor such that the longwave radiation emission level to space were lowered about 6 mb (~120 m) there would be an enhancement of OLR (with no change of temperature) sufficient to balance the suppression of OLR from a doubling of CO_2. The 1.3 W m^{-2} reduction in OLR that we have experienced since the mid-19th century (about one-third of the way to a doubling of CO_2) is very small when compared with the overall 399 W m^{-2} of solar energy impinging on the top of the tropical atmosphere and the mostly compensating 356 W m^{-2} of OLR and albedo energy going back to space. This 1.3 W m^{-2} energy gain (0.37% of the net energy returning to space) is much too small to ever allow a determination of its possible influence on TC activity. Any such potential CO_2 influence on TC activity is deeply buried as turbulence within the tropical atmospheres' many other energy components. It is possible that future higher atmospheric CO_2 levels may cause a small influence on global TC activity. But any such potential influence would likely never be able to be detected, given that our current measurement capabilities only allow us to assess TC intensity to within about 5 mph.

4. CONTRAST OF THEORIES OF HURRICANE ACTIVITY CHANGES

4.1. Theory of Human-Induced Increases Due to Rising CO_2 Levels

Those who think CO_2 increases have in the past and will in the future cause significant increases in hurricane activity believe that the physics of the CO_2-hurricane association is directly related to radiation changes as indicated in Fig. 8. They view CO_2 as blocking OLR to space. This acts to warm SSTs and add moisture to the boundary layer just above the ocean surface. These changes cause an increase in lapse rates (the lower levels warm while upper levels do not change much) which lead to more deep cumulonimbus (Cb) convection. More Cb convection leads to a higher percentage of tropical disturbances forming into tropical cyclones and a greater spin-up of the inner-core of those systems which do form.

This physical argument is too simplistic. It has no empirical verification in any other global TC basin except for the Atlantic where SST changes are primarily a result of ocean circulation changes. Table 2 shows the correlation of ACE with late summer-early fall SSTs in the Main Development Regions of the Northeast Pacific, the Northwest Pacific and the Southern Hemisphere. Note the low (or even negative) correlations between ACE and SST in each of these three

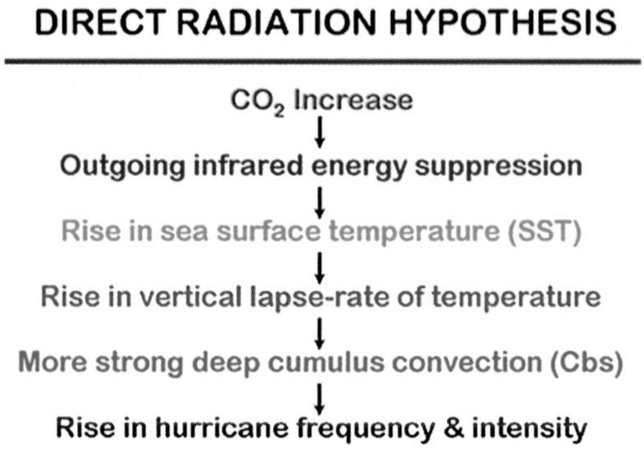

FIGURE 8 Physical linkage of those who believe that increases in CO_2 are making hurricanes more frequent and/or more intense.

DIRECT RADIATION HYPOTHESIS

CO_2 Increase
↓
Outgoing infrared energy suppression
↓
Rise in sea surface temperature (SST)
↓
Rise in vertical lapse-rate of temperature
↓
More strong deep cumulus convection (Cbs)
↓
Rise in hurricane frequency & intensity

TC basins. It is obvious that other physical processes besides SST are primarily responsible for differences in hurricane activity in these basins.

4.2. Theory of the THC (OR AMO)

We do not view seasonal hurricane variability in the Atlantic as being directly related to changes in CO_2-induced radiation forcing or to SST changes by themselves. For the Atlantic, we view long-period tropical cyclone variability primarily as a result of changes in the strength of the THC (or AMO). We hypothesize that these changes act as shown in Fig. 9. THC changes result in alterations of tropospheric vertical wind shear, trade-wind strength, and SSTs in the Main Development Region (MDR) of $10-20°N$; $20-70°W$ in the tropical Atlantic. A large component of the SST increase in this area is not a direct result of radiation differences but rather the combination of the effects of reduced southward advection of colder water in the east Atlantic and reduced trade-wind strength. Weaker trade winds reduce upwelling and evaporation and typically act to increase SST.

TABLE 2   Correlation of ACE with late summer-early fall SSTs in three TC basins from 1980 to 2009

	Yearly mean ACE	ACE vs. SST correlation (r)
Northeast Pacific	134	0.01
Northwest Pacific	310	−0.30
Southern Hemisphere	205	0.23
Globe (SST 20°N−20°S)	769	−0.08

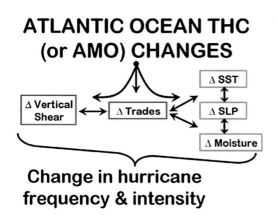

ATLANTIC OCEAN THC (or AMO) CHANGES

Change in hurricane frequency & intensity

FIGURE 9 Idealized portrayal of how changes in the Atlantic THC bring about various parameter changes in the Atlantic's MDR. Vertical shear, trade-wind strength, and SST are the key parameters which respond to THC changes. Favorable SLP and mid-level moisture changes occur in association with the shear, trade wind and SST changes. It is the combined package which is important, not the SST alone.

The influence of the warmer Atlantic SST, as previously discussed, is not primarily to enhance lapse rates and Cb convection but rather as a net overall positive influence on lowering the MDR's surface pressure and elevating mean upward tropospheric vertical motion and reducing vertical shear. This causes an increase in tropospheric moisture content. It is this combination of factors which brings about more TC activity.

5. DISCUSSION

In a global warming or global cooling world, the atmosphere's upper air temperatures will warm or cool in unison with the SSTs. Vertical lapse rates will not be significantly altered. We have no plausible physical reasons for believing that Atlantic hurricane frequency or intensity will change significantly if global ocean temperatures were to continue to rise. For instance, in the quarter-century period from 1945 to 1969 when the globe was undergoing a weak cooling trend, the Atlantic basin experienced 80 major (Category 3-4-5) hurricanes and 201 major hurricane days. By contrast, in a similar 25-year period from 1970 to 1994 when the globe was undergoing a general warming trend, there were only 38 Atlantic major hurricanes (48% as many) and 63 major hurricane days (31% as many) (Fig. 10). Atlantic SSTs and hurricane activity do not follow global mean temperature trends or amounts of CO_2.

5.1. US Landfall Observations

The most reliable long-period hurricane records we have are the measurements of US landfalling TCs since 1900 (Table 3). Although global mean ocean and Atlantic SSTs have increased by about 0.4 °C between two 55-year periods (1901−1955 compared with 1956−2010), the frequency of US landfall numbers actually shows a slight downward trend for the latter period. This

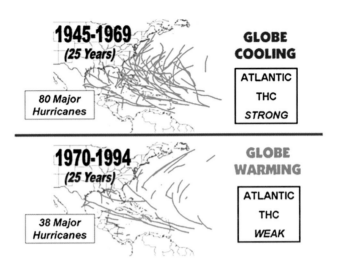

FIGURE 10   Tracks of major (Category 3-4-5) hurricanes during the 25-year period of 1945−1969 when the globe was undergoing a weak cooling vs. the 25-year period of 1970−1994 when the globe was undergoing a modest warming. CO_2 amounts in the later period were approximately 18% higher than in the earlier period. Major Atlantic hurricane activity was only about one-third as frequent during the latter period despite warmer global temperatures and higher CO_2 amounts.

downward trend is particularly noticeable for the US East Coast and Florida Peninsula where the difference in landfall of major (Category 3-4-5) hurricanes between the 45-year period of 1921−1965 (24 landfall events) and the 45-year period of 1966−2010 (7 landfall events) was especially large (Fig. 11). For the entire United States coastline, 39 major hurricanes made landfall during the earlier 45-year period (1921−1965) compared with only 26 major hurricanes for the latter 45-year period (1966−2010). This occurred despite the fact that CO_2 averaged approximately 365 ppm during the latter period compared with 310 ppm during the earlier period.

TABLE 3   U.S. landfalling tropical cyclones by intensity during two 55-year periods

Years	Named storms	Hurricanes	Major hurricanes (Cat. 3-4-5)	Global temperature increase
1901−1955 (55 years)	210	115	44	
1956−2010 (55 years)	180	87	34	+0.4°C

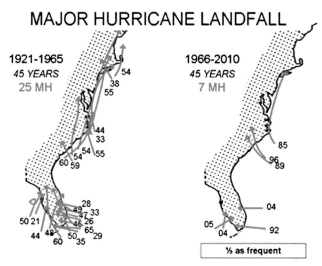

MAJOR HURRICANE LANDFALL

1921-1965
45 YEARS
25 MH

1966-2010
45 YEARS
7 MH

⅓ as frequent

FIGURE 11   Contrast of tracks of East Coast and Florida Peninsula major landfalling hurricanes during the 45-year period of 1921−1965 vs. the most recent 45-year period of 1966−2010.

Although 2005 had a record number of TCs (28 named storms), this should not be taken as an indication of something beyond natural processes. There have been several other years with comparable hurricane activity to 2005. For instance, 1933 had 21 named storms in a year when there was no satellite or aircraft data. Records of 1933 show all 21 named storms had tracks west of 60°W where surface observations were more plentiful. If we eliminate all of the named storms of 2005 whose tracks were entirely east of 60°W and therefore may have been missed given the technology available in 1933, we reduce the 2005 named storm total by seven (to 21) − the same number as was observed to occur in 1933.

Utilizing the National Hurricane Center's best track database of hurricane records back to 1875, six previous seasons had more hurricane days than the 2005 season. These years were 1878, 1893, 1926, 1933, 1950, and 1995. Also, five prior seasons (1893, 1926, 1950, 1961, and 2004) had more major hurricane days. Although the 2005 hurricane season was certainly one of the most active on record, it was not as much of an outlier as many have indicated.

We believe that the Atlantic basin remains in an active hurricane cycle associated with a strong THC. This active cycle is expected to continue for another decade or two at which time we should enter a quieter Atlantic major hurricane period like we experienced during the quarter-century periods of 1970−1994 and 1901−1925. Atlantic hurricanes go through multi-decadal cycles. Cycles in Atlantic major hurricanes have been observationally traced back to the mid-19th century. Changes in the THC have been inferred from Greenland paleo ice-core temperature measurements going back thousands of years. These changes are natural and have nothing to do with human activity.

6. IPCC-IV'S TROPICAL CYCLONE MIS-STATEMENTS

We completely disagree with the large number of papers written right after the time of the flurry of landfalling US major hurricanes during 2004–2005. We strongly disagree on how these authors interpreted the hurricane data to imply that rising levels of CO_2 were likely a significant contributing influence to the large amounts of hurricane destruction during those 2 years.

A number of these papers served as the basis for the IPCC-AR4 (2007) report concerning tropical cyclones of which one paragraph of the Executive Report (page 239) will be quoted:

"Intense tropical cyclone activity has increased since about 1970. Globally, estimates of the potential destructiveness of hurricanes show a significant upward trend since the mid-1970s, with a trend towards longer lifetimes and greater storm intensity, and such trends are strongly correlated with tropical SST. These relationships have been reinforced by findings of a large increase in numbers and proportion of hurricanes reaching categories 4 and 5 globally since 1970 even as total number of cyclones and cyclone days decreased slightly in most basins. The largest increase was in the North Pacific, Indian and southwest Pacific Oceans."

It is unfortunate indeed that, the IPCC-AR4 report, which shared a Noble Prize for science, would report this information which had already been rebutted by several studies at the time of its issue.

7. SPECIAL CHARACTERISTICS OF THE ATLANTIC OCEAN HURRICANE BASIN

The Atlantic Ocean basin is largely land-locked except for on its far southern margin. It has the highest salinity of any of the ocean basins and the highest upper-ocean density (due to the combination of high salinity and cold ocean temperatures) at its far northern latitudes (Fig. 12). Saline water has a higher density than freshwater at temperatures near freezing. The North Atlantic is the primary global location for the subsidence of upper-ocean water to deep levels in a process known as deep water formation. This can occur only if the upper-ocean water of the North Atlantic becomes denser than the water at deep levels. The global ocean fully ventilates itself every 1–2 thousand years through polar region (mainly the North Atlantic) subsidence to bottom levels of cold, dense, salty water and a compensating upwelling of less dense water in the Southern Hemisphere tropics (Figs. 13, 14). Ocean bio-life depends on this deep ocean ventilation.

The Atlantic Ocean is unique for having a continuous northward flow of upper-level water that moves into the polar region, cools, and then sinks because of its higher salinity-induced density. The deep water that is formed then returns to the Atlantic's southern fringes and mixes with the higher latitude water of the Southern Hemisphere. The Atlantic portion of this circulation feature has

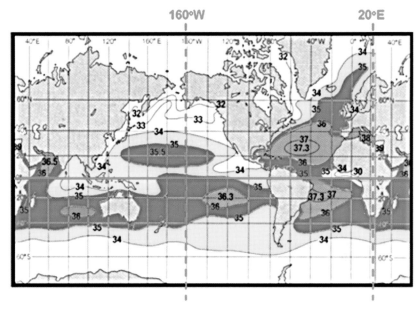

FIGURE 12   Average surface salinity of the global oceans. Notice the North Atlantic's high salinity values at northern latitudes.

GREAT OCEAN CONVEYOR BELT

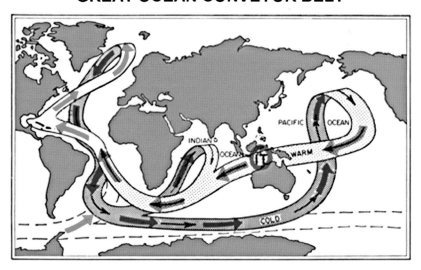

FIGURE 13   Modified view of the global ocean's deep circulation as originally portrayed by Wally Broecker, personal communication. Red arrows show upper-ocean circulation and blue arrows show deep ocean circulation.

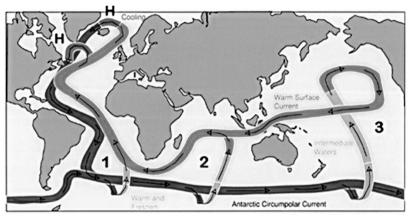

Courtesy of John Marshall (MIT)

FIGURE 14   A more recent portrayal of the ocean's deep water circulation Figure courtesy of John Marshall. Note upwelling of water in the tropics south of the equator as a response to the North Atlantic and Antarctic deep water subsidence.

been designated the Atlantic Thermohaline Circulation (THC). It is part of the Great Ocean Conveyor Belt or the Meridional Overturning Circulation (MOC).

The strength of the THC varies on multi-decadal time scales due to the nature of the Atlantic's naturally occurring multi-decadal salinity variations. When the THC is stronger than normal, the North Atlantic's upper-ocean water becomes warmer than usual and other tropical Atlantic meteorological parameter changes occur to cause more hurricane activity. The opposite occurs when the THC is weaker than average.

We have diagnosed that the THC has been significantly stronger than average since 1995 and during the period of the 1930s through 1960s. It was distinctly weaker than average in the quarter-century periods between 1970−1994 and 1900−1925.

The THC appears to be a product of the unique geometry of the Atlantic basin. The earth's most recent period of ice ages commenced about 2−3 million years ago and is associated with the time of Central American plate tectonic changes which lead to a rise of the Isthmus of Panama and the isolation of the Atlantic Ocean from the Pacific except at its southern margin. The Atlantic has since been mostly a closed ocean basin.

This sequestering has brought about special Atlantic climate conditions that were not present when the Atlantic and Pacific Oceans were connected. Before the filling in of the Isthmus of Panama, it was possible for ocean waters to flow freely between the Atlantic and Pacific. The isolation of the Atlantic together with the net energy deficit of the Western Hemisphere (in comparison with the Eastern Hemisphere) has acted to cause the development of especially large and strong surface high (or anticyclones) pressure systems in the Atlantic subtropics. These high pressure systems cause a strong suppression of Atlantic Basin sub-tropical

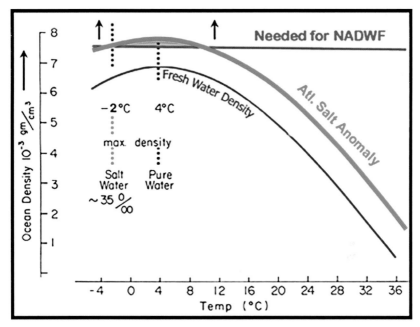

FIGURE 15   Portrayal of upper-level freshwater ocean density as a function of ocean temperature (blue line) and the typical variation of the combined fresh and salt water density as a function of ocean temperature (red line). When salt content is high enough and water has been cooled to near 4 °C, upper-ocean water is dense enough to sink to deep water levels forming NADWF.

rainfall. However, sub-tropical surface winds and evaporation rates (~1.2 m of water/year) remain quite high. From 30°S to 40°N, the Atlantic Ocean's net rate of surface evaporation (E) is substantially larger than the net rate of precipitation (P) by amounts as high as 30–40%. Such large positive values of evaporation minus precipitation ($E - P$) are not found in any other large areas of the global oceans. The Atlantic's unique geometry allows for the development and sustenance of high salinity conditions not experienced by the other ocean basins.

Salinity has a strong positive influence on water density. The higher salinity content of the cold water in the North Atlantic upper water enables it to sink to deep (or bottom) ocean levels where it then flows southward to the south Atlantic. Upper-level poleward moving North Atlantic water that is able to retain high salinity values and cool to temperature values near 4 °C (maximum density of freshwater) is dense enough to sink to deep levels. High salinity North Atlantic upper-level cold water is some of the globe's densest water. Such high density North Atlantic upper water is able to sink to deep levels forming NADWF. This deep water then flows southward to mix with the circumpolar vortex of the Southern Hemisphere and eventually to upwell in the Southern Hemisphere tropics. Figures 15 and 16 display some of the unique characteristics of the North Atlantic.

FIGURE 16   Portrayal of the typical latitudinal variation of North Atlantic upper water density due to variations in both ocean temperature (blue) and salinity (red). Their combined effect is given in purple. It is only in the high latitudes from approximately 55–70°N of the Atlantic that upper-ocean water is able to become dense enough to sink to bottom ocean levels and form NADW.

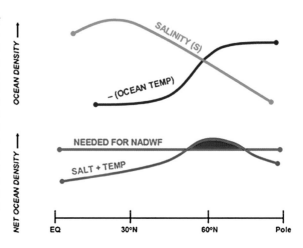

It is estimated (Schmitz, 1996) that the average strength of the Atlantic THC or the amount of water which sinks from upper to lower levels in the North Atlantic is about 14 Sverdrup (Sv) [$1\,Sv = 10^6\,m^3\,s^{-1}$]. This is equivalent to a yearly mass evacuation of Atlantic upper water occupying a volume of 1,000 km by 5,000 km to a depth of 100 m. The poleward advecting and cooling of this water are accomplished by ocean to atmosphere energy transfer and by the mixing of colder high latitude water. This causes a significant warming of the high latitude Arctic atmosphere and a decrease in high latitude westerly winds. It is estimated that the strength of the THC can undergo significant yearly, decadal, and century changes.

8. VARIATIONS IN STRENGTH OF THE THC

The Atlantic THC can vary due to a number of factors, such as the rate of buildup of Atlantic sub-tropical salinity, the salinity content of the South Atlantic water flowing into the North Atlantic, the rate of Atlantic evaporation over precipitation $(E - P)$, the amount and salinity reduction of upper water flowing into the North Atlantic due to Arctic and Labrador Sea current mixing, the amount of freshwater carried by rivers emptying into the North Atlantic, and the amount of high latitude Atlantic rainfall (Figs. 17–19). There must not be too much salinity diminution during the upper-ocean's advection from sub-tropical to high latitudes if a strong THC is to be maintained. If salinity is reduced too much, the upper ocean cannot maintain a high enough density to be able to sink in large quantities to deep levels.

When the THC is stronger than normal and higher amounts of saline water are being exported out of the sub-tropical gyre that are beyond what the sub-tropical gyre can replace through positive amounts of evaporation minus

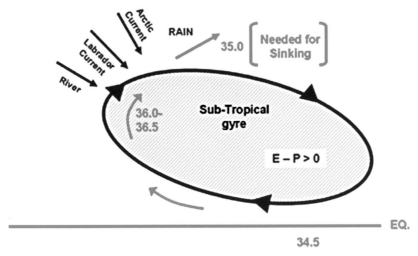

FIGURE 17 Idealized portrayal of North Atlantic sub-tropical upper-ocean gyre circulation conditions where salinity is continuously accumulating on the west side of the gyre. This high salinity water is then advected poleward where salinity diminution occurs due to its mixing with lower salinity water as well as due to rainfall. If salinity can be maintained at values as high or higher than about ~35 g kg^{-1}, then NADWF will occur.

precipitation $(E - P)$, then there will, in time, be a gradual reduction in the strength of the THC.

In the opposite sense, when the THC is weaker than normal, there is more time for salinity within the sub-tropical ocean gyre to gradually increase

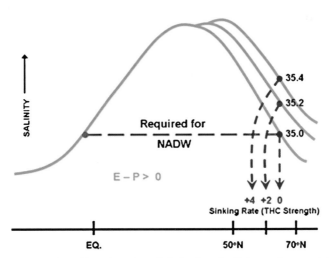

FIGURE 18 Illustration of how the poleward diminution of salinity can be a crucial factor in determining the strength of the THC. Variations of just a few units of 0.1 g kg^{-1} can determine whether the THC will be strong or weak.

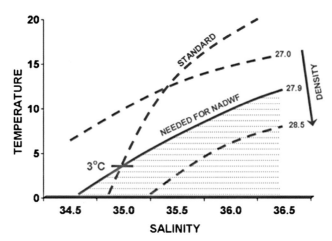

FIGURE 19   Illustration of the usual temperature and salinity changes that are required of the poleward flowing upper-level Atlantic water in order to be able to form NADW (red area). The standard blue curve shows the typical simultaneous changes in temperature and salinity that occur with poleward moving upper-level water in order that it has sufficient density to be able to sink to deep levels.

through evaporation being greater than rainfall. This allows salinity to gradually increase. There will then come a time (a decade or two later) when salinity has increased to the point where the THC becomes strong again. Atlantic salinity and the strength of the THC thus tend to vary inversely with each other as shown in Figs. 20, 21. There are also periods when Arctic ice flow and/or enhanced Labrador currents may so dilute the THC with freshwater that it is weakened.

When the THC is weak, the high latitude Atlantic and the atmosphere above it receive significantly less ocean-induced thermal energy than when the THC is strong. A weak THC causes the high latitude Atlantic ocean and atmosphere to cool and the westerly winds to strengthen (e.g., an increase in polarity of the North Atlantic Oscillation (NAO) and Arctic Oscillation (AO) indices). A strong THC typically brings about a warmer high latitude North Atlantic Ocean and atmosphere and weaker westerly winds (i.e., the NAO and AO decrease). It must be remembered that in a mechanical sense the atmosphere dissipates its kinetic energy at a rate of about 10% per day. The maintenance of the THC for multi-decadal periods, in a primary way, cannot be thought of as a consequence of the wind fields. The THC is a more dominant feature than the atmospheric wind circulation. The THC is a steady feature that once established, can maintain itself for years at a time. By contrast, the westerly wind currents, due to their rapid dissipation, must be fully regenerated on a time scale of 8−10 days.

The higher the salinity values coming out of the Atlantic subtropics, the greater is the amount of freshwater (from lower salinity water melting, rain,

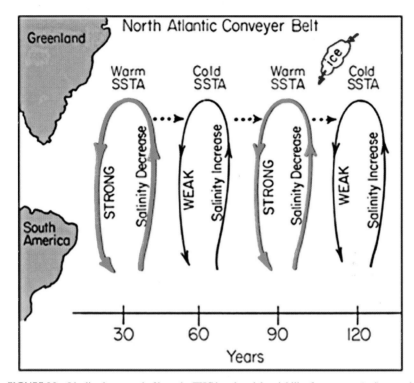

FIGURE 20 Idealized portrayal of how the THC has decadal variability from strong (red) to weak (black) and back to strong again in response to altering salinity patterns and deep water formation. Ice flows from the Arctic (upper right of figure) can sometimes disrupt or greatly alter these multi-decadal THC patterns.

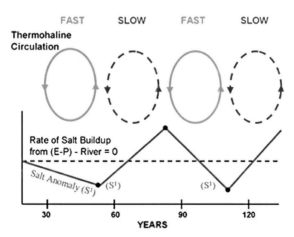

FIGURE 21 Illustration of how North Atlantic salinity values (green curve at bottom) are hypothesized to go up-and-down in response to the varying strength of the THC.

and river run-off) which the poleward moving ocean current can ingest and still be dense enough to produce NADW. The strength of the THC is thus related to the amount of poleward salinity advection from the south minus the amount of salinity diminution which the poleward moving water ingests from mixing with less saline ocean water and from river and rainfall freshwater inputs. The temperature of the poleward flowing water is less of a factor for density variation than is salinity. Variations of upper-ocean temperature around 4 °C cause very little variations in density as compared with salinity.

9. STORAGE OF WEST-ATLANTIC SUB-TROPICAL SALINITY

There can be a substantial storage of salinity in the sub-tropical Atlantic down to 500−600 m depth (Fig. 22). This depth of high salinity water is aided by the sub-tropical anticyclonic gyre having high values of $E - P$ that continually raises salt content. Also, the sub-tropical gyre winds cause a strong Ekman type of mechanically-forced ocean subsidence. This gradually drives upper ocean high salinity water to even deeper levels.

It is possible for the salt content of the subtropics to maintain a strong THC even when the rate of salt buildup is less than the rate of salt being advected poleward. It is necessary, however, that the THC not be too strong for too long a period so that it depletes too much of the sub-tropical salt gain from evaporation. The sub-tropical buildup of salt by evaporation over precipitation is a steadier feature than the strength of the THC.

Assuming that salinity is reduced by about $1-2$ g kg^{-1}, as the Atlantic THC water moves from the high salinity region of the west-Atlantic anticyclone to the North Atlantic where it sinks, then with an estimated THC of 14 Sv it is possible to continually maintain the THC and the high values of salinity within the sub-tropical high. On the long-period average, salinity buildup from evaporation minus precipitation $(E - P)$ must be balanced by salinity loss through poleward advection of higher salinity water.

If the THC is stronger or weaker than its average value of about 14 Sv, then salinity in the sub-tropical Atlantic would be decreased or increased by an amount approximately equivalent to the percentage alteration of the THC strength from 14 Sv. Thus, if the THC were 20 Sv in strength rather than 14 Sv there would be a gradual reduction in the rate of the Atlantic's sub-tropical gyre salinity buildup equivalent to about 25% of that required for steady state. Salt content in the sub-tropical gyre would thus be gradually reduced and some years later (if this reduction continues), the THC would begin to weaken. If the THC were, by contrast, to have a strength of 10 Sv rather than 14 Sv, there would be a rate of salinity increase within the Atlantic sub-tropical gyre that would be about 30% greater than would be required by salinity maintenance. Salt content would then gradually increase within the gyre, and some years later the THC would begin increasing in strength.

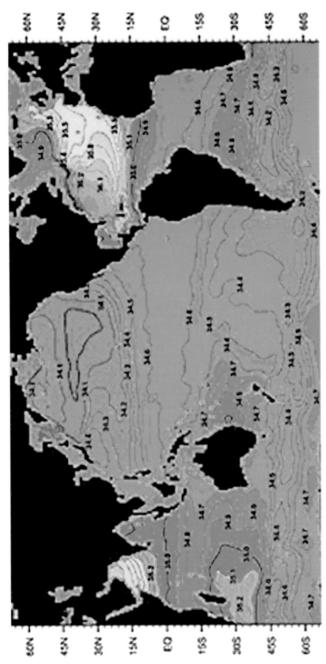

FIGURE 22 Salinity content of the global oceans at 500 m depth. Note the very high salinity contents of the western Atlantic sub-tropical region.

Such salinity variations and THC strength changes would be hard to detect within the Atlantic subtropics unless they persisted for a number of years. This is because the west-Atlantic subtropics maintain such a massive amount of high salinity which is continuously stored to deep levels. THC changes required to raise or lower significant amounts of salinity are, by comparison, small.

This large reservoir of high salinity water residing within the western Atlantic subtropics would require 20−30 years to deplete if the THC were to flow at its average strength with no replacement of salt from evaporation minus precipitation.

It is thus possible to store large amounts of salinity in the western North Atlantic subtropics for the maintenance of a strong THC for many years beyond the rate of its buildup replacement. Or oppositely, it is possible to maintain a weakened THC for many years despite steady salt buildup.

Another factor in determining the strength of the THC are the conditions on the opposite side of the globe in the Southern Hemisphere which allow for the mass compensating upwelling water to rise. It is required that there be an equivalent amount of upwelling water to balance NADW formation. This is possible through deep, high salinity, water mixing with less dense water and the development of positive upwelling buoyancy. There are likely times when favorable upwelling conditions in the Indian Ocean and Pacific basins are not present. The status of upwelling conditions is likely also a fundamental component of the MOC. They likely feed back to cause the strength of the THC to be altered from that specified solely by North Atlantic water density conditions alone.

10. SUMMARY

It is not possible to directly measure the strength of the THC. We think we can infer its strength from proxy measurements of the North Atlantic SST and salinity anomalies (which are directly related to each other) minus the SLPA over the broad Atlantic (0−50°N; 70°W−10°W). When the THC is strong the Atlantic atmospheric and oceanic sub-tropical gyres are weaker than normal. When the Atlantic THC is weaker than average, the gyres are stronger than normal (Figs. 23, 24).

With regards to multi-decadal variations of Atlantic major hurricane activity, it is possible to give a sequence of physical arguments (Fig. 25) for how a year or a multi-decadal period with a stronger than normal THC will have tropical Atlantic conditions associated which are more favorable for Atlantic basin major hurricane activity. Among these conditions are positive tropical Atlantic SSTAs, lower tropical Atlantic SLPAs, weaker trade winds, and smaller values of tropospheric vertical wind shear.

There is no evidence that Atlantic hurricane activity is significantly impacted by CO_2 increases or by global mean surface temperature changes. This myth should be put to rest. It is the natural variability of the Atlantic's meteorological parameters that we must be most concerned about.

FIGURE 23 Atlantic regions from which we diagnose the strength of the THC. We assume that the higher the SSTA plus salinity anomaly (SA) and the lower the SLPA, the stronger the THC. Our proxy for the THC is thus THC = [(SSTA + SA) − SLPA]. This proxy deviation is composed of the North Atlantic SSTA plus SA in the regions (50−65°N; 50°W−10°W) minus the SLPA anomaly in the region (0−50°N; 70°W−10°W). Positive anomalies indicate a stronger than normal THC and negative values indicate a weaker than average THC.

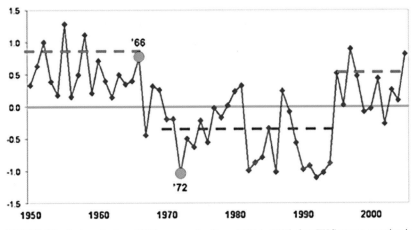

FIGURE 24 Portrayal of our THC proxy value from 1950 to 2006. Our THC proxy equation is given in Fig. 23.

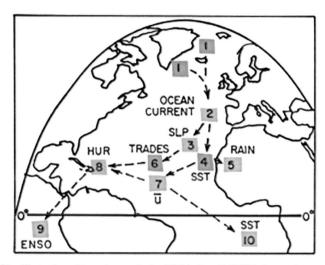

FIGURE 25   Illustration of how changes in the THC induce NADWF changes in area 1, causing ocean current changes in area 2 which lead to SLP (3), SST (4), and rain (5) changes which, in turn, cause changes in the strength of the trade winds (6), upper-tropospheric westerly winds (7), and other factors which lead to more or fewer hurricanes (8).

REFERENCES

Bjerknes, J., 1964. Atlantic air–sea interaction: Advances in Geophysics 10, 1–82.

Broecker, W.S., 1991. The great ocean conveyor: Oceanography 4 (7), 9–89.

Curry, R.G., McCartney, M.S., 2001. Ocean gyre circulation changes associated with the North Atlantic Oscillation. Journal of Physical Oceanography 31, 3374–3400.

Goldenberg, S.B., Landsea, C.W., Mestas-Nunez, A.M., Gray, W.M., 2001. The recent increase in Atlantic hurricane activity: causes and implications: Science 293, 474–479.

Gray, W.M., Sheaffer, J.D., Landsea, C.W., 1996. Climate trends associated with multi-decadal variability of intense Atlantic hurricane activity: Chapter 2 "Hurricanes, Climatic Change and Socioeconomic Impacts: A Current Perspective". In: Diaz, H.F., Pulwarty, R.S. (Eds.), 1996. Westview Press, p. 49.

Grossmann, I., Klotzbach, P.J., 2009. A review of North Atlantic modes of natural variability and their driving mechanisms: Journal of Geophysical Research 114.

Hurrell, J.W., 1995. Decadal trends in the North Atlantic Oscillation: regional temperatures and precipitation. Science 269, 676–679.

Klotzbach, P.J., 2006. Trends in global tropical cyclone activity over the past twenty years (1986–2005). Geophysical Research Letters 33.

Klotzbach, P.J., Gray, W.M., 2008. Multidecadal variability in North Atlantic tropical cyclone activity. Journal of Climate 21, 3929–3935.

Pielke Jr., R.A., Landsea, C.W., 1998. Normalized Atlantic hurricane damage, 1925–1995. Weather Forecasting 13, 621–631.

Pielke Jr., R.A., Gratz, J., Landsea, C.W., Collins, D., Masulin, R., 2008. Normalized hurricane damage in the United States: 1900–2005. Natural Hazard Revue 9, 29–42.

Schmitz, W.J., 1996. On the World Ocean Circulation: Volume I. Some Global Features/North Atlantic Circulation: Woods Hole Oceanographic Institute Technical Report, WHOI-96—03.

van Loon, H., Rogers, J.C., 1978. The seesaw in winter temperatures between Greenland and Northern Europe. Part I: General description: Monthly Weather Revue 106, 296—310.

Part IV

Solar Activity

Chapter 10

Solar Changes and the Climate

Joseph D'Aleo

Icecap US, 18 Glen Drive, Hudson, NH 03051, USA

Chapter Outline

1. INTRODUCTION

The IPCC AR4 discussed at length the varied research on the direct solar irradiance variance and the uncertainties related to indirect solar influences through variance through the solar cycles of ultraviolet and solar wind/geomagnetic activity. They admit that ultraviolet radiation by warming through ozone chemistry and geomagnetic activity through the reduction of cosmic rays and through that low clouds could have an effect on climate but in the end chose to ignore the indirect effect. They stated:

"Since TAR, new studies have confirmed and advanced the plausibility of indirect effects involving the modification of the stratosphere by solar UV irradiance variations (and possibly by solar-induced variations in the overlying mesosphere and lower thermosphere), with subsequent dynamical and radiative coupling to the troposphere. Whether solar wind fluctuations (Boberg and Lundstedt, 2002) or solar-induced heliospheric

Evidence-Based Climate Science. DOI: 10.1016/B978-0-12-385956-3.10010-5

modulation of galactic cosmic rays (Marsh and Svensmark, 2000b) also contribute indirect forcings remains ambiguous." (2.7.1.3)

For the total solar forcing, in the end the AR4 chose to ignore the considerable recent peer review in favor of Wang et al. (2005) who used an untested flux transport model with variable meridional flow hypothesis and reduced the net long-term variance of direct solar irradiance since the mini-ice age around 1,750 by up to a factor of 7. This may ultimately prove to be AR4's version of the AR3's "hockey stick" debacle.

2. EARTH–SUN CONNECTION

The sun is the ultimate source of all the energy on Earth; its rays heat the planet and drive the churning motions of its atmosphere. The brightness (irradiance) of the sun has been measured during recent 11 year solar cycles to vary just 0.1%. A conundrum for meteorologists was explaining whether and how such a small variation could drive major changes in weather patterns on Earth.

Though the sun's brightness or irradiance changes only slightly with the solar cycles, the indirect effects of enhanced solar activity including warming of the atmosphere in low and mid-latitudes by ozone reactions due to increased ultraviolet radiation, in higher latitudes by geomagnetic activity and generally by increased solar radiative forcing due to less clouds caused by cosmic ray reduction may greatly magnify the total solar effect on temperatures.

The following is an assessment of the ways the sun may influence weather and climate on short and long time scales.

2.1. The Sun Plays a Role in Our Climate in Direct and Indirect Ways

The sun changes in its activity on time scales that vary from 27 days to 11, 22, 80, 180 years and more. A more active sun is brighter due to the dominance of faculae over cooler sunspots with the result that the irradiance emitted by the sun and received by the earth is higher during active solar periods than during quiet solar periods. The amount of change of the solar irradiance based on satellite measurements since 1978 during the course of the 11-year-cycle just 0.1% (Willson and Hudson 1988) has caused many to conclude that the solar effect is negligible especially in recent years. Over the ultra-long cycles (since the Maunder Minimum), irradiance changes are estimated to be as high as 0.4% (Hoyt and Schatten, 1997; Lean et al., 1995; Lean, 2000; Lockwood and Stamper, 1999; Fligge and Solanki, 2000). This current cycle has seen a decline of 0.15%.

However, this does not take into account the sun's eruptional activity (flares, solar wind bursts from coronal mass ejections and solar wind bursts from coronal holes) which may have a much greater effect. This takes on more importance since Lockwood et al. (1999) showed how the total magnetic flux leaving the sun has increased by a factor of 2.3 since 1901. This eruptional activity may enhance warming through ultraviolet-induced ozone chemical reactions in the high atmosphere or ionization in higher latitudes during solar-induced geomagnetic storms. In addition, the work of Svensmark (2007), Palle Bago and Butler (2000), and Tinsley and Yu (2002) have documented the possible effects of the solar cycle on cosmic rays and through them the amount of low cloudiness. It may be that through these other indirect factors, solar variance is a much more important driver for climate change than currently assumed. Because, it is more easily measured and generally we find eruptional activity tracking well with the solar irradiance, we may utilize solar irradiance measurements as a surrogate or proxy for the total solar effect.

2.2. Correlations with Total Solar Irradiance

Studies vary on the importance of direct solar irradiance especially in recent decades. Lockwood and Stamper (GRL 1999) estimated that changes in solar luminosity can account for 52% of the change in temperatures from 1910 to 1960 but just 31% of the change from 1970 to 1999.

N. Scafetta and B. J. West of Duke University, in *"Phenomenological Solar Signature in 400 years of Reconstructed Northern Hemisphere Temperature Record"* (GRL 2006 and b0) showed how total solar irradiance accounted for up to 50% of the warming since 1900 and 25—35% since 1980. The authors noted the recent departures may result "from spurious non-climatic contamination of the surface observations such as heat-island and land-use effects [Pielke et al., 2002; Kalnay and Cai, 2003]". There analysis was done using the global databases which may also suffer from station dropout and improper adjustment for missing data which increased in the 1990s. In 2007, in their follow-up paper in the GRL, they noted the sun could account for as much as 69% of the changes since 1900.

This USHCN database though regional in nature would have been a better station database to use for analysis of change as it is more stable, has less missing data, and a better scheme for adjusting for missing data, as well as some adjustments for changes to siting and urbanization.

An independent analysis was conducted using the USHCN data and TSI data obtained from Hoyt and Schatten. The annual TSI composite record was constructed by Hoyt and Schatten (1993) (and updated in 2005) utilizing all five historical proxies of solar irradiance including sunspot cycle amplitude, sunspot cycle length, solar equatorial rotation rate, fraction of penumbral spots, and decay rate of the 11-year sunspot cycle.

The following includes a plot of this latest 11-year running mean solar irradiance vs. a similar 11-year running mean of NCDC annual mean US temperatures. It confirms this moderately strong correlation (r-squared of 0.59).

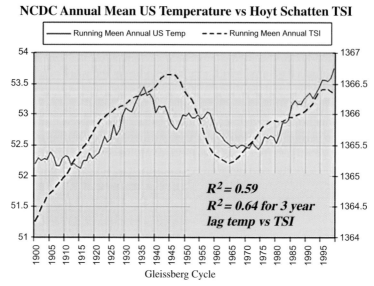

FIGURE 1   USHCN annual mean temperature (11-year running mean) correlated with Hoyt-Schatten-Willson total solar irradiance (also 11-year running mean).

The correlation increases to an *r*-squared value of 0.654 if you introduce a lag of 3 years for the mean USHCN data to the mean TSI. This is close to the 5-year lag suggested by Wigley and used by Scafetta and West. The highest correlation occurred with a 3-year lag (Fig. 1).

In recent years, satellite missions designed to measure changes in solar irradiance though promising have produced there own set of problems. As Judith Lean noted the problems is that no one sensor collected data over the entire time period from 1979 "forcing a splicing of from different instruments, each with their own accuracy and reliability issues, only some of which we are able to account for". Fröhlich and Lean in their 1998 GRL paper gave their assessment which suggested no increase in solar irradiance in the 1980s and 1990s.

Richard Willson, principal investigator of NASA's ACRIM experiments though in the GRL in 2003 was able to find specific errors in the data set used by Lean and Froelich used to bridge the gap between the ACRIM satellites and when the more accurate data set was used a trend of 0.05% per decade was seen which could account for warming since 1979 (Fig. 2).

Two other recent studies that have drawn clear connections between solar changes and the Earth's climate are Soon (2005) and Kärner (2002). Soon (2005 GRL) showed how the arctic temperatures (the arctic of course has no urbanization contamination) correlated with solar irradiance far better than with the greenhouse gases over the last century (see Fig. 3). For the 10-year running mean of total solar irradiance (TSI) vs. Arctic-wide air temperature

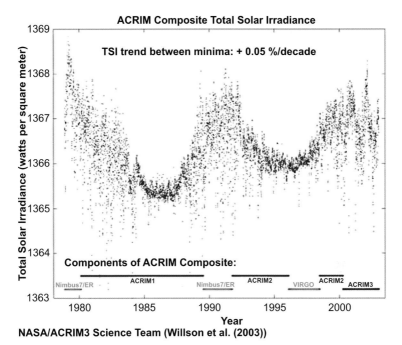

FIGURE 2 Richard Willson (ACRIMSAT) composite TSI showing trend of +0.05%/decade from successive solar minima.

anomalies (Polyokov), he found a strong correlation of (r-squared of 0.79) compared to a correlation vs. greenhouse gases of just 0.22.

2.3. Warming Due to Ultraviolet Effects Through Ozone Chemistry

Though solar irradiance varies slightly over the 11-year cycle, radiation at longer UV wavelengths is known to increase by several (6−8% or more) percent with still larger changes (factor of two or more) at extremely short UV and X-ray wavelengths (Baldwin and Dunkerton, JAS 2004).

Energetic flares increase the UV radiation by 16%. Ozone in the stratosphere absorbs this excess energy and this heat has been shown to propagate downward and affect the general circulation in the troposphere. Shindell et al. (1999) used a climate model that included ozone chemistry to reproduce this warming during high flux (high UV) years. Labitzke and Van Loon (1988) and later Labitzke in numerous papers have shown that high flux (which correlates very well with UV) produces a warming in low and middle latitudes in winter in the stratosphere with subsequent dynamical and radiative coupling to the troposphere. The winter of 2001/02, when cycle 23 had a very strong high flux

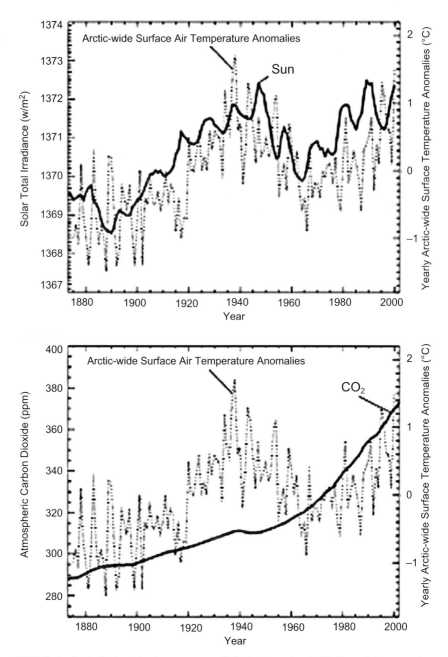

FIGURE 3 Arctic Basin wide air temperatures (Polyokov) correlated with Hoyt–Schatten total solar irradiance (TSI) and with annual average CO_2 (Soon 2005).

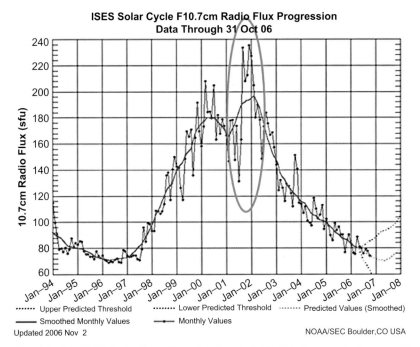

FIGURE 4 NOAA SEC solar flux (10.7 cm) during Cycle 23. Note the second solar max with extremely high flux from September 2001 to April 2002.

second maxima (Fig. 4) provided a perfect verification of Shindell and Labitzke and Van Loon's work.

The warming that took place with the high flux from September 2001 to April 2002 caused the northern winter polar vortex to shrink (Fig. 5) and the southern summer vortex to break into two centers for the first time ever observed (Fig. 7). This disrupted the flow patterns and may have contributed to the brief summer breakup (Fig. 8) of the Larsen ice sheet.

NASA reported on the use of the Shindell Ozone Chemistry Climate Model to explain the Maunder Minimum (Little Ice Age).

Their model showed (Fig. 9) when the sun was quiet in 1680, it was much colder than when it became active again 100 years later. "During this period, very few sunspots appeared on the surface of the Sun, and the overall brightness of the Sun decreased slightly. Already in the midst of a colder-than-average period called the Little Ice Age, Europe and North America went into a deep freeze: alpine glaciers extended over valley farmland; sea ice crept south from the Arctic; and the famous canals in the Netherlands froze regularly — an event that is rare today."

Writing in Environmental Research Letters (2010), Mike Lockwood et al. (2010) verified that solar activity does seem to have a direct correlation with Earth's climate by influencing North Atlantic blocking (NAO) as Shindell has

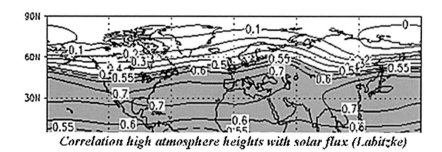

Correlation high atmosphere heights with solar flux (Labitzke)

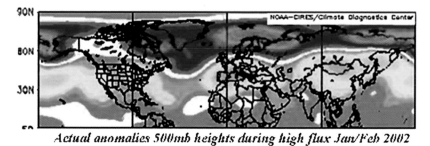

Actual anomalies 500mb heights during high flux Jan/Feb 2002

FIGURE 5   Labitzke correlated stratospheric heights with solar flux and actual height anomalies in the mid-troposphere during the high flux mode of the second solar max in early 2002.

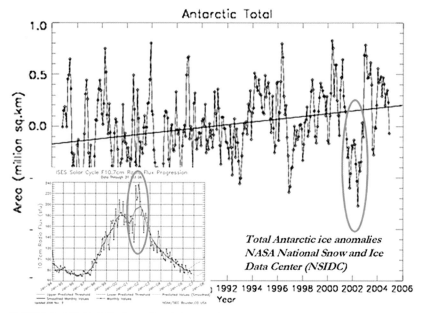

FIGURE 6   NASA NSIDC satellite derived total Antarctic ice extent anomalies from 1979 to 2005. Note the dropoff with the Larsen ice sheet break up in the summer of 2002 corresponding to major atmospheric changes during the high flux second solar maximum.

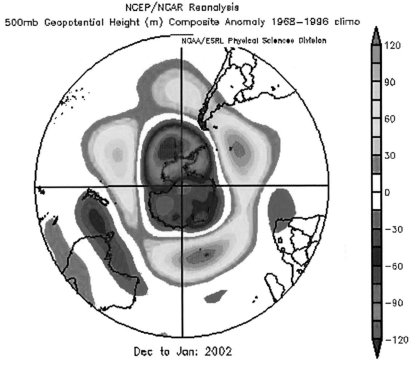

FIGURE 7 December 2001 to January 2002 500 mb height anomalies for Southern Hemisphere. Note the ring of warming with the high flux induced UV ozone chemistry as a ring surrounding a shrunken polar vortex as seen in the Northern Hemisphere in Fig. 2. Note how the vortex actually became a dipole with weakness in center. The changing winds and currents very likely contributed to the ice break of the Larsen ice sheet.

FIGURE 8 Larsen ice sheet break up late summer 2000 following strong solar flux break-up of southern polar vortex.

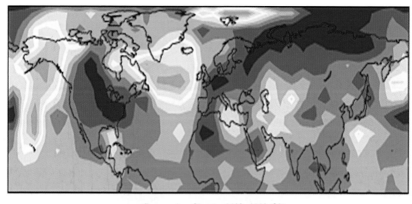

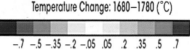

Temperature Change: 1680–1780 (°C)

−.7 −.5 −.35 −.2 −.05 .05 .2 .35 .5 .7

FIGURE 9 Shindell ozone chemistry model forecast of the difference between the quiet solar period of the Maunder Minimum and the active late 18ᵗʰ century. *Shindell NASA Observatory and Shindell 2001 http://earthobservatory.nasa.gov/IOTD/view.php?id=7122*

shown. The reason that the scope of the study is limited to that area, or at most Europe, is that it is one of the few regions that there is a reliable, continuous temperature record going back to the Little Ice Age.

They noted further "solar activity during the current sunspot minimum has fallen to levels unknown since the start of the 20ᵗʰ century. The Maunder Minimum (about 1650−1700) was a prolonged episode of low solar activity which coincided with more severe winters in the United Kingdom and continental Europe. Motivated by recent relatively cold winters in the UK, we investigate the possible connection with solar activity. We identify regionally anomalous cold winters by detrending the Central England temperature (CET) record using reconstructions of the northern hemisphere mean temperature.

We show that cold winter excursions from the hemispheric trend occur more commonly in the UK during low solar activity, consistent with the solar influence on the occurrence of persistent blocking events in the eastern Atlantic. We stress that this is a regional and seasonal effect relating to European winters and not a global effect. Average solar activity has declined rapidly since 1985 and cosmogenic isotopes suggest an 8% chance of a return to Maunder Minimum conditions within the next 50 years (Lockwood, 2010 *Proc. R. Soc.* A 466 303−29): the results presented here indicate that, despite hemispheric warming, the UK and Europe could experience more cold winters than during recent decades."

We note the NAO has ties to the Arctic Oscillation (AO) and together they have far greater influence on hemisphere than implied. This past winter with a record negative arctic oscillation and persistent negative NAO was the coldest in the UK and the southeastern United States since 1977/78, coldest in Scotland since 1962/63, coldest ever recorded in parts of Siberia. Coldest weather since 1971/72 was reported in parts of North China.

2.4. Tropical Effects

Elsner et al. (2010) found the probability of three or more hurricanes hitting the United States goes up drastically during low points of the 11-year sunspot cycle. Years with few sunspots and above-normal ocean temperatures spawn a less stable atmosphere and, consequently, more hurricanes, according to the researchers. Years with more sunspots and above-normal ocean temperatures yield a more stable atmosphere and thus fewer hurricanes. Elsner has found that between the high and low of the sunspot cycle, radiation can vary more than 10% in parts of the ultraviolet range. When there are more sunspots and therefore ultraviolet radiation, the warmer ozone layer heats the atmosphere below.

Elsner and Hodges found the probability of three or more hurricanes hitting the United States goes up drastically during low points of the 11-year sunspot cycle. Years with few sunspots and above-normal ocean temperatures spawn a less stable atmosphere and, consequently, more hurricanes, according to the researchers. Years with more sunspots and above-normal ocean temperatures yield a more stable atmosphere and thus fewer hurricanes.

The sun's yearly average radiance during its 11-year cycle only changes about one-tenth of 1%, according to NASA's Earth Observatory. But the warming in the ozone layer can be much more profound, because ozone absorbs ultraviolet radiation. Elsner has found that between the high and low of the sunspot cycle, radiation can vary more than 10% in parts of the ultraviolet range. When there are more sunspots and therefore ultraviolet radiation, the warmer ozone layer heats the atmosphere below. Their latest paper shows evidence that increased UV light from solar activity can influence a hurricane's power even on a daily basis (Fig. 10).

Meehl et al. (2009) found that chemicals (ozone) in the stratosphere and sea surface temperatures in the Pacific Ocean respond during solar maximum in a way that amplifies the sun's influence on some aspects of air movement. This can intensify winds and rainfall, change sea surface temperatures, and cloud cover over certain tropical and subtropical regions, and ultimately influence global weather. An international team of scientists led by the National Center for Atmospheric Research (NCAR) used more than a century of weather observations and three powerful computer models to tackle this question.

The answer, the new study found, has to do with the Sun's impact on two seemingly unrelated regions: water in the tropical Pacific Ocean and air in the stratosphere, the layer of the atmosphere that runs from around 6 miles (10 km) above Earth's surface to about 31 miles (50 km).

Hurricanes and the sunspot theory

Increased solar activity such as sunspots can warm upper layers of Earth's atmosphere, making the atmosphere more stable and decreasing hurricanes. Sunspot activity varies on an 11-year cycle. Researchers at Florida State University theorize that hurricane activity may increase as sunspots decrease. **Here's how:**

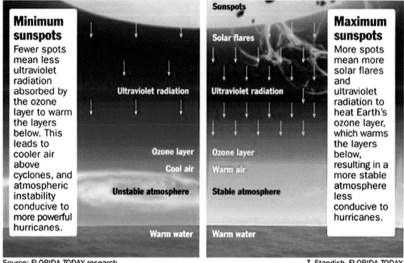

Source: FLORIDA TODAY research   T. Standish, FLORIDA TODAY

FIGURE 10   Research by Robert Hodges and Jim Elsner of Florida State University found the probability of three or more hurricanes hitting the United States goes up drastically during low points of the 11-year sunspot cycle, such as we're in now. Years with few sunspots and above-normal ocean temperatures spawn a less stable atmosphere and, consequently, more hurricanes, according to the researchers. Years with more sunspots and above-normal ocean temperatures yield a more stable atmosphere and thus fewer hurricanes.

The study found that chemicals in the stratosphere and sea surface temperatures in the Pacific Ocean respond during solar maximum in a way that amplifies the sun's influence on some aspects of air movement. This can intensify winds and rainfall, change sea surface temperatures, and cloud cover over certain tropical and subtropical regions, and ultimately influence global weather.

"The sun, the stratosphere, and the oceans are connected in ways that can influence events such as winter rainfall in North America," said lead author of the study, Gerald Meehl of NCAR. "Understanding the role of the solar cycle can provide added insight as scientists work toward predicting regional weather patterns for the next couple of decades."

The new study finds that weather patterns across the globe are partly affected by connections between the 11-year solar cycle of activity, Earth's stratosphere, and the tropical Pacific Ocean. The study could help scientists get an edge on eventually predicting the intensity of certain climate phenomena, such as the Indian monsoon and tropical Pacific rainfall, years in advance.

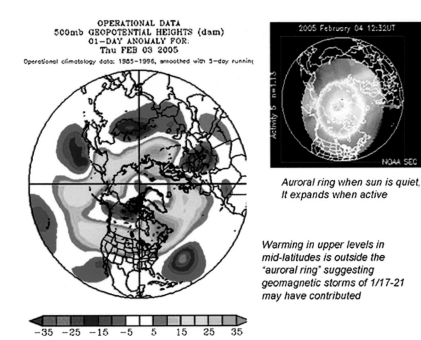

FIGURE 11 Anomaly at 500 mb 2 weeks after a major geomagnetic storm in 2005. Warmth seen in approximate location and shape of auroral ring.

2.5. Geomagnetic Storms and High Latitude Warming

When major eruptive activity (i.e., coronal mass ejections, major flares) takes place and the charged particles encounter the earth, ionization in the high atmosphere leads to the familiar and beautiful aurora phenomenon. This ionization leads to warming of the high atmosphere which like ultraviolet warming of the stratosphere works its way down into the middle troposphere with time.

Here is an example of an upper level chart two weeks after a major geomagnetic storm. Note the ring of warmth (higher than normal mid-tropospheric heights) surrounding the magnetic pole (Fig. 11).

2.6. Solar Winds, Cosmic Rays and Clouds

A key aspect of the sun's effect on climate is the indirect effect on the flux of Galactic Cosmic Rays (GCR) into the atmosphere. GCR is an ionizing radiation that supports low cloud formation. As the sun's output increases the solar wind shields the atmosphere from GCR flux. Consequently the increased solar irradiance is accompanied by reduced low cloud cover, amplifying the climatic effect. Likewise when solar output declines, increased GCR flux enters the

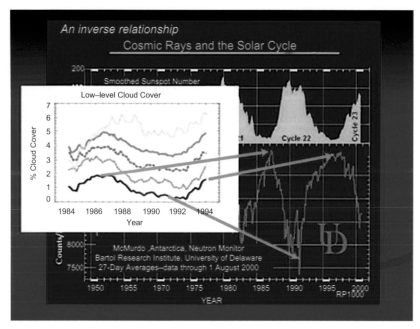

FIGURE 12 Cosmic ray neutrons are inversely proportional to solar activity and directly proportional to low cloudiness (from Palle Bago and Butler, 2001).

atmosphere, increasing low cloudiness and adding to the cooling effect associated with the diminished solar energy.

The conjectured mechanism connecting GCR flux to cloud formation received experimental confirmation in the recent laboratory experiments of Svensmark (Proceedings of the Royal Society, Series A, October 2006), in which he demonstrated exactly how cosmic rays could make water droplet clouds.

Palle Bago and Butler showed in 2002 (Int. J. Climate.) how the low clouds in all global regions changed with the 11-year cycle in inverse relation to the solar activity. Changes of 1−2% in low cloudiness could have a significant effect on temperatures through changes in albedo (Fig. 12).

Recently, Henrik Svensmark and Eigil Friis-Christensen published a reply to Lockwood and Fröhlich − The persistent role of the Sun in climate forcing, rebutting Mike Lockwood's Recent oppositely directed trends in solar climate forcings and the global mean surface air temperature. In it, they correlated tropospheric temperature with cosmic rays. Figure 13 features two graphs. The first graph compares tropospheric temperature (blue) to cosmic rays (red). The second graph removes El Nino, volcanoes, and a linear warming trend of 0.14 °C per decade.

Jasper Kirkby of CERN as an introduction to CERN's CLOUD experiment which "aims to study and quantify the cosmic ray-cloud mechanism in

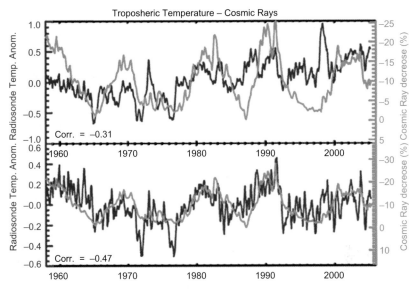

FIGURE 13 Tropospheric cosmic rays vs. radiosonde temperature anomalies raw and bottom filtered and detrended (Henrik Svensmark and Eigil Friis-Christensen).

a controlled laboratory experiment" and answer "the question of whether—and to what extent—the climate is influenced by solar/cosmic ray variability" found compelling evidence that indeed there could be such a connection (Figs. 14, 15).

Cosmic ray influence appears on the extremely long geologic time scales. Shaviv (JGR 2005 estimated from the combination of increased radiative forcing through cosmic ray reduction and the estimated changes in total solar luminosity (irradiance) over the last century that the sun could be responsible for up to 77% of the temperature changes over the 20th century with 23% for the anthropogenic. He also found the correlation extended back in the ice core data 500 million years (Fig. 16).

The degree to which cosmic rays is a climate driver is yet to be determined. It could still be that the cosmic rays are a proxy for one or more of the other solar factors, one more easily measured. There appears little doubt that the sun through many ways is the main driver for the earth's climate over time.

2.6.1. Throwback Solar Cycle 23

In NASA's David Hathaway's own words, Cycle 23 has been a cycle like we have not seen in century or more. The irradiance dropped 50% more than recent cycles, the solar wind was at the lowest levels of the satellite age. There were over 800 sunspotless days, well more than double those of the recent cycles. The cycle lasted 3 years longer than Cycle 22, longest since Cycle 6 that peaked in 1810 (Figs. 17, 18).

Cloud observations

- Original GCR-cloud correlation made by Svensmark & Friis-Christensen,1997
- Many studies since then supporting or disputing solar/GCR - cloud correlation
- Not independent - most use the same ISCCP satellite cloud dataset
- No firm conclusion yet - requries more data - but, if there is an effect, it is likely to be restricted to certain regions of globe and at certain altitudes & conditions
- Eg.correlation (>90% sig.) of low cloud amount and solar UV/GCR,1984-2004:

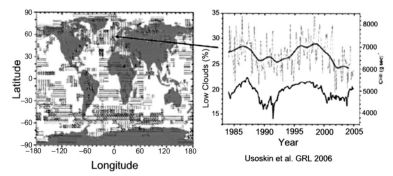

Usoskin et al. GRL 2006

FIGURE 14　Jasper Kirkby of CERN as an introduction to CERN's CLOUD experiment summarized the state of the understanding.

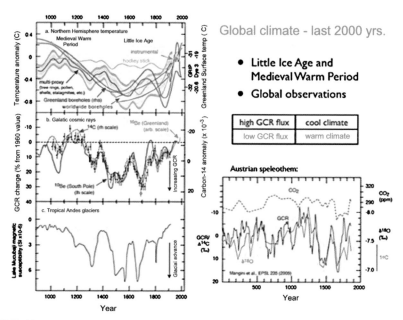

FIGURE 15　Jasper Kirkby of CERN shows excellent correlations of galactic cosmic rays and temperatures for the Northern Hemisphere, Greenland, Tropical Andes, and Austria.

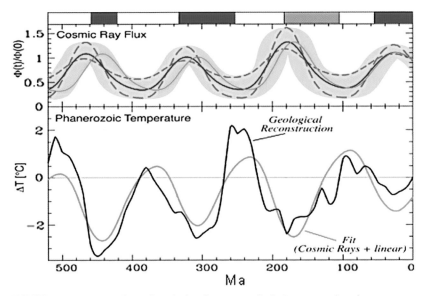

FIGURE 16   Shaviv cosmic ray flux plus irradiance vs. geological reconstruction of temperatures.

Historically, cycle length has correlated well with temperatures (Figs. 19, 20).

Cosmic rays according to NASA recently reached a space age high (Figs. 21, 22).

The Russian Pulkovo Observatory believes a Maunder like minimum is possible (Fig. 23).

2.7. Secular Cycles — Combined Natural Factors

Temperature trends coincides with the ocean and solar TSI cyclical trends as can be seen in this diagram that overlays standardized ocean temperature

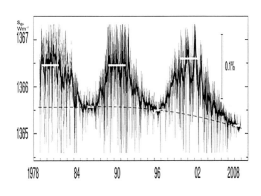

FIGURE 17   The irradiance in Cycle 23 dropped about 50% more than prior minima (with a change of near 0.15% from maxima).

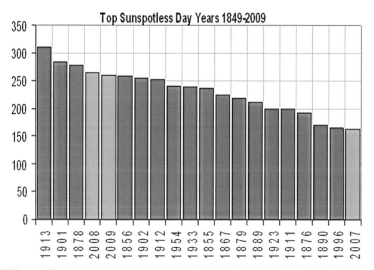

FIGURE 18   The number of sunspotless days per year since 1849. *Source: Daily Wolfnumbers from SIDC, RWC Belgium.*

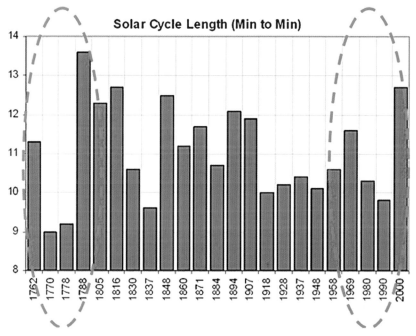

FIGURE 19   Solar cycle lengths (years) from minimum to minimum for cycles 1 through 23. Year of maximum shown as label.

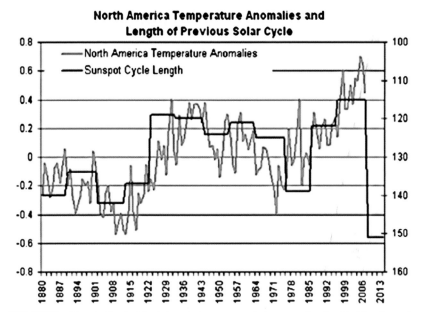

FIGURE 20 Historically, the North American temperatures have correlated well with solar cycle length. Note the rapid increase in length for Cycle 23, implying an upcoming cooling.

configuration indices (PDO + AMO and Hoyt Schatten/Willson TSI and USHCN version 2 temperatures). The 60-year cycle clearly emerges including that observed warming trend. The similarity with the ocean multidecadal cycle phases also suggest the sun play a role in their oscillatory behavior. Scafetta (2010) presents compelling evidence for this 60-year cyclical behavior (Fig. 24).

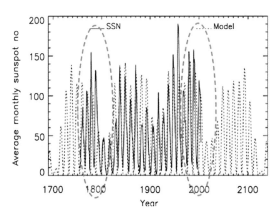

FIGURE 21 Clilverd et al. (2006) did a regression analysis of the various solar cycles and built a model that showed skill in predicting past cycles. The model suggests a Dalton like minimum is coming.

FIGURE 22 NASA depicted cosmic ray monitoring showing that the number of particles was approximately 19.4% higher than any other time since 1951.

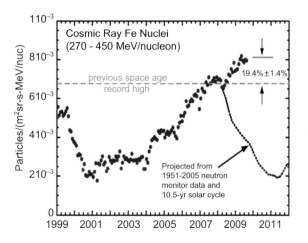

Dr. Don Easterbrook has used the various options of a 60-year repeat of the mid-20th century solar/ocean induced cooling, Dalton Minimum, and a Maunder Minimum scenarios to present this empirical forecast range of options (Fig. 25).

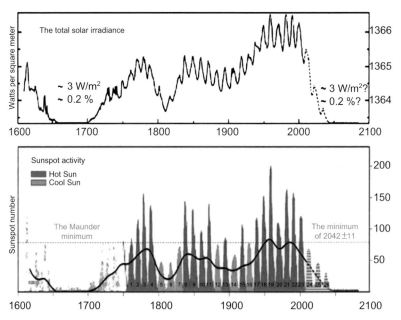

FIGURE 23   Habibullo Abdussamatov, Dr. Sc. Head of the Pulkovo Observatory has projected a decline in upcoming cycles to Maunder Minima levels.

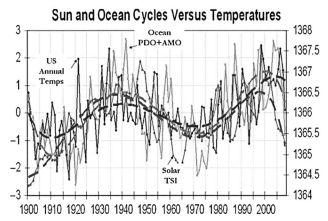

FIGURE 24   Solar TSI (Hoyt/Schatten TSI calibrated to Willson AMCRIMSAT TSI) and PDO + AMO (STD) vs. the USHCN annual plots with polynomial smoothing.

3. SUMMARY

Though the sun's brightness or irradiance changes only slightly with the solar cycles, the indirect effects of enhanced solar activity including warming of the atmosphere in low and mid-latitudes by ozone reactions due to increased ultraviolet radiation, in higher latitudes by geomagnetic activity and generally by increased radiative forcing due to less clouds caused by cosmic ray reduction may greatly magnify the total solar effect on temperatures. The sun appears to be the primary driver right up to the current time.

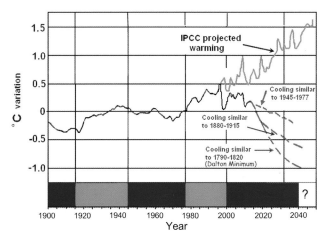

FIGURE 25   Projected future temperatures from the IPCC, to ocean/solar 60-year cycle cooling, to a Dalton minima to a Maunder Minima (D'Aleo and Easterbrook, 2010).

REFERENCES

Baldwin, M.P., Dunkerton, T.J., 2004, The solar cycle and stratosphere—troposphere dynamical coupling: JAS.

Boberg, F., Lundstedt, H., 2002. Solar wind variations related to fluctuations of the North Atlantic Oscillation. Geophysical Research Letters 29.

Clilverd, M.A., Clarke, E., Ulich, T., Rishbeth, H., Jarvis, M.J., 2006. Predicting solar cycle 24 and beyond: Space Weather 4.

D'Aleo, J., Easterbrook, D., 2010. Multidecadal Tendencies in ENSO and Global Temperatures Related to Multidecadal Oscillations. Energy & Environment 21 (5), 437—460.

Elsner, J.B., Jagger, T.H., Hodges, R.E., 2010. Daily tropical cyclone intensity response to solar ultraviolet radiation: Geophysical Research Letters 37.

Fröhlich, C., Lean, J., 1998. The sun's total irradiance: cycles, trends, and related climate change uncertainties since 1976. Geophysical Research Letters 25, 4377—4380.

Gleissberg, W., 1958. The 80-year sunspot cycle. Journal of British Astronomy Association 68, 150.

Hoyt, D.V., 1979. Variations in sunspot structure and climate: Climate Change 2, pp. 79—92.

Hoyt, D.V., Schatten, K.H., 1993. A discussion of plausible solar irradiance variations, 1700—1992. J. Geophys. Res. 98, 18895—18906.

Hoyt, D.V., Schatten, K.H., 1997. The Role of the Sun in Climate Change. Oxford University Press, New York.

Kärner, O., 2002. On non-stationarity and anti-persistency in global temperature series. J. Geophys. Res. 107 (D20) doi:10.1029/2001JD002024.

Labitzke, K., The global signal of the 11-year sunspot cycle in the stratosphere: differences between solar maxima and minima. Meteorology Zeitschrift 10, 83—90.

Labitzke, K., van Loon, H., 1988. Associations between the 11-year solar cycle, the QBO and the atmosphere. Part I: The troposphere and stratosphere in the northern hemisphere winter. J. Atmos. Terr. Phys. 50, 197—206.

Landscheidt, T., 2000. Solar wind near earth, indicator if variations in global temperatures. In: Vazquez, M., Schmiedere, E. (Eds.), The Solar Cycle and Terrestrial Climate. European Space Agency, Special Publication, 463, pp. 497—500.

Lean, J., 2000. Evolutiom of the Sun's spectral irradiance since the Maunder Minimum. Geophys. Res. Lett. 27, 2425—2428.

Lean, J., Beer, J., Bradley, R., 1995. Reconstruction of solar irradiance since 1610: Implications for climate change. Geophys. Res. Lett. 22, 3195—3198.

Lockwood, M., Stamper, R., 1999. Long-term drift of the coronal source magnetic flux and the total solar irradiance. Geophysical Res. Ltrs. GL900485 26 (16), 2461.

Lockwood, M., Stamper, R., Wild, M.N., 1999. A doubling of the sun's coronal magnetic field during the last 100 years. Nature 399, 437—439. (doi:10.1038/20867).

Lockwood, M., 2010. Solar change and climate: an update in the light of the current exceptional solar minimum. Proceedings of the Royal Society A 466, 303—329.

Lockwood, M., Harrison, R.G., Woollings, T., Solanki, S.K., 2010. Are cold winters in Europe associated with low solar activity? Environmental Research Letters 024001. doi:10.1088/1748—9326/5/2/024001.

Meehl, G.A., Arblaster, J.M., Matthes, K., Sassi, F., van Loon, H., 2009. Amplifying the Pacific climate system response to a small 11 year solar cycle forcing. Science 325, 1114—1118.

Moberg, A., et al., 2005. Highly variable Northern Hemisphere temperatures reconstructed from low- and high-resolution proxy data. Nature 433, 613—617.

Palle Bago, E., Butler, C.J., 2000. The influence of cosmic rays on terrestrial clouds and global warming. Astronomical Geophysics 41, 4.18–4.22.

Palle Bago, E., Butler, C.J., 2001. Sunshine records from Ireland: cloud factors and possible links to solar activity and cosmic rays. International Journal of Climatology 21, 709–729.

Scafetta, N., West, B.J., 2005. Estimated solar contribution to the global surface warming using the ACRIM TSI satellite composite. Geophysical Research Letters 32, L18713. doi:10.1029/2005GL023849.

Scafetta, N., West, B.J., 2006a. Phenomenological solar contribution to the 1900–2000 global surface warming. Geophysical Research Letters 33, L05708. doi:10.1029/2005GL025539.

Scafetta, N., West, B.J., 2006b. Reply to comment by J.L. Lean on "Estimated solar contribution to the global surface warming using the ACRIM TSI satellite composite". Geophysical Research Letters 33, L15702. doi:10.1029/2006GL025668.

Scafetta, N., West, B.J., 2006c. Phenomenological solar signature in 400 years of reconstructed Northern Hemisphere temperature. Geophysical Research Letters 33, L17718. doi:10.1029/2006GL027142.

Scafetta, N., West, B.J., 2007. Phenomenological reconstructions of the solar signature in the Northern Hemisphere surface temperature records since 1600. Journal of Geophysical Research 112, D24S03. doi:10.1029/2007JD008437.

Scafetta, N., 2010. Empirical evidence for a celestial origin of the climate oscillations and its implications. Journal of Atmospheric and Solar-Terrestrial Physics 10.1016/j.jastp.2010.04.015.

Singer, S.F., 2008. ed., Nature, Not Human Activity, Rules the Climate: Summary for Policy-makers of the Report of the Nongovernmental International Panel on Climate Change. The Heartland Institute, Chicago, IL, p. 855.

Shaviv, N.J., 2002. The spiral structure of the Milky Way, cosmic rays, and Ice Age epochs on Earth. New Astronomy 8, 39–77.

Shaviv, N.J., 2005. On climate response to changes in cosmic ray flux and radiative budget. Journal of Geophysical Research 110.

Shindell, D.T., Rind, D., Balachandran, N., Lean, J., Lonergan, P., 1999. Solar cycle variability, ozone, and climate. Science 284, 305–308.

Shindell, D.T., Schmidt, G.A., Mann, M.E., Rind, D., Waple, A., 2001. Solar Forcing of Regional Climate Change During the Maunder Minimum, Science 7 December, 2149–2152. doi:10.1126/science.1064363

Soon, W.H., 2005. Variable solar irradiance as a plausible agent for multidecadal variations in the Arctic-wide surface air temperature record of the past 130 years. Geophysical Research Letters 32, doi:10.1029/2005GL023429.

Soon, W.H., Posmentier, E., Baliunas, S.L., 1996. Inference of solar irradiance variability from terrestrial temperature changes: 1880–1993: an astrophysical application of the sun–climate relationship. Astrophysical Journal 472, 891–902.

Svenmark, H., Friis-Christensen, E., 1997. Variation of cosmic ray flux and global cloud cover—a missing link in solar–climate relationships. Journal of Atmospheric and Solar-Terrestrial Physics 59, 1125–1132.

Svensmark, H., 2007. Cosmoclimatology. Astronomical Geophysics 48, 18–24.

Theijl, P., Lassen, K., 2000. Solar forcing of the northern hemisphere land air temperature. Journal Atmospheric Solar Terrestrial Physics 62, 1207–1213.

Tinsley, B.A., Yu, F., 2002. Atmospheric ionization and clouds as links between solar activity and climate: in American Geophysical Union Monograph, Solar Variability and Its Effects on the Earth's Atmosphere and Climate System.

Veizer, J., Godderis, Y., François, L.M., 2000. Evidence for decoupling of atmospheric CO_2 and global climate during the Phanerozoic eon. Nature 408.

Wang, Y.-M., Lean, J.L., Sheeley Jr., N.R., 2005b. Modeling the Sun's magnetic field and irradiance since 1713. Ap. J. 625, 522−538. (doi:10.1086/429689).

Willson, R., 1997. Total solar irradiance trend during solar cycles 21 and 22. Science 277, 1963−1965.

Willson, R.C., Mordvinov, A.V., 2003. Secular total solar irradiance trend during solar cycles 21−23: Geophysical Research Letters 30, 1199.

Willson, R.C., Hudson, H.S., 1988. Solar luminosity variations in Solar Cycle 21. Nature, 810−812. U332U.

Willson, R.C., Mordvinov, A.V., 2003. Secular total solar irradiance trend during solar cycles 21−23. GRL 30 (5), 1199.

Chapter 11

The Current Solar Minimum and Its Consequences for Climate

David Archibald

Summa Development Limited

1. INTRODUCTION

A number of cycles in solar activity have been recognized, including the Schwabe (11 years), Hale (22 years), Gleissberg (88 years), de Vries (210 years), and Bond (1,470 years) cycles. There is nothing to suggest that cyclic behavior in solar activity has ceased for any reason. Therefore, predicting when the next minimum should occur should be as simple as counting forward from the last one. The last major minimum, the Dalton Minimum from 1798 to 1822, was two solar cycles long — Solar Cycles 5 and 6. Recent Gleissberg minima appear to be the decade 1690–1700, Solar Cycle 13 from 1889 to 1901, and Solar Cycle 20 from 1964 to 1976.

A de Vries cycle event, herein termed the Eddy Minimum, has started exactly 210 years after the start of the Dalton Minimum.

Friis-Christensen and Lassen theory, using methodology pioneered by Butler and Johnson at Armagh, can be used to predict the temperature response to the Eddy Minimum for individual climate stations with a high degree of confidence. The latitude of the US-Canadian border is expected to lose a month from its growing season with the potential for un-seasonal frosts to further reduce agricultural productivity.

Evidence-Based Climate Science. DOI: 10.1016/B978-0-12-385956-3.10011-7

2. THE CURRENT MINIMUM

As recently as 2008, there was a wide range in estimated amplitudes for Solar Cycle 24, from Dikpati at 190 and Hathaway at 170 respectively to Clilverd (2005) at 42 and Badalyan et al. (2001) at 50. This enormous divergence in projections of solar activity generated very little interest from the climate science community, despite the large impact it would have on climate (Archibald, 2006, 2007). The basis of Clilverd's prediction was a model for sunspot number using low-frequency solar oscillations, with periods of 22, 53, 88, 106, 213, and 420 years modulating the 11-year Schwabe cycle. The model predicts a period of quiet solar activity lasting until approximately 2030 followed by a recovery during the middle of the century to more typical solar activity cycles with peak sunspot numbers around 120.

The graphs in Figs. 1−15 show data related to solar activity. Additional data may be found in Archibald (2010).

The Eddy Minimum (Fig. 1) has started 210 years after the start of the Dalton Minimum, consistent with it being a de Vries Cycle event.

The graph in Fig. 2 shows that Solar Cycles 3 and 4, leading up to the Dalton Minimum, are very similar in amplitude and morphology to Solar Cycles 22 and 23, leading up to the current minimum. The two data sets are aligned on the month of transition between Solar Cycles 4 and 5 and between Solar Cycles 23 and 24.

In the absence of a significant change in Total Solar Irradiance over the solar cycle, modulation of the Earth's climate by the changing flux in galactic cosmic

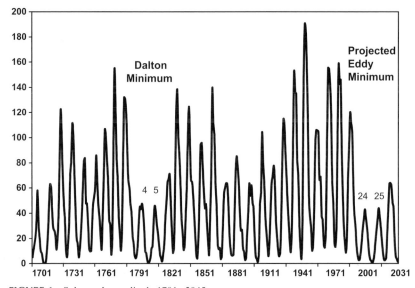

FIGURE 1 Solar cycle amplitude 1701−2045.

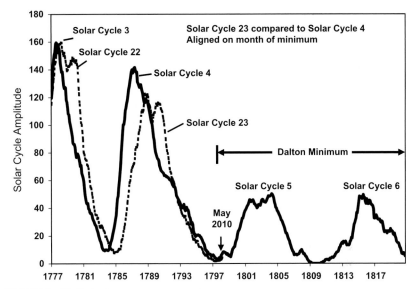

FIGURE 2   Similarity between precursor cycles.

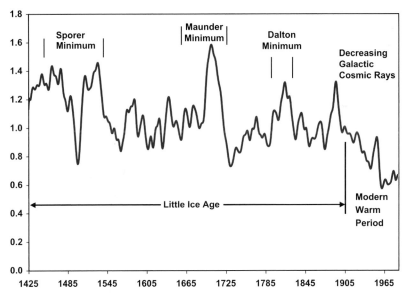

FIGURE 3   Be^{10} from the Dye 3 ice core, Greenland Plateau.

rays was proposed by Svensmark and Friis-Christensen (1997). The Dye 3 Be^{10} record (Fig.3) shows a correlation between spikes in Be^{10} and cold periods for the last 600 years. It also shows a steep decline in Be^{10} in the Modern Warm Period, suggesting a solar origin for this warming. Usokin et al. (2005) found

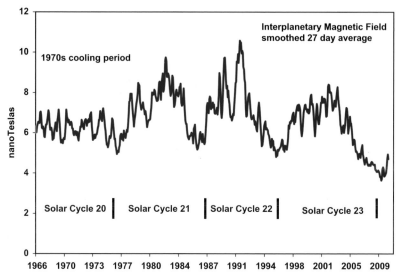

FIGURE 4 Interplanetary Magnetic Field 1966−2010.

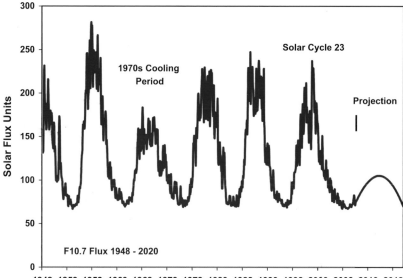

FIGURE 5 F10.7 flux 1948−2020.

that the level of solar activity during the past 70 years is exceptional, and the previous period of equally high activity occurred more than 8,000 years ago.

The strength of the Interplanetary Magnetic Field (Fig. 4) has fallen to levels below that of previous solar cycle transitions. What is also interesting in

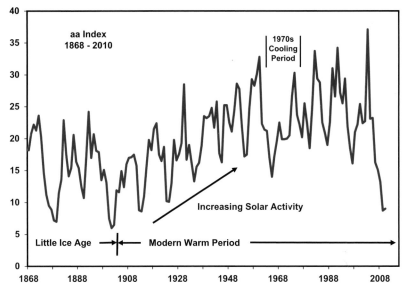

FIGURE 6   The aa Index 1868–2010.

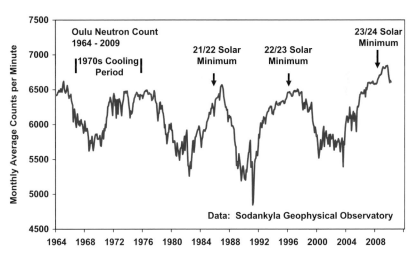

FIGURE 7   Oulu, Finland neutron monitor count 1960–2010.

this data is the flatness of this solar magnetic indicator during the 1970s cooling period.

The F10.7 index (Fig. 5) is a measure of the solar radio flux near the peak of the observed solar radio emission. Emission from the Sun at radio wavelengths

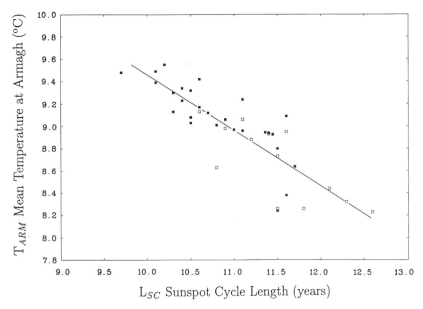

FIGURE 8   The correlation between solar cycle length and mean annual temperature at Armagh, Northern Ireland.

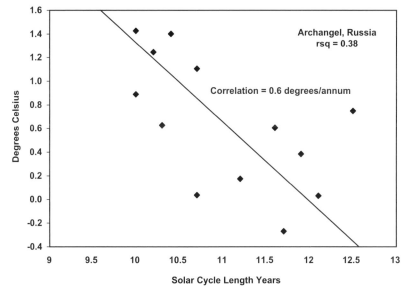

FIGURE 9   Archangel, Russia — solar cycle length relative to average annual temperature.

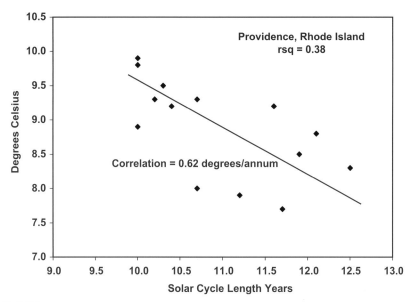

FIGURE 10   Providence, Rhode Island — solar cycle length relative to average annual temperature.

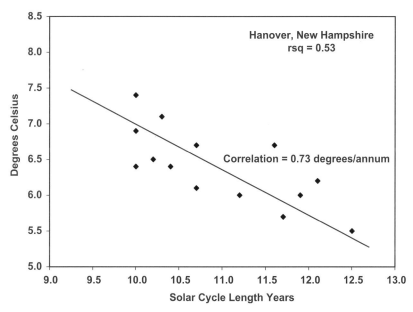

FIGURE 11   Hanover, New Hampshire — solar cycle length relative to average annual temperature.

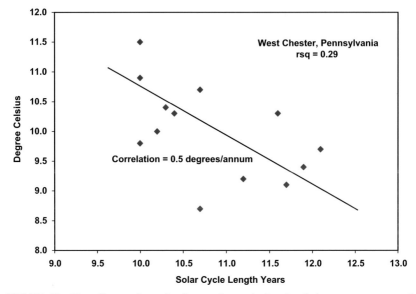

FIGURE 12 West Chester, Pennsylvania — solar cycle length relative to average annual temperature.

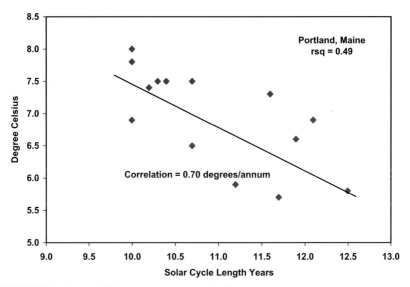

FIGURE 13 Portland, Maine — solar cycle length relative to average annual temperature.

is due primarily to diffuse, non-radiative heating of coronal plasma trapped in the magnetic fields overlying active regions. It is the best indicator of overall solar activity levels and is not subject to observer bias in the way that the counting of sunspots is. The graph above shows the F10.7 flux from 1948 with

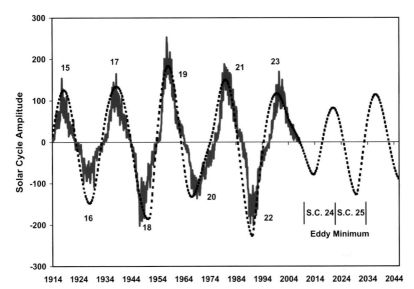

FIGURE 14   The ability to look forward using a model of solar activity.

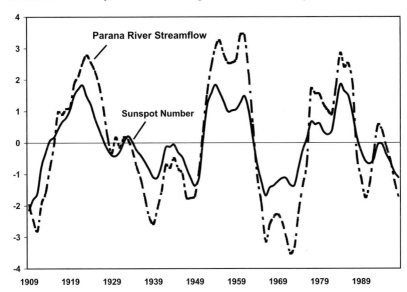

FIGURE 15   The correlation between the de-trended time series for the Parana River stream flow and sunspot number.

a projection to 2020. Note the lower activity of the 1970s cooling period. Activity over the next 10 years is projected to be much lower again.

The aa Index (Fig. 6) is a geomagnetic activity index which is driven by the solar coronal magnetic field strength. The strength of the solar coronal magnetic field doubled over the 20th century. At the same time, the Earth came out of the Little Ice Age. There was a dip in the aa Index associated with the 1970s cooling period. The aa Index has now fallen back to levels last seen in the Little Ice Age in the late 19th century.

A weaker Interplanetary Magnetic Field results in more galactic cosmic rays reaching the inner planets of the solar system, seen in Fig. 7 of the neutron count of the Oulu station in Finland. The peak neutron count can be more than a year later than the month of solar minimum. This is due to the time the solar wind takes to reach the heliopause, which is the boundary of the solar atmosphere with interstellar space. Note the much higher average neutron count during the 1970s cooling period associated with Solar Cycle 20. The increased galactic cosmic ray flux expected over Solar Cycle 24 will cause increased cloudiness, which will in turn increase the Earth's albedo, and the world will then cool in search of a new equilibrium temperature.

Friis-Christensen and Lassen (1991) demonstrated that global temperature over a solar cycle is better correlated with the length of the previous solar cycle than with solar cycle amplitude. In 1996, Butler and Johnson at the Armagh Observatory applied that theory to the temperature record of the observatory and produced the graph shown in Fig. 8. This graph can be considered as the Rosetta Stone of solar–climate studies; in that it has significant predictive power. Simply, Solar Cycle 22 was 9.6 years long, equating to a temperature of about 9.6 °C at Armagh. Solar Cycle 23 was 12.5 years long, equating to a temperature of 8.2 °C at Armagh. The difference is 1.4 °C which is the temperature fall, on average, predicted over Solar Cycle 24. There is not much scatter on this graph and therefore this result is almost certain.

A number of European temperature records show a correlation between solar cycle length and temperature, including the Central England Temperature record and de Bilt in the Netherlands. Generally, the more northerly the location, the better the correlation. The graph in Fig. 9 shows the correlation for Archangel in Russia.

Providence, Rhode Island will be 1.8 °C colder over Solar Cycle 24 relative to its average temperature over Solar Cycle 23 (Fig. 10).

Hanover, New Hampshire will be 2.2 °C colder over Solar Cycle 24 relative to its average temperature over Solar Cycle 23 (Fig. 11).

West Chester, Pennsylvania will be 1.5 °C colder over Solar Cycle 24 relative to its average temperature over Solar Cycle 23 (Fig. 12).

Portland, Maine will be 2.1 °C colder over Solar Cycle 24 relative to its average temperature over Solar Cycle 23 (Fig. 13).

The graph in Fig. 14 is from a model provided by Ed Fix (this volume). The notion that the orbits of the planets, particularly Jupiter, are responsible for generating the sunspot cycle has been with us since the discovery of the sunspot cycle by Samuel Schwabe in 1843. The model is based on changes in the Sun's orbit about the barycenter of the solar system as the driver of the sunspot cycle. It is a simple oscillatory model driven by the acceleration of the radial component of the barycenter's position relative to the Sun. The model has a very good hindcast match. At face value, it is predicting two very short and weak cycles. What is more likely is that there will be phase destruction over Solar Cycle 24, including the possibility that the solar magnetic poles will not reverse at solar maximum, predicted by other methods to be in 2015.

The Parana River, in central South America, runs through Brazil, Paraguay, and Argentina for nearly 4,000 km and is the second largest river in South America after the Amazon. Its outlet is the River Plata a few kilometers north of Buenos Aires. In 2010, three Argentinean researchers, Pablo Mauas, Andrea Buccino, and Eduardo Flamenco, published a paper showing the very strong correlation between sunspot activity and stream flow of the Parana River (Fig. 15). The relationship demonstrated has predictive power, and points to future drought conditions in the Amazon region as a consequence of the weak activity of Solar Cycle 24.

3. SUMMARY

The world has entered a de Vries cycle event in solar activity which will produce a decline in temperature in the range of 1.2−2.2 °C in the mid-latitude regions, with a consequent impact on agricultural productivity.

REFERENCES

Archibald, D., 2006. Solar cycles 24 and 25 and predicted climate response: Energy and Environment 17, 29−38.

Archibald, D., 2007. Climate outlook to 2030: Energy and Environment 18, 615−619.

Archibald, D., 2010. The Past and Future of Climate. Rhaetian Management Pty Ltd, p. 142.

Badalyan, O.G., Obridko, V.N., Sykora, J., 2001. Brightness of the coronal green line and prediction for activity cycles 23 and 24: Solar Physics 199, 421−435.

Butler, C.J., Johnston, D.J., 1996. A provisional long mean air temperature series for Armagh Observatory. Journal of Atmospheric and Terrestrial Physics 58, 1657−1672.

Clilverd, M., 2005. Prediction of solar activity the next 100 years. Solar Activity: Exploration, Understanding and Prediction, Workshop in Lund, Sweden.

Friis-Christensen, E., Lassen, K., 1991. Length of the solar cycle: an indicator of solar activity closely associated with climate: Science 254, 698−700.

Solheim, E., 2010. Solen varsler et kaldere tiar: Astronomi 4/10, 4−6.

Svensmark, H., Friis-Christensen, E., 1997. Variation in cosmic ray flux and global cloud coverage—a missing link in solar-climate relationships: Journal of Atmospheric and Solar Terrestrial Physics 59, 1225.

Usokin, I.G., Schuessler, M., Solanki, S.K., Mursula, K., 2005. Solar activity, cosmic rays, and the Earth's temperature: a millennium-scale comparison. Journal of Geophysical Research 110, A10102.

Chapter 12

Total Solar Irradiance Satellite Composites and their Phenomenological Effect on Climate

Nicola Scafetta

Active Cavity Radiometer Irradiance Monitor (ACRIM) Lab, Coronado, CA 92118, USA,
Duke University, Durham, NC 27708, USA

Chapter Outline

1. INTRODUCTION

A contiguous total solar irradiance (TSI) database of satellite observations extends from late 1978 to the present, covering 30 years, that is, almost three sunspot 11-year cycles. This database comprises the observations of seven independent experiments: NIMBUS7/ERB (Hoyt et al., 1992), SMM/ACRIM1 (Willson and Hudson, 1991), ERBS/ERBE (Lee et al., 1995), UARS/ACRIM2 (Willson, 1994, 1997), SOHO/VIRGO (Crommelynck and Dewitte, 1997; Fröhlich et al., 1997), and ACRIMSAT/ACRIM3 (Willson, 2001). Another TSI satellite record,

Evidence-Based Climate Science. DOI: 10.1016/B978-0-12-385956-3.10012-9

289

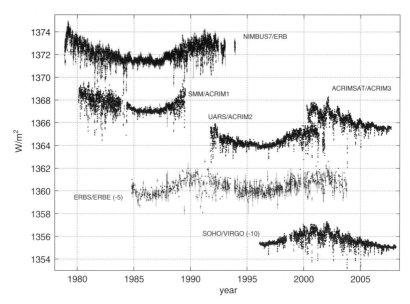

NIMBUS7/ERB

ACRIMSAT/ACRIM3

SMM/ACRIM1

UARS/ACRIM2

ERBS/ERBE (-5)

SOHO/VIRGO (-10)

FIGURE 1 TSI satellite records. ERBS/ERBE and SOHO/VIRGO TSI records are shifted by -5 W/m^2 and -10 W/m^2, respectively, from the 'native scale' for visual convenience. (Units in Watts/meter^2 at 1 A.U.)

SORCE/TIM (Kopp et al., 2003), exists but it is not studied here because its mission started in February 2003 and this TSI satellite record is still too short for our purpose. None of these independent datasets cover the entire period of observation; thus, a composite of the database is necessary to obtain a consistent picture about the TSI variation. In this paper we use the TSI records depicted in Fig. 1 as available by 2008.

Three TSI satellite composites are currently available: ACRIM composite (Willson and Mordvinov, 2003), PMOD composite (Fröhlich and Lean, 1998; Fröhlich, 2000, 2006), and IRMB composite (Dewitte et al., 2004), respectively (see Fig. 2). Each composite is compiled by using different models corresponding to different mathematical philosophies and different combinations of data.

For example, one of the most prominent differences between ACRIM and PMOD composites is due to the different way of how the two teams use the NIMBUS7/ERB record to fill the period 1989.53−1991.75, the so-called ACRIM-gap between ACRIM1 and ACRIM2 records. Consequently, these two composites significantly differ from each other, in particular about whether the TSI minimum during solar Cycles 22−23 (1995/96) is approximately 0.45 W/m^2 higher (ACRIM composite) or approximately at the same level (PMOD) as the TSI minimum during the previous solar Cycles 21−22 (1985/96).

The difference among the TSI satellite composites has significant implications not only for solar physics, where the correctness of the theoretical

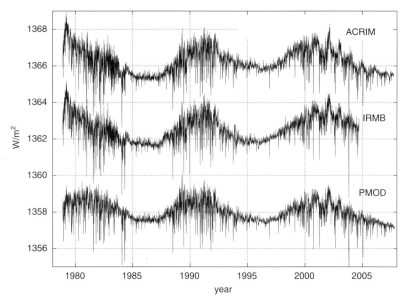

FIGURE 2 ACRIM, IRMB, and PMOD TSI satellite composites.

models has to be tested against the actual observations and not vice versa, but also for the current global warming debate. Phenomenological analyses (Scafetta and West, 2007, 2008; Scafetta, 2009) using TSI proxy, satellite composites, and global surface temperature records of the past 400 years show that solar variation has been a dominant forcing for climate change during both the pre- and industrial periods. According to these analyses, the sun will likely be a dominant contributor to climate change in the future. However, the solar contribution to the global warming during the last three decades remains severely uncertain due mostly to the difference between the TSI satellite composites.

The phenomenological solar signature on the global temperature is found to match quite well 400 years of global climate reconstructions since 1600 (Scafetta and West, 2007, 2008; Scafetta, 2009), but such an almost continuous matching would be abruptly interrupted since 1975 if the PMOD composite is adopted. Instead, by adopting the ACRIM composite it is still possible to notice a significant correlation between temperature data and the reconstruction of the solar effect on climate. Thus, a significant fraction of the +0.4 K warming observed since 1980 can be ascribed to an increase of solar activity if ACRIM composite is adopted, but almost none of it would be linked to solar activity if PMOD composite is adopted. Evidently, if the solar contribution is uncertain, the anthropogenic contribution to the global warming observed during the last three–four decades is uncertain as well. Hence, determining the correct TSI composite during the last three decades should be considered of crucial importance.

The general circulation climate models adopted by the Intergovernmental Panel on Climate Change (IPCC, 2007) do not agree with the above phenomenological findings and predict a minor solar contribution to climate change during the last century and, in particular, during the last 30–40 years. However, such climate models assume that the TSI forcing is the only solar forcing of the climate system and usually use TSI proxy records such as those proposed by Lean (Lean, 2000; Wang et al., 2005), which are compatible with the PMOD TSI composite since 1978. This choice would be evidently problematic and misleading if PMOD TSI composite is flawed.

In any case, the small climate sensitivity to solar changes predicted by the current climate models is also believed to be due to the absence in the models of several climate feedback mechanisms that may be quite sensitive to several solar related changes, in addition to TSI changes alone. Some of these phenomena include, for example, the UV modulation of ozone concentration that would influence the stratosphere water-vapor feedback and the modulation of the cloud cover due to the variation of cosmic ray flux that is linked to changes of the magnetic solar activity (Pap et al., 2004; Kirkby, 2007). These climate mechanisms are expected to magnify the influence of solar activity changes on climate. Because all solar related observables have similar geometrical patterns, TSI satellite records and proxy reconstructions can also be used as geometrical solar proxies in phenomenological and holistic models, instead of just using them as radiative forcings alone in analytical computer climate models. Indeed, this empirical methodology may circumvent the current limitation of our knowledge in the microscopic mechanisms involved in the climatic phenomena.

The original ACRIM composite (Willson and Mordvinov, 2003) has been constructed by simply calibrating the three ACRIM datasets and the NIMBUS7/ERB record on the basis of a direct comparison of the *entire* overlapping region between two contiguous satellite records. This composite uses the actual observations as published by the original experimental groups without any alteration. However, if some degradation or glitches do exist in the data, this composite is flawed for at least two reasons: (1) the mathematical methodology used for merging two contiguous satellite records, which uses just the average estimate of the residual during the entire overlapping regions between two records, may easily give biased estimates; (2) if the NIMBUS7/ERB record presents some glitches, or degradation did occur during the ACRIM-gap, the relative position of ACRIM1 and ACRIM2 would be erroneous.

The IRMB composite (Dewitte et al., 2004) is constructed by first referring all datasets to space absolute radiometric references, and then the actual value for each day is obtained by averaging all available satellite observations for that day. Thus, IRMB composite adopts a statistical average approach among all available observations; evidently, because the daily average estimate is based on a small set of data (1, 2, or in a few cases 3 data per day), it is not statistically robust, and this may easily produce artificial slips every time data from a specific record are missing or added.

The PMOD composite (Fröhlich and Lean, 1998; Fröhlich, 2000, 2004, 2006) is constructed with altered published experimental TSI satellite data. In fact, the PMOD team claims that the published TSI data are corrupted. Thus, PMOD claims that the published TSI satellite records need to be "corrected" before merging them into a TSI composite. Data corruption is claimed to be caused by sudden glitches due to changes in the orientation of the spacecraft and/or to switch-offs of the sensors, or because of some kind of instrumental degradation. Some TSI theoretical model predictions (Lee et al., 1995; Fröhlich and Lean, 1998; Chapman et al., 1996) have been heavily used by the PMOD team to identify, correct and evaluate these presumed errors in the published records, and these models have been changed constantly during the last 10 years. PMOD composite is claimed to be consistent with some TSI theoretical proxy models (Wenzler et al., 2006; Krivova et al., 2007). However, differences between the model and the PMOD TSI composite can be easily recognized: for example, Wenzler et al. (2006) needed to calibrate the model on the PMOD composite itself to improve the matching, and several details are not reproduced. Also, it cannot be excluded that an alternative calibration of the parameters of these TSI proxy models may better fit the ACRIM TSI satellite composite. Evidently, if the above theoretical models and/or the corrections of the satellite records implemented by the PMOD team are flawed, PMOD would be flawed as well. In any case, an apparent agreement between some theoretical TSI model, which depends on several calibration parameters and a TSI satellite composite does not necessarily indicate the correctness of the latter because in science theoretical models should be tested and evaluated against the actual observations, and not vice versa.

In this paper, alternative TSI satellite composites are constructed using an approach similar to that adopted by the ACRIM team, that is, we do not alter the published satellite data by using predetermined theoretical models that may bias the composite. However, contrary to the original ACRIM team's approach we use a methodology that takes into account the evident statistical relative differences that are found in the published satellite records. The three ACRIM records are preferred and the ACRIM-gap is filled by using the measurements from the NIMBUS7/ERB and the ERBS/ERBE satellite experiments. Finally, we use these alternative TSI satellite composites in conjunction with a recent TSI proxy reconstruction proposed by Solanki's team (Wenzler et al., 2006; Krivova et al., 2007) to reconstruct the signature of solar variation on global climate using a phenomenological model (Scafetta and West, 2007, 2008; Scafetta, 2009).

2. SMM/ACRIM1 vs. NIMBUS7/ERB

The first step is to merge SMM/ACRIM1 and NIMBUS7/ERB. Note that the relative accuracy, precision, and traceability of these two databases are radically different. In particular, the average error for the NIMBUS7/ERB measurements is ± 0.16 W/m^2 while the average error for the SMM/ACRIM1 measurements is ±0.04 W/m²: thus, SMM/ACRIM1 is significantly more precise than NIMBUS7/ERB. Moreover, NIMBUS7/ERB instrumentations were not able to self-calibrate their sensor degradations as well as ACRIM1. NIMBUS7/ERB radiometer was calibrated electrically every 12 days. For the above reasons, ACRIM1 measurements are supposed to be more accurate than the NIMBUS7/ERB ones and, when available, they are always preferred to the NIMBUS7/ERB values.

It is necessary to adopt the NIMBUS7/ERB record for reconstructing the TSI record for a few days and during three prolonged periods: (1) before 17/02/1980, (2) from 04/11/1983 to 03/05/1984, and (3) after 14/07/1989. To accomplish this goal the position of NIMBUS7/ERB is evaluated relative to SMM/ACRIM1 and plotted in Fig. 3. The black smooth curve is a 91-day moving average.

The data shown in Fig. 3 have an average of 4.3 W/m². The original ACRIM TSI composite (Willson and Mordvinov, 2003) is constructed using the above average value to merge SMM/ACRIM1 and NIMBUS7/ERB. However, if the differences between SMM/ACRIM1 and the NIMBUS7/ERB were only due to random fluctuations around an average value, the error associated to the 91-day moving average values would have been about ±0.02−0.03 W/m², as calculated from the measurement uncertainties. Because the standard deviation of the smooth data shown in Fig. 3 is significantly larger, ±0.17 W/m², the difference between SMM/ACRIM1 and NIMBUS7/ERB measurements is not due just to random fluctuations, but it is due to biases and trends in the data, which are probably produced by the poorer sensor calibration of NIMBUS7/ERB.

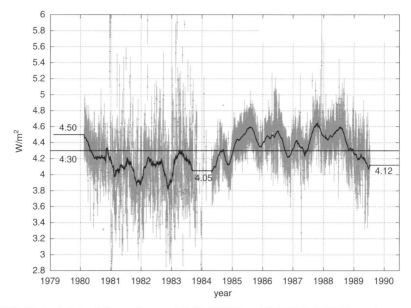

FIGURE 3   Relative difference between NIMBUS7/ERB and SMM/ACRIM1 TSI records.

Under the theoretical assumption that SMM/ACRIM1 measurements are more accurate than NIMBUS7/ERB ones, the black 91-days moving average smooth curve depicted in Fig. 3 suggests that NIMBUS7/ERB measurements could gradually shift during a relatively short period of time, a few months, by an amount that on average is about 0.2 W/m^2. In a few cases, this shift can also be as large as 0.5 W/m^2. In fact, irregular large oscillations with periods ranging from 5 to 12 months are clearly visible in Fig. 3.

The above finding suggests that the original methodology adopted by the ACRIM team to merge SMM/ACRIM1 and NIMBUS7/ERB records may be inappropriate because it assumes that NIMBUS7/ERB data are statistically stationary relative to the ACRIM1 measurements, while this is not what the data show. By not taking into account this problem the ACRIM team's methodology can introduce significant artificial slips in the TSI satellite composite at the chosen merging day.

To reduce the errors due to the above irregular large oscillations of NIMBUS7/ERB measurements, we use the black 91-days moving average smooth curve shown in Fig. 3 to reduce NIMBUS7/ERB record to the level of SMM/ACRIM1 during the overlapping period. Then, we use the NIMBUS7/ ERB corrected record to fill all days that SMM/ACRIM1 record misses. Before 17/02/1980 NIMBUS7/ERB record is shifted by -4.5 W/m^2, while after 14/07/1989 it is shifted by -4.12 W/m^2 as indicated in the figure. This means that relative to the original ACRIM composite, our composite is 0.2 W/m^2 lower before 17/02/1980, and 0.18 W/m^2 higher after 14/07/1989.

However, these values do depend on the adopted moving average window. In fact, by increasing the window the two above levels approach the average level at -4.3 W/m^2, which is the value used by the ACRIM team to merge the two records. Thus, our proposed merging methodology can have an error as large as 0.2 W/m^2 (Fig. 4).

On the contrary, the PMOD team significantly alters both SMM/ACRIM1 and NIMBUS7/ERB records before 1986 (see Fig. 4). These corrections are not justified by the data themselves, but by theoretical models, which may be erroneous and/or have their own large uncertainties. About the SMM/ACRIM1 data, PMOD team assumes that the SMM/ACRIM1 record from 1984 to 1986 significantly degraded. In fact, as Fig. 4 shows, the position of NIMBUS7/ERB relative to SMM/ACRIM1 gradually increases from 1984 to 1986. However, this pattern could have been caused by a degradation of SMM/ACRIM1, as the PMOD team interprets, or by an increase of sensitivity of NIMBUS7/ERB sensors due to undetermined factors. From 1988 to 1989.5, the position of NIMBUS7/ERB relative to SMM/ACRIM1 gradually decreased of the same amount of the gradual increase observed from 1984 to 1986. This suggests that NIMBUS7/ERB record can gradually vary by these large amounts.

In any case, ACRIM team has never published an update of their SMM/ ACRIM1 record and, they have publicly disagreed with the PMOD team on numerous occasions (Willson and Mordvinov, 2003; Scafetta and Willson, 2009)

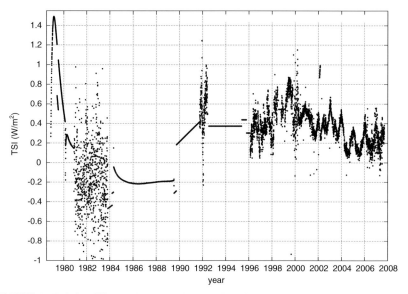

FIGURE 4   Relative difference between ACRIM and PMOD TSI satellite composites.

arguing that no physical explanation could explain the PMOD's claimed degradation of the SMM/ACRIM1 record. Herein, we believe that the ACRIM team's opinion cannot be just ignored and lightly dismissed. In fact, they are the authors of the data. The correction implemented by the PMOD team on the SMM/ACRIM1 record should be considered just as based on a hypothesis not taken for granted. In any case, this proposed correction would not alter the position of the TSI minimum in 1985/1986 relative to the minimum in 1996, which herein is a more important issue. PMOD team's correction would only lower the TSI maximum in 1981/1982 by about 0.2 W/m^2 relative to the TSI satellite composite after 1986.

About the NIMBUS7/ERB record before 1980, although NIMBUS7/ERB trend appears to be quite uncertain, the PMOD team's proposed correction should be considered hypothetical as well. In particular, PMOD team believes that the large NIMBUS7/ERB peak occurred during the first months of 1979 (see Figs. 1 and 2) is an artifact due to changes in the orientation of the spacecraft. However, we observe that the TSI theoretical reconstruction proposed by Solanki (Wenzler et al., 2006) shows that a large TSI peak occurred during the first months of 1979. In fact, a careful look at their Figs. 14 and 15, where the TSI proxy reconstruction is compared against the PMOD composite reveals that during the first months of 1979 there is a discrepancy of about 1 W/m^2 between the proposed TSI proxy records and the PMDO TSI composite. This is a sufficient evidence for considering the PMOD team's corrections of NIMBUS7/ERB questionable. Moreover, from 2000 to 2004 large peaks in TSI have been

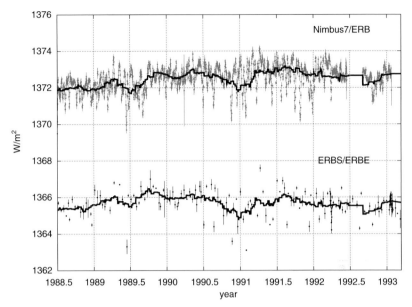

FIGURE 5 NIMBUS7/ERB and ERBS/ERBE TSI records during the ACRIM-gap. The smooth curves are 91-day moving averages calculated on the TSI values only for those days available in both records.

recorded by all available instruments: see Figs. 1 and 2. These TSI peaks, in particular the one occurred in 2002, do not seem particularly different from the peak recorded by NIMBUS7 during 1979.

In any case, the exact TSI patterns before 17/02/1980 and during the ACRIM-gap are highly uncertain because they are obtained from lower quality satellite measurements.

3. THE ACRIM-GAP: 15/07/1989−03/10/1991

SMM/ACRIM1 and UARS/ACRIM2 records can only be bridged by using two low quality satellite records: NIMBUS7/ERB and ERBS/ERBE. Figure 5 shows the two records from 1988.5 to 1993. Note that NIMBUS7/ERB and ERBS/ERBE present opposite trend. From 1990 to 1991.5, NIMBUS7/ERB record shows an increasing trend while ERBS/ERBE record shows a decreasing trend (Willson and Mordvinov, 2003). Thus, these two satellite records are not compatible with each other, and at least one of the two may be corrupted. Note that ERBS/ERBE too was unable to properly calibrate its sensor degradations and a direct comparison with the ACRIM records reveals that the discrepancy between local ACRIM smooth trends and ERBS/ERBE smooth trends could be as large as ±0.2 W/m^2. The amplitude of these non-stationary biases is smaller than that observed in the NIMBUS7/ERB measurements, but they are still

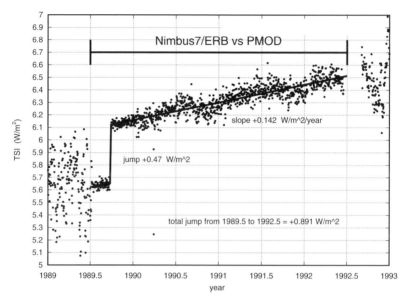

FIGURE 6   Relative difference between NIMBUS7/ERB original record and PMOD TSI satellite composite.

significant. Moreover, the average error of ERBS/ERBE's measurements is the largest among all satellite observations: ±0.26 W/m².

The PMOD team claims that NIMBUS7/ERB record must be severely corrected during the ACRIM-gap. The justifications for the proposed corrections can be found in the scientific literature. Lee et al. (1995) compared the NIMBUS7/ERB dataset with a TSI model based on a multi-regression analysis of March 1985 to August 1989 ERBS/ERBE irradiance measurements, and they concluded that after September 1989 NIMBUS7/ERB time series appeared to, and have abruptly increased by +0.4 W/m² after a switch-off of NIMBUS7/ERB for 4 days. Another +0.4 W/m² upward shift appeared to have occurred on April 1990. Thus, this study suggested a two-step shift correction that, once combined, moves down the NIMBUS7/ERB record by 0.8 W/m² at the end of April 1990.

Later Chapman et al. (1996) reviewed the finding by Lee et al. (1995) and concluded that on 29/09/1989 NIMBUS7 experienced a sudden increase by +0.31 W/m² and on 9/05/1990 there was another upward shift by +0.37 W/m²: the combined two-step correction shift had to be −0.68 W/m² after 9/05/1990. Afterward, Fröhlich and Lean (1998) suggested a different two-step shift correction, that is, the NIMBUS7/ERB record had to be adjusted by −0.26 W/m² and −0.32 W/m² near October 1, 1989 and May 8, 1990, respectively: the combined correction shift of NIMBUS7/ERB during the ACRIM-gap was estimated to be −0.58 W/m².

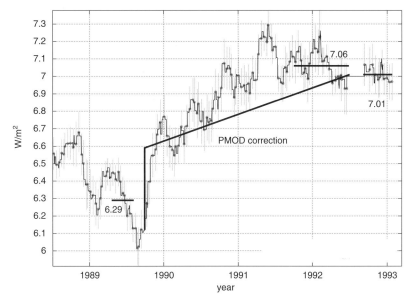

FIGURE 7   It is depicted the 91-day moving average of the relative difference between NIMBUS7/ERB original record and ERBS/ERBE TSI satellite composite during the ACRIM-gap. The figure shows the correction (solid line) of NIMBUS7/ERB record implemented by PMOD team shown in Fig. 6. Note the rapid, but gradual divergence between NIMBUS7 and ERBS during October and November 1989, that contrasts with the sudden one-day glitch shift claimed by PMOD on September 29.

Finally, Fröhlich (2000, 2004) revised significantly his previous model correction of NIMBUS7/ERB data. He first acknowledged that the supposed slip on May 1990 was indeed difficult to identify. Then, he substituted the two-step correction model with a new model in which there was only one sudden slip on September 29, 1989 followed by a upward linear trend. Figure 6 shows our analysis of these new corrections in their latest version: on 29/09/1989 there is a sudden jump of $+0.47$ W/m^2, which is significantly larger than what was previously estimated, which is followed by an upward trend of $+0.142$ W/m^2/ year. The total downward shift forced on NIMBUS7/ERB record from 1989.5 to 1992.5 is about -0.89 W/m^2, which is significantly larger than what Fröhlich himself and the other groups had previously estimated.

From the above studies, it is evident that several opinions have been formulated to solve the ACRIM-gap, even by the same research team, and they quantitatively disagree with each other. The above conflicting solutions indicate that it is not so certain how NIMBUS7/ERB should be corrected, if some corrections were truly needed.

Figure 7 shows our analysis of the comparison between NIMBUS7/ERB and ERBS/ERBE. The 91-day moving average curve of the relative difference between NIMBUS7/ERB and ERBS/ERBE decreases until August 1989 around the time when SMM/ACRIM1 merges with NIMBUS7/ERB at the level 6.29 W/m², as shown in the graph. Since the beginning of September 1989 to the beginning of 1990 the curve rises rapidly, but, in contrast with the PMOD's claims, no sudden one-day jump by about 0.65 W/m² due to an instrumental glitch is observed. From 1990 to 1991.5 the curve rises by about 0.40 W/m². Finally, from 1991.5 to 1993 the curve decreases slightly by about 0.05 W/m².

The total shift from 1989.5 to 1993 is about 0.72 W/m². Note that the error related to the single measurements is about ±0.11 W/m². Thus, the observed difference between NIMBUS7/ERB and ERBS/ERBE is significant and must be interpreted as due to biases in the data. These are likely due to uncorrected problems in one or both the satellite instrumentations.

Figure 7 shows also the correction implemented by the PMOD team on NIMBUS/ERB record (Fröhlich, 2000, 2004), which approximately reproduce the divergence between NIMBUS7 and ERBS. It is evident from the figure that the PMOD team believes that the observed difference between the two TSI satellite records is due to uncorrected problems occurring only on NIMBUS7/ERB's sensors. Even so, the correction of NIMBUS/ERB record implemented by the PMOD team (0.89 W/m² from 1989.5 to 1992.5) appears to be over-estimated at least by about 0.12 W/m² because the total shift observed during the ACRIM-gap period is about 0.77 W/m². The difference seems to be due to the fact that PMOD team did not take into account that the real comparison must be done with the level when SMM/ACRIM1 merges with NIMBUS7/ERB around the middle of 1989, and the level during this period, as indicated in the figure, is about 6.29 W/m². Thus, if on 29/09/1989 a shift really occurred in the NIMBUS7/ERB record, its magnitude has to be by about 0.30 W/m², as previously estimated by Chapman et al. (1996) and Fröhlich and Lean (1998). Finally, the PMOD team's correction with a linear increase from 29/09/1989 to 1992.5 is also poorly observed in data depicted in Fig. 7; indeed, the PMOD linear correction during this period appears to be just a linear simplification of the complex pattern observed in the figure.

However, the theoretical studies (Lee et al., 1995; Fröhlich and Lean, 1998; Fröhlich, 2000, 2004; Chapman et al., 1996) claiming that NIMBUS7/ERB is erroneous during the ACRIM-gap may be questionable. In fact, although these authors did notice a difference between NIMBUS7/ERB and ERBS/ERBE records, they have interpreted such a difference as only due to a corruption of the NIMBUS7/ERB record. This interpretation was preferred despite the fact that ERBS/ERBE too was unable to continuously calibrate its sensor degra-dations and its data had larger uncertainties than NIMBUS7/ERB data. Indeed, the increase observed in Fig. 7 during the ACRIM-gap could result from increased ERBS/ERBE degradation relative to NIMBUS7/ERB, a relative increase in the sensitivity of the NIMBUS7/ERB sensor, or both (5).

It is important to stress that in 1992, the experimental team responsible of the NIMBUS7/ERB record (Hoyt et al., 1992) corrected all biases in the data they could find and afterward they ever come up with a physical theory for the instrument that could cause it to become more sensitive. The NIMBUS7/ERB calibrations before and after the September 1989 shutdown gave no indication of any change in the sensitivity of the radiometer. When Lee et al. (1995) of the ERBS team claimed that there was an increase in NIMBUS7/ERB sensitivity, the NIMBUS7 team examined the issue and concluded there was no internal evidence in the NIMBUS7/ERB record to warrant the correction that the latter team was proposing. In a personal communication on September 16, 2008 with Dr. Hoyt, which was published in Scafetta and Willson (2009), Dr. Hoyt stated:

Concerning the supposed increase in Nimbus7 sensitivity at the end of September 1989 and other matters as proposed by Frohlich's PMOD TSI composite: 1. There is no known physical change in the electrically calibrated Nimbus7 radiometer or its electronics that could have caused it to become more sensitive. At least neither Lee Kyle nor I could never imagine how such a thing could happen and no one else has ever come up with a physical theory for the instrument that could cause it to become more sensitive. 2. The Nimbus7 radiometer was calibrated electrically every 12 days. The calibrations before and after the September shutdown gave no indication of any change in the sensitivity of the radiometer. Thus, when Bob Lee of the ERBS team originally claimed there was a change in Nimbus7 sensitivity, we examined the issue and concluded there was no internal evidence in the Nimbus7 records to warrant the correction that he was proposing. Since the result was a null one, no publication was thought necessary. 3. Thus, Frohlich's PMOD TSI composite is not consistent with the internal data or physics of the Nimbus7 cavity radiometer. 4. The correction of the Nimbus7 TSI values for 1979—1980 proposed by Frohlich is also puzzling. The raw data was run through the same algorithm for these early years and the subsequent years and there is no justifi-cation for Frohlich's adjustment in my opinion.

Indeed, it is not possible to exclude that the TSI increase between 1989 and 1991, or part of it, observed in Fig. 7 is an indication of ERBS losing sensitivity rather than NIMBUS7 gaining sensitivity. In fact, a careful look at the pattern depicted in Fig. 7 reveals a rapid but gradual increase immediately after September 1989, which lasted for two months. This pattern suggests that the discrepancy between NIMBUS7 and ERBS during that period is not due to a one-day glitch event on 29/9/1989 causing a sudden upward increase in the sensitivity of NIMBUS7's sensors, as claimed by PMOD. There are several physical reasons to believe that ERBS/ERBE could degrade more likely than NIMBUS7/ERB, in particular during the ACRIM-gap. For example: a) The NIMBUS7/ERB cavity radiometer was in a relatively high altitude (about 900 km), while ERBS/ERBE was in a low earth orbit (ca. 200 km). It is possible that ERBS would degrade much faster than NIMBUS7/ERB due to more atmospheric bombardment of its sensor. b) During the ACRIM-gap ERBS/ERBE was experiencing for the first time the enhanced solar UV radi-ation, which occurs during solar maxima, and this too may have caused a much faster degradation of the cavity coating of ERBS than of NIMBUS7/ERB

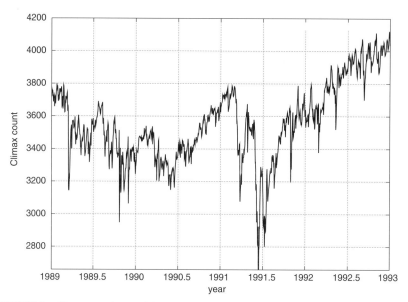

FIGURE 8   Climax cosmic ray of the University of New Hampshire (Version 4.8 21 December 2006). Data from http://ulysses.sr.unh.edu/NeutronMonitor/.

because NIMBUS7 already experienced such degradation during the previous solar maximum. c) From spring 1990 to May/June 1991, when according to Fig. 7 the difference between NIMBUS7/ERB and ERBS/ERBE increased by about 0.40 W/m², there was a rapid increase of cosmic ray flux, as Fig. 8 shows. In addition, the above phenomena could have more likely damaged ERBS/ERBE's sensors than NIMBUS7/ERB's ones, which already experienced a solar maximum 10 years earlier.

The cosmic ray count is negative-correlated to TSI and magnetic flux, and its minima correspond to solar activity maxima. Fig. 8 shows that the minimum around 1991.5 was lower than the minimum around 1989.8−1990.5. This implies that according to this record, the solar activity was likely higher around 1991.5 than around 1989.8−1990.5. This contradicts the pattern observed in ERBS/ERBE and confirms the NIMBUS7/ERB pattern, as Fig. 5 shows. However, other solar indexes, such as the sunspot number index, present the opposite scenario. It is unlikely that solar proxy indexes can be used to definitely and precisely solve this issue.

We believe that unless the experimental teams find a physical theory for explaining the divergences observed in their own instrumental measurements and definitely solve the problem, there exists only a statistical way to address the ACRIM-gap problem by using the published data themselves. This requires just the acknowledgment of the existence of a still unresolved uncertainty in the TSI satellite data. This can be done in the following way:

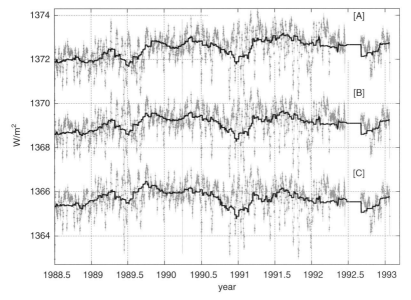

FIGURE 9   Reconstructions of the NIMBUS7/ERB record during the ACRIM-gap in agreement with three alternative scenarios. (A) NIMBUS7/ERB data are unaltered; (C) the data are adapted in such a way that their smooth component matches exactly the smooth component of ERBS/ERBE; (B) the data are adapted in such a way that their smooth component matches exactly the average between the smooth components in (A) and (C). The smooth component is a 91-day moving average.

(1) Assuming that NIMBUS7/ERB is correct and ERBS/ERBE is erroneous; this would imply that during the ACRIM-gap, ERBS/ERBE record degraded and should be shifted upward by $0.72-0.77$ W/m^2.

(2) Assuming that ERBS/ERBE is correct and NIMBUS7/ERB is erroneous; this would imply that during the ACRIM-gap NIMBUS7/ERB increased its sensitivity to TSI and should be shifted downward by $0.72-0.77$ W/m^2.

(3) Assuming that both ERBS/ERBE and NIMBUS7/ERB records need some correction.

Note that there is no objective way to implement method # 3 and infinitely different solutions may be proposed, as those proposed by Fröhlich and other authors. Herein, we propose that all configurations between case # 1 and case # 2 may be possible, and for case # 3 we just propose an average between methods 1 and 2 stressing that this arithmetic average should not be interpreted as a better physical solution for the ACRIM-gap problem.

Fig. 9 depicts the three reconstructions of NIMBUS7/ERB record in agreement with the above three scenarios. (A) NIMBUS7/ERB data are unaltered. (C) NIMBUS7/ERB data are altered in such a way that their

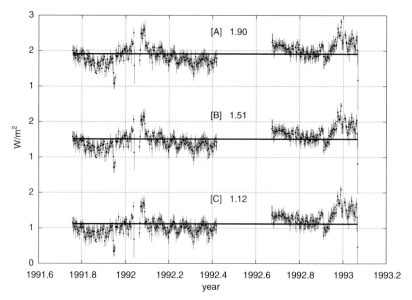

FIGURE 10 Relative difference between NIMBUS7/ERB and UARS/ACRIM2 TSI satellite records according to the three hypotheses discussed in the article. Note that NIMBUS7/ERB record has been already merged with SMM/ACRIM1 record. Thus, the value reported in the figure refers to the position SMM/ACRIM1 relative to UARS/ACRIM2 in the three alternative cases.

91-day moving average curve in Fig. 9 matches exactly the 91-day moving average curve of ERBS/ERBE shown in Fig. 5. (B) NIMBUS7/ERB data are altered in such a way that their 91-day moving average curve matches exactly the average between the two 91-day moving average curves of NIMBUS7/ERB and ERBS/ERBE shown in Fig. 5.

4. SMM/ACRIM1 vs. UARS/ACRIM2

To align SMM/ACRIM1 and UARS/ACRIM2 records we proceed as follows. First, we merge NIMBUS7/ERB record and its two alternative records shown in Fig. 9 with the SMM/ACRIM1 record. The merging is done by guaranteeing a continuity of the 91-day moving average curves at the merging day, 03/10/1991.

Second, we use the finding shown in Fig. 10. This figure depicts the overlapping period between NIMBUS7/ERB and UARS/ACRIM2 records. This interval is quite short, and it is made of two separated intervals during which both satellite measurements were interrupted for several months. Note that the two intervals are not aligned: there is a difference of about 0.2 W/m^2 between the two levels. Because the standard deviation of the data is about 0.26 W/m^2, which is significantly larger than the statistical error of measure 0.16 W/m^2, the figure indicates that the data are not statistically stationary.

However, it is not evident which record is performing poorly: NIMBUS7/ ERB and ERBS/ERBE or UARS/ACRIM2. Because the difference observed between NIMBUS7/ERB and UARS/ACRIM2 records in (A), and between the adapted NIMBUS7/ERB and UARS/ACRIM2 records in (C) (where NIMBUS7/ERB record is adapted to reproduce ERBS/ERBE smooth trending) are almost equal, the first impression is that UARS/ACRIM2 sensors experienced a downward slip between the two intervals by about 0.2 W/m^2. However, because both NIMBUS7/ERB and ERBS/ERBE were less able to calibrate their sensor degradation, it is still uncertain whether it is UARS/ ACRIM2 record that has to be corrected and, if so, how large this correction should be. Indeed, given the short time period and that both NIMBUS7/ERB and ERBS/ERBE are characterized by non-stationary biases as large as ±0.2 W/m^2, it is possible that during 1992 the two latter records experienced a similar upward bias. Thus, here we decided to keep UARS/ACRIM2 record unaltered and merge the two sequences using the average of the relative differences during the entire overlapping period in all three cases, as shown in Fig. 10. The error associated with this merging is about ±0.1 W/m^2. However, if UARS/ACRIM2 record needs a correction, its global implication would be that the TSI satellite composite before 1992.5 should be shifted downward by about 0.1 W/m^2 in all three cases.

5. UARS/ACRIM2 vs. ACRIMSAT/ACRIM3

The merging between UARS/ACRIM2 and ACRIMSAT/ACRIM3 is done by using the information shown in Fig. 11 that depicts the relative difference between ACRIMSAT/ACRIM3 and UARS/ACRIM2, and for comparison, the relative difference between SOHO/VIRGO and UARS/ACRIM2. Note that UARS/ACRIM2 measurements were interrupted from 05/06/2001 to 08/16/ 2001.

The latter comparison is necessary for studying the discrepancy observed between ACRIMSAT/ACRIM3 and UARS/ACRIM2, which is significantly larger than the statistical error of the measurements. In fact, the average statistical error of UARS/ACRIM2 data is 0.01 W/m^2, while the average statistical error of ACRIMSAT/ACRIM3 data is 0.008 W/m^2. The average statistical error of the relative difference between ACRIMSAT/ACRIM3 and UARS/ACRIM2 is 0.018 W/m^2. However, the data in the figure have a standard deviation of 0.3 W/m^2, which is significantly larger than their statistical errors. Thus, the observed difference between ACRIMSAT/ACRIM3 and UARS/ ACRIM2 records is not due to random fluctuations, but to non-stationary trends in the data.

Because a similar pattern appears when UARS/ACRIM2 is compared against both SOHO/VIRGO and ACRIMSAT/ACRIM3 records, it appears that UARS/ACRIM2 sensors may have experienced some problem. Perhaps an annual cycle has been filtered off in some way. However, these problems

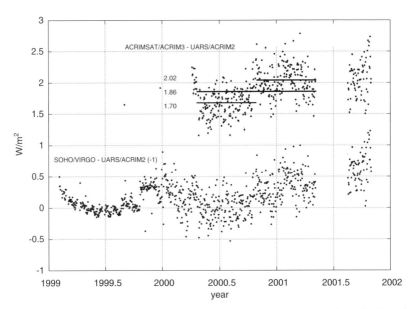

FIGURE 11 Relative difference between ACRIMSAT/ACRIM3 and UARS/ACRIM2 TSI satellite records (top) and between SOHO/VIRGO and ACRIM2 TSI satellite records (bottom). The latter is shifted down by 1 W/m² for visual convenience.

appear to have significantly modified a natural variation in the TSI data characterized by a time scale close to 1 year. Because the difference between ACRIMSAT/ACRIM3 and UARS/ACRIM2 appears to present a cyclical pattern, an accurate way to merge the two sequences is to evaluate the average during an entire period of oscillation. The period from 04/05/2000 to 05/06/2001 covers approximately one period of oscillation, and during this period the average difference between ACRIMSAT/ACRIM3 and UARS/ACRIM2 is 1.86 W/m²: we use this value for the merging. As the figure shows, the averages during the first and the second half of the cycle are 1.70 W/m² and 2.02 W/m², respectively. This suggests that our merging has an uncertainty of ±0.18 W/m².

6. THREE UPDATED ACRIM TSI COMPOSITES

The satellites records are merged and our three TSI composites are shown in Fig. 12. Table 1 summarizes how SMM/ACRIM1 and UARS/ACRIM2 records are adjusted and aligned with ACRIMSAT/ACRIM3.

The composite (A) shows that the 1996 minimum is about 0.67 ± 0.1 W/m² higher than the 1986 minimum. The composite (B) shows that the 1996 minimum is about 0.28 ± 0.1 W/m² higher than the minimum in 1986. The composite (C) shows that the 1996 minimum is about 0.11 ± 0.1 W/m² lower

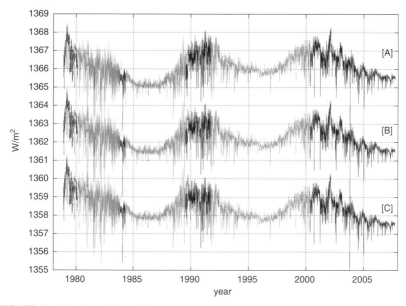

FIGURE 12 The three TSI satellite composites: see text for details. The composites (B) and (C) are shifted by -4 W/m^2 and -8 W/m^2, respectively, for visual convenience.

than the minimum in 1986. Thus, only in the eventuality that during the ACRIM-gap ERBS/ERBE record is uncorrupted the two solar minima would almost coincide. However, on average the composites indicate that the TSI minimum in 1996 is 0.28 ± 0.4 W/m^2 higher than the minimum in 1985/6.

TABLE 1 Position in W/m^2 of SMM/ACRIM1 (A1) and UARS/ACRIM2 (A2) relative to ACRIMSAT/ACRIM3 (A3) in the three scenarios (A), (B), and (C) as discussed in the test. The errors are calculated by taking into account the highest uncertainty due to the statistical non-stationarity of NIMBUS7/ERB and UARS/ACRIM2 when they merge and when UARS/ACRIM2 merges ACRIMSAT/ACRIM3. Note that the error of the position of SMM/ACRIM1 compared to UARS/ACRIM2 is ± 0.1 W/m^2. The error associated to the non-stationarity of NIMBUS7/ERB during the ACRIM-gap is described by the three scenarios (A), (B), and (C).

	(A)	(B)	(C)
A1	-1.90 ± 0.19	-1.51 ± 0.19	-1.12 ± 0.19
A2	$+1.86 \pm 0.16$	$+1.86 \pm 0.16$	$+1.86 \pm 0.16$
A3	0	0	0

Note that if UARS/ACRIM2 record needs to be corrected during its merging with NIMBUS7/ERB, as explained above, the TSI 1996 minimum relative to the TSI 1986 minimum would be about 0.1 W/m^2 higher than the above three estimates. In this case, according to the satellites data the difference between the two minima would be about 0.38 ± 0.4 W/m^2. This would further stress that the TSI satellite measurements indicate on average that TSI likely increased during solar cycles 21−23 (1980−2002).

7. TSI PROXY SECULAR RECONSTRUCTIONS

It is necessary to use reconstructions of the solar activity as long as possible, at least one century, for determining the effect of solar variations on climate. The TSI record that is possible to obtain from direct TSI satellite measurements covers the period since 1978, and this period is far too short to correctly estimate how the sun may have altered climate. The reason is because the climate system is characterized by a slow characteristic time response to external forcing that is estimated to be about 8 years (which, theoretically, can be as large as 12 years) (Scafetta, 2008; Schwartz, 2008). This decadal time response of the climate requires several decade long records for a correct evaluation of the climatic effect of an external forcing on climate. Thus, it is necessary to merge the TSI satellite composites together with long TSI secular reconstructions, which are quite uncertain because they are necessarily based on proxy data, and not direct TSI measurements.

Long-term TSI changes over the past 400 years since the 17th-century Maunder minimum have been reconstructed by several authors, for example: Hoyt and Schatten (1997), Lean (2000), Wang et al. (2005), and Krivova et al. (2007). These TSI proxy reconstructions are based on the sunspot number record, the long-term trend in geomagnetic activity, the solar modulation of cosmogenic isotopes such as ^{14}C and ^{10}Be records, and other solar related records. These observables are used because they are supposed to be linked to TSI variations. However, it is not known exactly how the TSI can be reconstructed from these historical records, nor whether these records are sufficient to faithfully reconstruct TSI changes. In fact, the proposed TSI secular proxy reconstructions are quite different from each other and show different patterns, trends and maxima, as depicted in Fig. 13. Nevertheless, they reproduce similar patterns: in particular, note the minima during the Maunder Minimum (1645−1715) and the Dalton Minimum (1790−1820), and the TSI increase during the first half of the 20th century.

The TSI increase during the first half of the 20th century is particularly important. In fact, because the characteristic time response of the climate to external forcing is about 8−12 years, an increase of TSI during the first half of the 20th century would have induced a warming also during the second half of the 20th century, even if the TSI have remained almost constant during the second half of the 20th century (16).

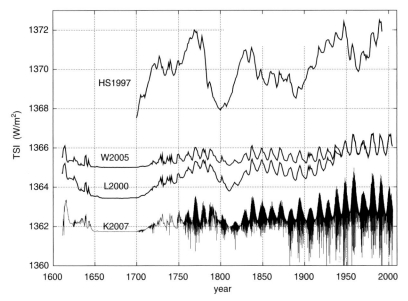

FIGURE 13   Secular TSI proxy reconstructions by Hoyt and Schatten (30) (HS1997), Lean (20) (L2000), Wang et al. (21) (W2005), and Krivova et al. (26) (K2007). K2007 has been shifted by -3 W/m^2 for visual convenience.

The four TSI proxy reconstructions shown in Fig. 13 present different trends since 1975. The TSI reconstruction proposed by Hoyt and Schatten (1997) suggests that TSI increased during this period, as shown in our TSI satellite composites (A) and (B), and in the original ACRIM TSI satellite composite. However, the other three TSI proxy reconstructions (Lean, 2000; Wang et al., 2005; Krivova et al., 2007) suggest that TSI did not change on average since 1978, as shown in our TSI satellite composite (C) and in the PMOD TSI satellite composite. Thus, the uncertainty that we have found in composing the TSI satellite records appears unresolved also by using the TSI proxy reconstructions because different solar proxies do suggest different TSI patterns as well.

Because TSI satellite composites refer to actual TSI measurements, we propose their merging with TSI proxy reconstructions for obtaining a TSI secular record. Here, we chose a recent TSI proxy reconstruction (Krivova et al., 2007), which has a daily resolution, and merge it to the TSI satellite composites in such a way that their 1980–1990 average coincides. Other choices and their implications by using the original ACRIM and the PMOD TSI satellite composites with the TSI proxy reconstructions of Lean (2000) and Wang et al. (2005) can be found in Scafetta and West (2007). The three TSI merged records herein proposed are depicted in Fig. 14. The figure shows that during the last decades the TSI has been at its highest value since the 17th century in all three cases.

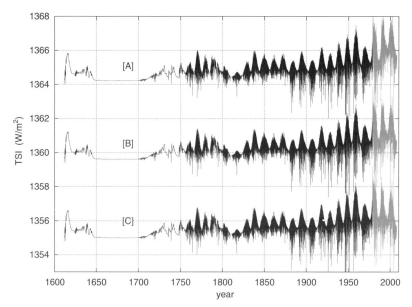

FIGURE 14 Merging of the secular TSI proxy reconstruction by Krivova et al. (26) (black) with the three TSI satellite composites proposed in Fig. 12 (gray). The TSI reconstructions (B) and (C) are shifted by -5 W/m^2 and -10 W/m^2, respectively, for visual convenience. The merging is made by shifting the original secular TSI proxy reconstruction by Krivova et al. (26) by -0.5174 W/m^2, -0.1295 W/m^2, and $+0.2584$ W/m^2 in the cases (A), (B), and (C) respectively.

8. PHENOMENOLOGICAL SOLAR SIGNATURE ON CLIMATE

The phenomenological solar signature on climate can be estimated with a phenomenological energy balance model (PEBM) (Scafetta and West, 2007; Scafetta, 2009). PEBM assumes that the climate system, to the lowest-order approximation, responds to an external radiative forcing as a simple thermodynamical system, which is characterized by a given relaxation time response τ with a sensitivity α. The physical meaning of it is that a small anomaly (with respect to the TSI average value) of the solar input, measured by ΔI, would force the climate to reach a new thermodynamic equilibrium at the asymptotic temperature value $\alpha \Delta I$ (with respect to a given temperature average value). Thus, if $\Delta I(t)$ is a small variation (with respect to a fixed average) of an external forcing and $\Delta T_s(t)$ is the Earth's average temperature anomaly induced by $\Delta I(t)$, $\Delta T_s(t)$ evolves in time as:

$$\frac{\mathrm{d}\Delta T_S(t)}{\mathrm{d}t} = \frac{\alpha \Delta I(t) - \Delta T_S(t)}{\tau}. \tag{1}$$

A simple thermal model equivalent to (1) has been used as a basic energy balance model (Douglass and Knox, 2005; North et al., 1981), but in this paper we use TSI records as a proxy forcing. We implement the PEBM by imposing that the global peak-to-trough amplitude of the 11-year solar cycle signature on the surface temperature is about 0.1 K from 1980 to 2002, as found by several authors (see IPCC, 2007, page 674 for details). This implies that the climate sensitivity Z_{11} to the 11-year solar cycle is $Z_{11} = 0.11 \pm 0.02$ K/W/m^2, as found by Douglass and Clader (2002) and Scafetta and West (2005).

In addition, the characteristic time response to external forcing has been phenomenologically estimated to be $\tau = 8 \pm 2$ years (28, 29). Note that Scafetta (28) has also found that climate is characterized by two characteristic time constants $\tau_1 = 0.40 \pm 0.1$ year and $\tau_2 = 8 \pm 2$ years, with the latter estimate that may be a lower limit (the upper limit being $\tau_2 = 12 \pm 3$ years), but an extensive discussion about the consequences of this finding is present in another study (18).

The value of the parameter α is not calculated theoretically by using the TSI as a climate forcing as usually done in the traditional climate models. The value of α is calculated by using the phenomenological climate sensitivity to the 11-year solar cycle found in Eq. (2) by means of the following equation

$$\alpha(\tau) = \sqrt{1 + \left(\frac{2\pi\tau}{11}\right)^2}, \tag{2}$$

which solves Eq. (1). Thus, we find that the phenomenological climate sensitivity to TSI changes is $\alpha = 0.51 \pm 0.15$ K/W/m^2. This value is the phenomenological climate sensitivity to irradiance changes where the TSI record is used only as a geometrical proxy of the total solar forcing on climate. Thus, the parameter α has a very different meaning than the climate sensitivity to solar irradiance as estimated in the traditional physical models. In our case, for example, the parameter α would include also other effects such as, for example, that of a cloud modulation by means of cosmic ray flux, which is regulated by solar activity and presents a geometrical form approximately similar to that of the TSI reconstructions.

With the above value of τ and α, Eq. (1) can be numerically solved by using as input the TSI records shown in Fig. 14. The phenomenological solar signatures (PSSs) are shown in Fig. 15 where the three PSSs are plotted since 1600 against a paleoclimate Northern Hemisphere temperature reconstruction (Moberg et al., 2005) and since 1850 against the actual instrumental Northern Hemisphere surface record (Brohan et al., 2006).

The figure shows that there is a good agreement between the PSSs and the temperature record. The patterns between 1600 and 1900 are well recovered such as the cooling during the Maunder (1650−1730) and Dalton (1790−1840) minima. The warming during the first half of the 20th century is partially recovered. The observed discrepancy around 1940−1950 could be due to an error in the used TSI proxy model. A better matching would occur by using Hoyt TSI model because it peaks during 1940−1950. Finally, since 1978 the output strongly depends on the TSI behavior. If the TSI reconstruction (A) is adopted, a significant portion of the warming, about 66% observed since 70s

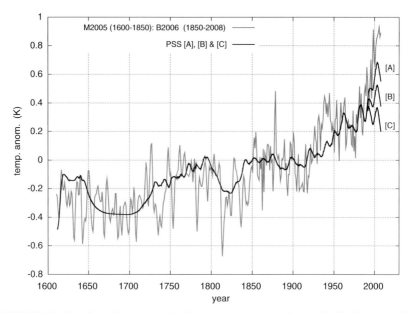

FIGURE 15 The three phenomenological solar signatures on climate (black) obtained with Eq. (1) forced with the three TSI records shown in Fig. 14 against a paleoclimate Northern Hemisphere temperature reconstruction (35) from 1600 to 1850, and since 1850 against the actual instrumental Northern Hemisphere surface record (36) (gray). The figure shows temperature anomalies relative to the 1895−1905 average.

has been induced by solar variations, while if the TSI reconstruction (C) is adopted, almost all warming, about 85% observed since 70s has been induced by factors alternative to solar variations. If the average TSI reconstruction (B) is adopted, about 50% of the warming observed since 70s has been induced by solar variations.

9. CONCLUSION

We have reconstructed new TSI satellite composites by using the three ACRIM records and have shown that different composites are possible, depending on how the ACRIM-gap problem from 1989.5 to 1992 is solved. Our three TSI composites indicate that the TSI minimum in 1996 is 0.30 ± 0.40 W/m^2 higher than the TSI minimum in 1986. On the contrary, the two TSI minima in 1986 and 1996 would be located at the same level only in the eventuality that the TSI ERBS/ERBE satellite record is uncorrupted during the ACRIM-gap, a fact that has been questioned by our analysis. For example, in Fig. 7 we did not notice a sudden one-day jump at the end of September 1989 in the sensitivity of NIMBUS7 sensors, as claimed by PMOD. On the contrary, we noticed a rapid, but gradual divergence between NIMBUS7 and ERBS that occurred during

October and November 1989. This gradual divergence may also imply a rapid degradation of ERBS sensors.

None of the TSI satellite composites proposed by the ACRIM, IRMB, and PMOD teams can be considered rigorously correct. All three teams have just adopted alternative methodologies that yield to different TSI composites, but these teams have also ignored the unresolved uncertainties in the data that yields to an unresolved uncertainty in the TSI composites as well.

The comparison with theoretical TSI proxy models, for example Wenzler et al. (2006) and Krivova et al. (2007), cannot be used to resolve the issue, as the PMOD team assumes. In fact: (1) in science, theoretical models have to be tested against the observations, not vice versa, (2) the TSI proxy models adopt a reductionistic scientific approach, that is, they assume that some given solar observable that refers to a particular solar measure (for example measurements from magnetograms or measurements of the intensity of a given frequency of the spectrum) can be used to faithfully reconstruct a global solar measure such as the TSI, and (3) the TSI proxy models do depend on parameters that opportunely calibrated give different outcomes that may, eventually, fit alternative satellite composites.

Because it is not possible to reconstruct with certainty the TSI behavior during the ACRIM-gap, the TSI decadal trend during the last three decades is unfortunately uncertain, and any discussion that needs to use the TSI record has to take into account this unresolved uncertainty.

However, because the uncertainty in the data indicates that the TSI minimum in 1996 is at least approximately 0.30 ± 0.40 W/m^2 higher than the TSI minimum in 1986, on average the satellite records do suggest that TSI may have increased from 1980 to 2000. Therefore, the sun may have significantly contributed to the warming observed during the last three decades, as suggested by the phenomenological model simulations herein proposed. This result has been further confirmed by Scafetta (2010) where it was shown that the warming observed from 1970 to 2000 has been mostly induced by a 60-year modulation of the climate during its warm phase. This 60-year modulation appears to be present in the climate records for centuries and it appears to be correlated to a correspondent and clear 60-year heliospheric cycle which is also revealed in several solar proxy records.

Note that a recent paper by Lockwood (2008) concluded that even with the adoption of the original ACRIM composite, the sun's contribution to the global surface warming would be negligible during the last three decades, in contrast with the findings of Scafetta and West (2007, 2008) and those presented here. However, Lockwood's findings derive from his evaluation of the characteristic time response of the climate to solar variation: $\tau = 0.8$ years. This value strongly differs from the value herein adopted of $\tau = 8$ years and recently measured by Scafetta (2008) and Schwartz (2008). The problem with Lockwood's short time constant is that according to the climate physics implemented in most climate models, the characteristic time response of the climate varies from a few months to several years and even decades, as Lockwood himself acknowledges in his own paper. For example, the linear upwelling/diffusion energy balance model adopted by Crowley (2000) uses about $\tau = 10$ years. In addition, Scafetta (2008) and Schwartz (2008) have found that climate is characterized by at least two characteristic time constants, one short with a time scale of several months and one long with a decadal time scale. The climate processes with a fast response are usually responsible for the fast fluctuations seen in the data. Instead, the climate processes with a slow response are those that drive the decadal and secular trends observed in the global temperature. This slow climate response derives from the fact that the processes that regulate the decadal and secular variation of the climate (most of all energy exchange with the deep ocean and changes of the albedo due to the melting of the glaciers and forestation and desertification processes) are very slow processes, and they work as powerful climate feedbacks. Thus, we believe that Lockwood's analysis is inappropriate because it failed to take into account the climate processes with a slow time response that would be responsible for a strong climate response to solar changes. However, a more detailed discussion about this issue, which would imply also an update of the PEBM presented here, is present in another dedicated study (2009).

REFERENCES

Brohan, P., Kennedy, J.J., Harris, I., Tett, S.F.B., Jones, P.D., 2006. Uncertainty estimates in regional and global observed temperature changes: a new dataset from 1850: Journal of Geophysical Research 111, D12106.

Chapman, G.A., Cookson, A.M., Dobias, J.J., 1996. Variations in total solar irradiance during solar cycle 22. Journal of Geophysical Research 101, 13541−13548.

Crommelynck, D., Dewitte, S., 1997. Solar constant temporal and frequency characteristics. Solar Physics 173, 177−191.

Crowley, T.J., 2000. Causes of climate change over the past 1000 years. Science 289, 270−277.

Dewitte, S., Crommelynck, D., Mekaoui, S., Joukoff, A., 2004. Measurement and uncertainty of the long-term total solar irradiance trend. Solar Physics 224, 209−216.

Douglass, D.H., Clader, B.D., 2002. Climate sensitivity of the Earth to solar irradiance. Geophysical Research Letters 29, 1786−1789.

Douglass, D.H., Knox, R.S., 2005. Climate forcing by the volcanic eruption of Mount Pinatubo. Geophysical Research Letters 32, L05710.

Fröhlich, C., Lean, J., 1998. The Sun's total irradiance: cycles, trends and related climate change uncertainties since 1978: Geophysical Research Letters 25, 4377−4380.

Fröhlich, C., 2000. Observations of irradiance variations. Space Science Reviews 94, 15−24.

Fröhlich, C., 2004. Solar irradiance variability. In: Geophysical Monograph 141: Solar Variability and Its Effect on Climate. American Geophysical Union, Washington, DC, USA, Chapt. 2: Solar Energy Flux Variations, pp. 97−110.

Fröhlich, C., 2006. Solar irradiance variability since 1978: revision of the PMOD composite during solar cycle 21. Space Science Reviews 125, 53−65.

Fröhlich, C., Crommelynck, D., Wehrli, C., Anklin, M., Dewitte, S., Fichot, A., Finsterle, W., Jimènez, A., Chevalier, A., Roth, H.J., 1997. In-flight performances of VIRGO solar irradiance instruments on SOHO. Solar Physics 175, 267–286.

Hoyt, D.V., Schatten, K.H., 1997. The Role of the Sun in the Climate Change. Oxford Univ. Press, New York.

Hoyt, D.V., Kyle, H.L., Hickey, J.R., Maschhoff, R.H., 1992. The Nimbus 7 solar total irradiance: a new algorithm for its derivation. Journal of Geophysical Research 97, 148–227.

Intergovernmental Panel on Climate Change, et al., 2007. Climate Change 2007: The Physical Science Basis. In: Solomon, S. (Ed.), Cambridge Univ. Press, New York.

Kirkby, J., 2007. Cosmic rays and climate. Surveys in Geophysics 28, 333–375.

Kopp, G., Lawrence, G., Rottman, G., 2003. Total irradiance monitor design and on-orbit functionality: Proceedings of SPIE 5171, 15–25.

Krivova, N.A., Balmaceda, L., Solanki, S.K., 2007. Reconstruction of solar total irradiance since 1700 from the surface magnetic flux. Astronomy and Astrophysics 467, 335–346.

Lean, J., 2000. Evolution of the Sun's spectral irradiance since the Maunder Minimum. Geophysical Research Letters 27, 2425–2428.

Lee III, R.B., Gibson, M.A., Wilson, R.S., Thomas, S., 1995. Long-term total solar irradiance variability during sunspot cycle 22. Journal of Geophysical Research 100, 1667–1675.

Lockwood, M., 2008. Recent changes in solar output and the global mean surface temperature. III. Analysis of the contributions to global mean air surface temperature rise. Proceedings of the Royal Society A 464, 1387–1404.

Moberg, A., Sonechkin, D.M., Holmgren, K., Datsenko, N.M., Karlén, W., 2005. Highly variable Northern Hemisphere temperatures reconstructed from low- and high-resolution proxy data. Nature 433, 613–617.

North, G.R., Cahalan, R.F., Coakley Jr., J.A., 1981. Energy balance climate models. Reviews in Geophysics 19, 91–121.

Pap J.M. et al., 2004 (Eds.), Solar Variability and Its Effects on Climate: Geophysical Monograph Series Volume 141 American Geophysical Union, Washington, DC.

Scafetta, N., West, B.J., 2005. Estimated solar contribution to the global surface warming using the ACRIM TSI satellite composite. Geophysical Research Letters 32, L18713.

Scafetta, N., Willson, R.C., 2009. ACRIM-gap and Total Solar Irradiance (TSI) trend issue resolved using a surface magnetic flux TSI proxy model. Geophysical Research Letters 36, L05701.

Scafetta, N., 2008. Comment on "Heat capacity, time constant, and sensitivity of Earth's climate system" by S.E. Schwartz. Journal of Geophysical Research 113, D15104.

Scafetta, N., 2009. Empirical analysis of the solar contribution to global mean air surface temperature change. Journal of Atmospheric and Solar-Terrestrial Physics 71, 1916–1923.

Scafetta, N., 2010. Empirical evidence for a celestial origin of the climate oscillations and its implications. Journal of Atmospheric and Solar-Terrestrial Physics 72, 951–970.

Scafetta, N., West, B.J., 2007. Phenomenological reconstructions of the solar signature in the NH surface temperature records since 1600. Journal of Geophysical Research 112, D24S03.

Scafetta, N., West, B.J., 2008. Is climate sensitive to solar variability? Physics Today 3, 50–51.

Schwartz, S.E., 2008. Reply to comments by G. Foster et al., R. Knutti et al., and N. Scafetta on "Heat capacity, time constant, and sensitivity of Earth's climate system". Journal of Geophysical Research 113, D15105.

Wang, Y., Lean, J.L., Sheeley Jr., N.R., 2005. Modeling the sun's magnetic field and irradiance since 1713. The Astrophysical Journal 625, 522–538.

Wenzler, T., Solanki, S.K., Krivova, N.A., Fröhlich, C., 2006. Reconstruction of solar irradiance variations in cycles 21–23 based on surface magnetic fields. Astronomy and Astrophysics 460, 583–595.

Willson, R.C., 2001. The ACRIMSAT/ACRIM3 experiment—extending the precision, long-term total solar irradiance climate database: The Earth Observer 13, 14–17.

Willson, R.C., Hudson, H.S., 1991. The Sun's luminosity over a complete solar cycle: Nature 351, 42–44.

Willson, R.C., Mordvinov, A.V., 2003. Secular total solar irradiance trend during solar cycles 21–23. Geophysical Research Letters 30, 1199–1202.

Willson, R.C., 1994. Irradiance observations of SMM, Spacelab 1, UARS and ATLAS experiments: The Sun as a Variable Star, Int. Astron. Union Colloq. 143 Proc., In: Pap, J. et al., (Eds.), pp. 54–62, Cambridge Univ. Press, New York.

Willson, R.C., 1997. Total solar irradiance trend during solar cycles 21 and 22. Science 277, 1963–1965.

Chapter 13

Global Brightening
and Climate Sensitivity

Christopher Monckton of Brenchley
Carie, Rannoch, PH17 2QJ, Scotland, UK

Chapter Outline

1. INTRODUCTION

The radiative-forcing method of the Intergovernmental Panel on Climate Change for determining climate sensitivity without resorting to atmosphere–ocean general-circulation models is deployed to determine the quantum of projected transient global warming from the global brightening of 1983–2001, as well as from anthropogenic influences over the same period and, separately, from 1950 to 2005. Results, compared with observations, shed light on the reliability of the IPCC's method.

2. THE DATA

The Hadley/CRU global mean surface temperature dataset (Jones et al., 1999, and Brohan et al., 2006, cited in Solomon et al., 2007) shows a rapid and sustained global warming from 1975 to 2001 at 0.16 K/decade, a rate identical,

Evidence-Based Climate Science. DOI: 10.1016/B978-0-12-385956-3.10013-0

within measurement error, to those observed from 1860 to 1880 and again from 1910 to 1940. Warming in the two earlier periods preceded any possible significant anthropogenic influence on the climate. Was the third and the most recent period of rapid warming also chiefly natural?

Satellites first measured global surface temperature, cloud cover, and radiant-energy flux leaving the top of the atmosphere in the early 1980s. Pinker et al. (2005), finding agreement between satellite and terrestrial records for the first time, reported a strong global brightening from 1983 to 2001, possibly caused by a decline in cloud cover with changes in water vapor and aerosols. Pinker, relying chiefly on the International Satellite Cloud Climatology Project (ISCCP: Rossow and Schiffer, 1991, 1999), applied linear and second-order least-squares fits to the satellite-derived time-series of globally-averaged short-wave anomalies in solar radiative flux at the Earth's surface from 1983 to 2001, after removing the annual cycle. The linear slope was positive at $0.16 \, \mathrm{W \, m^{-2} \, year^{-1}}$ (Fig. 1), close to the $0.143 \, \mathrm{W \, m^{-2} \, year^{-1}}$ found in a recent reanalysis by Boston (Appendix).

The Earth Radiation Budget Experiment satellites (ERBE: Barkstrom, 1984) also detected a reduction in short-wave radiation reflected from clouds to space during the same period, with a corresponding increase in long-wave radiation as more short-wave radiation reached the Earth's surface and was Wien-displaced to the near-infrared, consistent with a reduction in global cloud cover, especially at low altitudes and latitudes, and particularly in the 1990s.

At the time when Pinker reported, the ERBE outgoing-radiation data presented in Wielicki et al. (2002a,b) had not been corrected to allow for orbital decay. After allowing for adjustments published by Wong et al. (2006), Boston (Appendix) used the ERBE data as a validity check for the ISCCP data, concluding that there was reasonably close agreement with them after allowance for differences in spatial and temporal coverage, and that the annual net

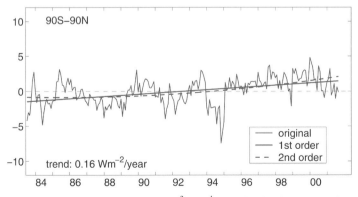

FIGURE 1 A globally-averaged $+0.16 \, \mathrm{W \, m^{-2} \, year^{-1}}$ trend in the short-wave solar surface radiative-flux anomaly, 1983–2001, after removal of the mean annual cycle. From Pinker, Fig. 1.

surface flux trend for 1983—2001 was +0.143 W m^{-2}. This value is taken as the starting-point for the calculations in the present paper.

Other observations confirm the fact of the global brightening, while disagreeing as to its magnitude. Solomon et al. (2007, at Table 3.5) compare tropical (20°S—20°N) top-of-atmosphere long-wave and short-wave radiative fluxes for 1994—1997 with the fluxes for 1985—1989, citing the ERBE satellite data, which showed outgoing long-wave radiation increasing by 0.7 W m^{-2} and outgoing short-wave radiation decreasing by 2.1 W m^{-2} over the period. Solomon also cites the ISCCP data as showing that outgoing long-wave radiation rose by 0.5 W m^{-2}, while outgoing short-wave radiation fell by 2.4 W m^{-2}. The AVHRR Pathfinder gave an opposite result: but, as Solomon notes, "Calibration issues, conversion from narrow to broadband, and satellite orbit changes are thought to render the AVHRR record less reliable for decadal changes compared to ERBS."

Wild et al. (2005), focusing on changes in mean surface temperature over the land, concluded that the global dimming up to the 1980s was offset by the period of brightening (or at least the absence of dimming) thereafter, and that the rapid warming that followed the transition was attributable almost entirely to anthropogenic influences. However, Pinker (at Fig. 5A) shows a very slight dimming over land only from the 1980s onward. Though Wild makes a passing reference to Pinker, he relied not upon satellite data but upon data from the Global Energy Balance archive and the Baseline Surface Radiation Network. Wild focused solely on land surfaces. The absence of solar dimming and the insignificant net brightening over land after 1980, deduced from surface-station measurements, is consistent with the analysis of land-only satellite data in Pinker.

Solomon et al. (2007, ch. 3) concludes the sub-chapter on clouds as follows:

"In summary, while there is some consistency between ISCCP, ERBS, SAGE II and surface observations of a reduction in high cloud cover during the 1990s relative to the 1980s, there are substantial uncertainties in decadal trends in all datasets and at present there is no clear consensus on changes in total cloudiness over decadal time-scales."

However, the data indicate that from 1983 to 2001 there was a significant decline in cloud cover generally, not merely in high cloud cover. It is the optically-dense clouds at low altitude and latitude that are most influential in global brightening or dimming.

3. ANALYSIS AND RESULTS

For clarity, this analysis is confined to central estimates. First, established methods are used to determine the surface pre-feedback warming ΔT_0 caused by global brightening. Next, simplified methods provided in Solomon are used to determine post-feedback surface climate sensitivity ΔT both to natural and to anthropogenic radiative forcings.

Boston's net global surface short-wave flux trend $+0.143$ W m^{-2} year^{-1} represents a flux increment $\Delta F_{nat} \approx 2.586$ W m^{-2} over the 18.083 years 1983–2001 (Appendix: Fig. A3). Warming before feedbacks is the product of ΔF_{nat} and the surface value of the Planck pre-feedback climate-sensitivity parameter κ_{sfc}, which, from data in Kiehl and Trenberth (2009) and methods in Kiehl and Trenberth (1997) and Kimoto (2009), is the first differential of the Stefan–Boltzmann equation where emissivity ε is unity, mean surface temperature T is 288 K, and radiative fluxes F_{sfc}, F_{eva}, F_{thm} from surface radiation, evapo-transpiration, and thermal convection are 390, 80, 17 W m^{-2} respectively:

$$\kappa_{sfc} = \Delta T/\Delta F_{sfc} = T/[4(F_{sfc} + F_{eva} + F_{thm})] \approx 0.1478 \text{ K W}^{-1} \text{ m}^2. \quad (1)$$

Accordingly, the naturally-occurring transient pre-feedbacks warming $\Delta T_{0,nat}$ that would be expected to have arisen from the global brightening mentioned in Pinker was:

$$\Delta T_{0,nat} = \Delta F_{nat}\kappa_{sfc} \approx 0.38 \text{ K.} \quad (2)$$

Allowance is then made for the amplifying influence of temperature feedbacks arising in response to the surface warming at Eq. (2). The feedback multiplier f, where b is the sum of all unamplified climate-relevant feedbacks and where, at the characteristic-emission altitude, the Planck parameter $\kappa_{cel} \approx 3.2^{-1} = 0.3125$ K W^{-1} m^2 (Solomon, ch. 10, p. 631, footnote), is given by the feedback-amplification function in Bode (1945):

$$f = (1 - b\kappa_{cel})^{-1}. \quad (3)$$

At a doubling of atmospheric CO_2 concentration, where the multi-model mean projected climate sensitivity $\Delta T_{2x} \approx 3.26$ K (Solomon, ch. 10, p. 798, box 10.2) and the anthropogenic CO_2 radiative forcing $\Delta F_{2x} = 5.35 \ln 2 \approx 3.71$ W m^{-2} (Myhre et al., 1998, and in Solomon), the IPCC's implicit central estimate is:

$$f_{man} = \Delta T_{2x}/\kappa_{cel}\Delta F_{2x} \approx 2.813. \quad (4)$$

With this central estimate of f_{man}, Eq. (3) is rearranged to derive a central estimate of the sum b_{man} of all unamplified temperature feedbacks that is implicit in Solomon:

$$b_{man} = (f_{man} - 1)/f_{man}\kappa_{cel} \approx 2.062 \text{ W m}^{-2} \text{ K}^{-1}. \quad (5)$$

However, it is necessary to deduct from b_{man} the cloud feedback of 0.69 W m^{-2} K^{-1} and the surface-albedo feedback of 0.26 W m^{-2} K^{-1} (Soden and Held, 2006, cited in Solomon), since the observed global brightening will have encompassed these feedbacks. Thus,

$$b_{nat} = b_{man} - 0.69 - 0.26 \approx 1.112 \text{ W m}^{-2} \text{ K}^{-1}. \quad (6)$$

Then
$$f_{\text{nat}} = (1 - b_{\text{nat}}\kappa_{\text{cel}})^{-1} \approx 1.533, \tag{7}$$
and

$$
\begin{aligned}
\Delta T_{\text{nat}} &= \Delta T_{0,\text{nat}} f_{\text{nat}} \\
&= \Delta F_{\text{nat}} k_{\text{sfc}} f_{\text{nat}} = \Delta F_{\text{nat}} \kappa_{\text{sfc}} (1 - b_{\text{nat}}\kappa_{\text{cel}})^{-1} \\
&\approx 0.59 \text{ K}.
\end{aligned} \tag{8}
$$

Accordingly, Solomon's radiative-forcing method implies that, in response to a natural global brightening of the magnitude determined by Boston, pre-feedback and post-feedback warmings would be 0.38 K (Eq. (2)) and 0.59 K (Eq. (8)), respectively.

In addition, anthropogenic warming would be expected to have occurred over the period. Table 1, again relying upon simplified methods described in Solomon, gives central estimates of the radiative forcings arising from observed increases in the concentrations of the major greenhouse gases from 1983 to 2001.

Since the forcings in Table 1 represent changes in net (down minus up) radiative flux at the tropopause rather than at the surface, the surface value κ_{sfc}

TABLE 1 Radiative forcings arising from changes in the atmospheric concentrations of key greenhouse gases from 1983 to 2001, determined from functions given in Myhre et al. (1998) and cited in Houghton et al. (2001) and in Solomon. CFC concentrations are from Hartley et al. (1996). The ozone forcing is an estimate derived from relative magnitudes of forcings in Solomon. The negative forcings from aerosol effects are omitted, since the measured surface radiative-flux anomaly occasioned by the global brightening of 1983–2001 implicitly takes them into account. Forcings from minor halocarbons, land-use changes, and aircraft contrails are omitted as *de minimis*.

Greenhouse gas	1983	2001	Radiative forcing
Carbon dioxide	342 ppmv	370 ppmv	0.421 W m^{-2}
CFC-12	343 pptv	541 pptv	0.065 W m^{-2}
Methane	1630 ppbv	1775 ppbv	0.054 W m^{-2}
Ozone	Estimate	Estimate	0.048 W m^{-2}
Nitrous oxide	304 ppbv	312 ppbv	0.024 W m^{-2}
CFC-11	183 pptv	256 pptv	0.018 W m^{-2}
		Total forcing	$\Delta F_{\text{man}} = 0.630$ W m^{-2}

of the Planck parameter is inapplicable. Instead, the substantially greater value $\kappa_{cel} \approx 3.2^{-1} \approx 0.3125$ K W^{-1} m^2 at the characteristic-emission altitude applies. However, the feedback factor f_{nat} remains appropriate thanks to the dominance of the global brightening over the period.

The warming ΔT_{nat} is *transient*, since — in the sufficiently short term, at any rate — global brightenings and dimmings come and go. However, in Solomon's methodology, the temperature change ΔT_{man} that is projected to arise in consequence of greenhouse-gas forcings is *equilibrium* warming: i.e., the global-temperature anomaly that would arise only after the climate had settled to a new equilibrium after the perturbation of the climate arising from the forcing. Thus, at equilibrium:

$$\Delta T_{man} = \Delta F_{man}\kappa_{cel}f_{nat} = \Delta F_{man}\kappa_{cel}(1 - b_{nat}\kappa_{cel})^{-1} \approx 0.30 \text{ K.} \quad (9)$$

To determine transient temperature change, we introduce an additional term r to represent the ratio of transient to equilibrium warming. In Solomon, the implicit ratio r for transient warming occurring in the medium term (i.e., in any period within the century and a half from 1950 to 2000) is approximately 0.57.

$$\Delta T_{man} = r\Delta F_{man}\kappa_{cel}f_{nat} \approx 0.17 \text{ K.} \quad (10)$$

Then

$$\Delta T = \Delta T_{nat} + \Delta T_{man} \approx 0.76 \text{ K.} \quad (11)$$

Of this 0.76 K warming over the 18-year period, >75% would be attributable to the naturally-occurring global brightening. However, the linear-regression trend on the HadCRUt monthly global land and sea surface temperature anomalies shows observed transient warming $\Delta T_{obs} = 0.34$ K. Solomon's methods appear to lead to an overstatement of climate sensitivity by a factor >2.2.

Even if the flux anomaly occasioned by the global brightening of 1983–2001 had been little more than half of that which Pinker had reported, it would have caused warming approximately equivalent to the observed warming, consequently leaving little room for any contribution from anthropogenic influences. Alternatively, even if temperature feedbacks were taken as net-zero rather than strongly positive, the warming arising from the global brightening over the period would be almost equal to the global warming that was actually observed, again leaving little room for any anthropogenic component.

It is possible that the global brightening from 1983 to 2001 was less than Pinker or Boston had found, or that the reported rise in radiative flux arising from the global brightening was to some extent offset by unreported natural influences. It is also possible that over so short a period the influence of long-acting feedbacks may not have made itself felt, though this possibility is to some extent accounted for in the transience ratio r. Besides, if the initial warming were small, long-acting feedbacks would also be small.

TABLE 2 Radiative forcings arising from changes in the atmospheric concentrations of greenhouse gases from 1950 to 2005, determined on the same basis as Table 1.

Greenhouse gas	1950	2005	Radiative forcing
Carbon dioxide	308 ppmv	378 ppmv	$1.095 \ \mathrm{W \ m^{-2}}$
Methane	1,100 ppbv	1,775 ppbv	$0.281 \ \mathrm{W \ m^{-2}}$
Ozone	Estimate	Estimate	$0.237 \ \mathrm{W \ m^{-2}}$
CFC-12	20 pptv	541 pptv	$0.172 \ \mathrm{W \ m^{-2}}$
Nitrous oxide	287 ppbv	319 ppbv	$0.102 \ \mathrm{W \ m^{-2}}$
CFC-11	10 pptv	248 pptv	$0.060 \ \mathrm{W \ m^{-2}}$
		Total forcing	$\Delta F_{man} = 1.948 \ \mathrm{W \ m^{-2}}$

Where $\Delta F_{man} \approx 1.948 \ \mathrm{W \ m^{-2}}$ (Table 2), and where the feedback multiplier $f_{man} \approx 2.813$ (Eqs. (3), (4)) encompasses cloud and surface-albedo feedbacks, Solomon's methods (Eq. (10)) would indicate transient warming of 0.98 K from 1950 to 2005 in the absence of any global brightening or dimming. However, observed warming ΔT_{obs}, taken as the least-squares linear-regression trend on the Had-CRUt global mean surface temperature anomalies from 1950 to 2005, was only 0.65 K, again suggesting that use of Solomon's methodology may lead to overstatement of climate sensitivity, this time by a factor 1.5.

In the sufficiently long term even a significant transient global brightening such as that of 1983–2001 might be canceled out. Solomon finds, with 90% confidence, that more than half of the global warming from 1950 to 2005 was of anthropogenic origin. On the assumption that all of it was anthropogenic, and that in particular no substantial net global brightening or dimming occurred from 1950 to 2005, radiative forcings from 1950 to 2005 were determined using the functions in Solomon, and are listed in Table 2.

To determine the warming that would be expected to have arisen over the 56-year period, it is again appropriate to omit Solomon's strongly-negative very-long-term forcing from anthropogenic particulate pollution, since no global brightening or dimming is assumed to have occurred over the period. However, over so long a period as half a century, the positive forcings from the cloud and surface-albedo feedbacks posited by Solomon would be expected to have operated.

4. DISCUSSION

The contribution of the naturally-occurring global brightening from 1983 to 2001 to warming over the period that was determined using Solomon's methods substantially exceeded observed warming. Even if the true global brightening had been only half of that reported by Pinker, its magnitude would have been sufficient to account for nearly all of the observed warming over the period, leaving little room for any anthropogenic contribution.

The surface brightening from 1983 to 2001 appears to have been real, substantial, and of natural origin. CO_2 concentration has continued to rise near-monotonically until the present, but the monotonicity of its increase, set against the stochasticity of the fluctuations in global brightening and dimming, implies an absence of correlation and hence of causation between the former and the latter, at least on decadal-to-subdecadal timescales. By contrast, at least from 1983 to 2001, there is some agreement between the global brightening and the observed warming, suggesting a perhaps causative and certainly far from counter-intuitive correlation between the two. The question arises whether Solomon was correct in listing direct and indirect aerosol forcings as being very strongly negative. If not, then the true magnitude of net anthropogenic forcings may have been considerably greater than the values given in Solomon. If so, Solomon's method may tend appreciably to overstate climate sensitivity.

The anthropogenic warming from 1950 to 2005 that was determined using Solomon's methods again substantially exceeded observation, raising the question whether there were any negative natural changes in net surface flux during that period. The natural cycles of brightening and dimming that may contribute to the quasi-periodicity of the global-temperature record appear to correspond to some extent with the warming and cooling phases of the Pacific Decadal Oscillation, but they may not necessarily cancel one another out even over long periods. However, in the absence of adequate satellite instrumentation before the early 1980s, that question cannot be definitively answered.

In future the deployment of standardized, automated, surface-mounted thermometers and pyranometers at locations all over the planet, reporting by satellite much as the Argo bathythermographs do for ocean temperature and salinity today, may provide better measurement of changes in cloud cover. Pyranometers deployed in Japan for >100 years show a remarkably close, and possibly causative, correlation between changes in surface solar flux (expressed as hours of sunlight) and changes in surface temperature in the region (W. W.-H. Soon, 2009, personal communication). However, data sources from many other regions do not show a similar correlation.

It is not likely that the net global brightening from 1983 to 2001 was caused in part by a systemic decline in particulate aerosols resulting from environmental measures in Western nations to improve the quality of the atmosphere: particulate aerosols continue to be emitted in increasing quantities by emerging nations such as China. If there had indeed been an enduring global clearing of the air sufficient to influence global temperatures, the warming from 1983 to 2001 might have been expected to continue thereafter: however, there has been a small cooling trend since late 2001.

5. A NOTE ON THE PLANCK PARAMETER

The Planck parameter κ is of particular significance in the determination of climate sensitivity because not only pre-feedback temperature change ΔT_0 but also the feedback factor f are dependent upon it, so that post-feedback temperature change ΔT is dependent upon κ^2:

$$
\begin{aligned}
\Delta T &= (\Delta F + b\Delta T)_\kappa \\
\Rightarrow \quad \Delta T(1 - b\kappa) &= \Delta F\kappa \\
\Rightarrow \quad \Delta T &= \Delta F\kappa(1 - b\kappa)^{-1} \\
\Rightarrow \quad \Delta T/\Delta F &= \kappa(1 - b\kappa)^{-1} \\
&= \kappa f \\
\Rightarrow \quad f &= (1 - b\kappa)^{-1}.
\end{aligned}
\tag{12}
$$

Equivalently, expressing the feedback loop as the sum of an infinite series,

$$
\begin{aligned}
\Delta T &= \Delta F\kappa + \Delta F\kappa^2 b + \Delta F\kappa^3 b^2 \\
&= \Delta F\kappa(1 + \kappa b + \kappa^2 b^2 + L) \\
&= \Delta F\kappa(1 - \kappa b)^{-1} \\
&= \Delta F\kappa f.
\end{aligned}
\tag{13}
$$

From Eqs. (12, 13) it is evident that it is inappropriate to take different values of κ when determining pre-feedback warming ΔT_0 and the feedback factor f. Yet that is what occurred in Eqs. (7, 8), where $\Delta T_{0,\text{nat}}$ was taken as the product of the naturally-occurring surface brightening ΔF_{nat} and the surface value κ_{sfc} of the Planck parameter, but $f_{\text{nat}} = (1 - b\kappa_{\text{cel}})^{-1}$ was determined using the characteristic-emission-altitude value κ_{cel}.

We acted thus because in Eq. (7) it was surely appropriate to convert a surface radiative-flux change ΔF_{nat} to pre-feedback surface temperature change $\Delta T_{0,\text{nat}}$ using the surface value κ_{sfc} of the Planck parameter, and because in Eq. (8) it was appropriate to adopt the characteristic-emission-altitude value κ_{cel}, which Solomon, following Houghton, uses to determine f in response to the pre-feedback surface temperature change ΔT_0.

The dilemma thus posed by the IPCC's treatment of the Planck parameter has received surprisingly little attention in the literature. However, Kimoto (2009) addresses it in some detail, and concludes that the surface value κ_{sfc} is preferable even for determination of climate sensitivity to radiative forcings at the tropopause. In that event, and on that ground alone, all of the IPCC's climate-sensitivity values would be overstated by $(\kappa_{\text{cel}}/\kappa_{\text{sfc}})^2$, approximately a factor 4.4.

If we had used κ_{sfc} not only for the initial surface warming caused by the net global brightening of 1983–2001 but also for the consequent feedbacks, on the IPCC's methodology the consequent natural warming ΔT_{nat} shown in Eq. (8) would have been 0.46 K rather than 0.59 K, and projected ΔT including anthropogenic influences would have been 0.63 K, greater than the observed 0.34 K by a factor 1.9. On the other hand, if we had used κ_{cel} throughout

Eqs. (7, 8), projected ΔT_{nat} would have been 1.24 K and projected ΔT 1.48 K over the 18-year period, above observed warming by a factor 4.4.

6. CONCLUSIONS

If the global brightening of 1983−2001 was as significant as Pinker reported and Boston broadly confirms, the fact that application of Solomon's method to the brightening flux implies a pre-feedback warming $\Delta T_0 \approx 0.38$ K from 1983 to 2001, when only 0.34 K global warming was observed, raises the question whether Solomon was correct in regarding short-acting feedbacks as strongly net-positive. Use of Solomon's method implies that the naturally-occurring global brightening of 1983−2001 should have caused a transient warming of 0.76 K, more than four times the 0.17 K anthropogenic warming over the period inferred by use of the same method, and more than twice the 0.34 K observed transient warming of those 18 years.

Likewise, even if no net global brightening were assumed to have occurred over the 56 years 1950−2005, use of Solomon's method would lead us to expect a warming of ~1 K, which is half as much again as the observed transient anthropogenic warming of 0.65 K. These discrepancies are too great to be attributable merely to the failure to use atmosphere−ocean general-circulation models when determining climate sensitivity. In any event, the IPCC's simplified methods used here were derived from the models.

If the data are in substance correct, the positive anthropogenic forcing over the period may have been offset to some extent by unidentified negative forcings or temperature feedbacks. For instance, a temporary reduction in relative humidity associated with a possibly cyclical reduction in cloud cover (in approximate alignment with the warming phases of the Pacific Decadal Oscillation) may have transiently altered the sign of the water-vapor feedback. The cloud feedback may also be strongly negative rather than strongly positive (Spencer et al., 2007). Temperature feedbacks in general may be somewhat net-negative, rather than strongly net-positive, as Solomon, citing Soden and Held (2006), finds them.

It is also possible that, in today's atmosphere, true climate sensitivity not only to atmospheric CO_2 enrichment but also to other anthropogenic and natural radiative forcings is less by a factor 1.5−2.2 than the estimates in Solomon.

It has been determined theoretically (e.g., Lindzen, 2007; Schwartz, 2007; Monckton of Brenchley, 2008) and confirmed empirically by direct measurement of outgoing radiation from the Earth's characteristic-emission level (e.g., Covey, 1995; Chen et al., 2002; Cess and Udelhofen, 2003; Hatzidimitriou et al., 2004; Clement and Soden, 2005, and by direct measurement of ocean temperatures in the mixed layer (Lyman et al., 2006 as amended, Gouretski and Koltermann, 2007; Willis, 2008; Willis et al., 2009), that the IPCC's current estimates of climate sensitivity to atmospheric CO_2 enrichment may be excessive.

It was perhaps no mere coincidence that the global brightening of 1983–2001 — a naturally-occurring event — coincided with the greater part of the only supra-decadal period of sustained and rapid warming observed since 1950.

ACKNOWLEDGMENTS

I am most grateful to Dr. Joseph Boston for having kindly re-evaluated (in the Appendix) the satellite data examined by Pinker, and for having made many other vital contributions to this work; to Dr. David Evans for having provided a helpful discussion of the Planck parameter; to Professor Larry Gould for having eliminated some errors; to Dr. Susan Solomon for some constructive comments; to Professor Richard Lindzen for having patiently answered many questions; to Dr. Willie Wei-Hock Soon for having drawn my attention to the literature on the global brightening of 1983–2001, and for having supplied much other useful material; and to Professor Antonino Zichichi and my fellow-participants in the 43rd annual Seminar of the World Federation of Scientists on Planetary Emergencies at the Ettore Majorana Center for Scientific Culture, Erice, Sicily, in the late summer of 2010 for several important observations on climate sensitivity. Any errors that remain are mine alone.

APPENDIX: REANALYSIS OF SATELLITE RADIATIVE-FLUX DATA

Joseph Boston j.boston@earthlink.net

Period of Study

The 18-years 1 month from mid-1983 to just after mid-2001 were selected as the period of study (Fig. A1) for three reasons: first, satellite radiative-flux datasets began in the early 1980s; secondly, global surface temperature rose rapidly until the end of the period, and global temperatures have exhibited no trend since then; and thirdly, the period of study is near-coincident with that of Pinker et al. (2005).

ISCCP FD and ERBE Satellite Radiative-flux Data Sources

Two sources of satellite-based radiation data were examined. The International Satellite Cloud Climatology Project (ISCCP) was established in 1982 to collect and analyze satellite radiance measurements.

Data were collected from a suite of weather satellites operated by several nations and processed by groups in several government agencies, laboratories and universities. Measurements were obtained every three hours over the whole globe on a 280 km equal-area global grid (equivalent to 2.5° latitude–longitude at the equator) from July 1983 to December 2007.

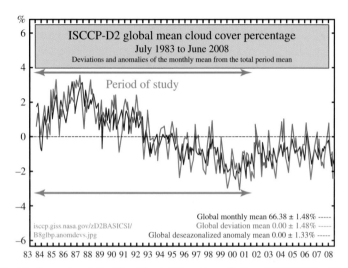

FIGURE A1 International Satellite Cloud Climatology Project (ISCCP-D2) global mean percentage cloud cover, July 1983 to June 2008, showing deviations (green) and deseasonalized anomalies (blue) of the monthly mean from the total period mean.

Results include global averages of upwelling and downwelling short-wave and long-wave radiative fluxes at top of atmosphere, in the atmosphere, and at the surface, for full, clear, and cloudy-sky conditions.

These results were compiled in the ISCCP FD datasets, which are available in specified latitude groupings. For this study separate FD datasets were obtained for five latitude groupings, all covering the entire 1983−2001 period. The principal groupings analyzed in the study were 20N−20S, 60N−60S, and 90N−90S.

The Earth Radiation Budget Experiment (ERBE) datasets are based upon non-scanner wide-field-of-view data obtained from three NASA satellites (ERBS, NOAA-9, and NOAA-10) from November 1984 to September 1999.

The instrument package included five active-cavity radiometers, two medium-field-of-view, two wide-field-of-view and a solar monitor, with a combined ability to measure incoming and earth-reflected solar irradiances in the short-wave (0.2−5 micron) spectral region, and earth-emitted long-wave radiances up to 100 microns.

The wide-field-of-view instruments observed irradiances from the entire earth disc enclosed by a 2-angular-degree space ring. The medium-field-of-view instruments viewed earth irradiances from regions with earth-centered angles of 10°.

The raw data were processed by the data management team in the Atmospheric Science Division of NASA to develop temporally and spatially averaged short-wave and long-wave fluxes at the top of the atmosphere, which the team compiled into a number of different data products.

The team discovered two sources of bias, which were quantified, and for which corrections were subsequently introduced. One of the biases was caused by slight satellite orbital decay, the other by a slow drift in the short-wave sensor.

The two datasets used in this study are edition 3, revision 1 of the 72-day mean datasets for the tropics (20N−20S) and near-global (60N−60S), both of which incorporate corrections for the two sources of bias. The datasets encompass the 14.5 years for which ERBE data (Barkstrom, 1984) are available, from 1985 to mid-1999, approximately centered on the 18.1-year period of the ISCCP FD datasets.

Determination of Short-wave Flux Trends

A central objective of this work was to examine the impact on the global warming trend of increasing solar flux to the Earth's surface and into the atmosphere. For this purpose it was essential to determine the underlying trends in the radiative fluxes.

Since the eruption of Mt. Pinatubo in mid-1991 caused a significant short-term distortion in the underlying trend, for both the ISCCP FD and ERBE datasets all of the flux data were pre-processed by removing the Pinatubo-affected data over 18 months from mid-1991 to end 1992. Remaining data were deseasonalized by determining the average value for a given calendar month over the whole period and subtracting that value from the individual yearly values for that calendar month. This procedure was repeated for each of the 12 months in turn. The Pinatubo-adjusted, deseasonalized monthly data were then regressed to obtain least-squares linear trends in $W\,m^{-2}\,year^{-1}$.

The key item of satellite data for this work was the net mean global surface short-wave flux trend from 1983 to 2001. The ERBE satellites did not produce data beyond 60N and 60S; they covered only a 14.5-year subset of the period of interest; and the data did not explicitly include the net short-wave surface flux. Therefore the ISCCP 90N−90S dataset was used as the basis for the study, and the ERBE data were used as a validity check with the ISCCP 60N−60S data shown in Fig. A2.

The ERBE data, covering 1985−1999, are divided into five 72-day periods for each year, a total of 75 periods. Owing to orbital characteristics, periods that are near-exact multiples of 36 days are essential for accurate analysis of the ERBE data. For seven of those periods (three each in 1993 and 1998 and one in 1999) the data are missing.

The analysis examined the possibility that the missing data might have a significant impact on deseasonalization. A quadratic regression was carried out over 1985−1999 for each 72-day period in turn, whereupon the five second-order fits were used to provide estimates of the missing values. There were two missing values for period 1, one missing value for each of periods 2−4, and two missing values for period 5. The deseasonalization and the overall regression

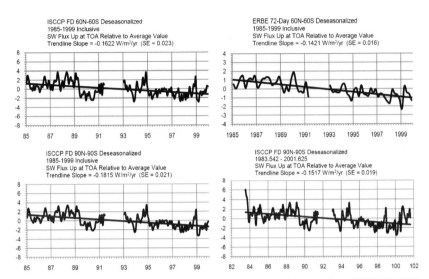

FIGURE A2 Deseasonalized east squares linear-regression trends on data from ISCCP FD 60N-60S, 1985−1999 (top left); ERBE 60N−60S, 1985−1999 (top right); ISCCP FD 90N−90S, 1985−1999 (bottom left); and ISCCP FD 90N−90S, 1983.542-2001.625 (bottom right, for comparison with Pinker).

were then repeated to obtain the least-squares linear-regression trend-line for upwelling short-wave flux at the top of the atmosphere. Agreement between the 60N−60S ISCCP and ERBE trends, and between the ISCCP results and those in Pinker (Table 3) was found to be close.

Figure A3 is a short-wave energy budget diagram showing how the average fluxes and flux trends are distributed among the significant flux components, including the surface, cloud, and aerosol albedo components. Effects that are *de minimis* are not shown. Net surface flux is found to be 0.143 W m^{-2}.

TABLE 3 Deseasonalized least-squares linear-regression trends for upwelling short-wave radiative flux at the top of the atmosphere, 1985−1999 (ISCCP FD and ERBE), and 1983.542-2001.625 (ISCCP FD and Pinker)

Flux (W m^{-2} year^{-1})	Latitudes	ISCCP	ERBE	Pinker	Variance
1985−1999	60N−60S	−0.162	−0.142		−12%
1985−1999	90N−90S	−0.182			
1983.542-2001.625	90N−90S	−0.152		−0.160	+5%

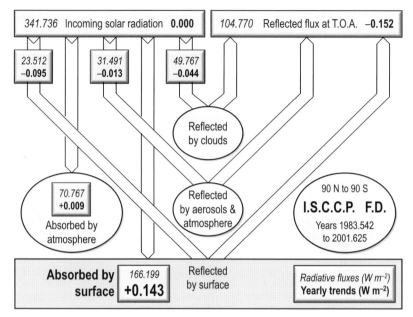

FIGURE A3 Simple radiation budget showing the net increase in surface solar radiance, the net brightening flux, for 1983.542-2001.625. Total brightening flux, the sum of the components 0.143 W m^{-2} absorbed by the surface and 0.009 W m^{-2} absorbed by the atmosphere, is 0.152 W m^{-2}.

REFERENCES

Barkstrom, B.R., 1984. The Earth Radiation Budget Experiment (ERBE). Bulletin of the American Meteorogical Society 65, 1170.

Bode, H.W, 1945. Network Analysis and Feedback Amplifier Design. p. 551 Van Nostrand, New York.

Brohan, P., Kennedy, J.J., Harris, I., Tett, S.F.B., Jones, P.D., 2006. Uncertainty estimates in regional and global observed temperature changes: a new dataset from 1850. Journal of Geophysical Research 111, D12106. 10.1029/2005JD006548.

Cess, R.D., Udelhofen, P.M., 2003. Climate change during 1985–1999: cloud interactions determined from satellite measurements. Geophysical Research Letters 30, 1. 1019, 10.1029/2002GL016128.

Chen, J., Carlson, B.E., Del Genio, A.D., 2002. Evidence for strengthening of the tropical general circulation in the 1990s. Science 295, 838–841.

Clement, A.C., Soden, B., 2005. The sensitivity of the tropical-mean radiation budget. Journal of Climate 18, 3189–3203.

Covey, C., 1995. Correlation Between Outgoing Long-wave Radiation and Surface Temperature in the Tropical Pacific: A Model Interpretation. Lawrence Livermore National Laboratory, Livermore, CA. 94551, November. UCRL-ID-122565.

Gouretski, V., Koltermann, K.P., 2007. How much is the ocean really warming? Geophysical Research Letters 34, 10. 1029/2006GL027834.

Hartley, D.E.T., Kindler, D.E., Cunnold, Prinn, R.G., 1996. Evaluating chemical transport models: comparison of effects of different CFC-11 emission scenarios. Journal of Geophysical Research 101, 14381−14385.

Hatzidimitriou, D., Vardavas, I., Pavlakis, K.G., Hatzianastassiou, N., Matsoukas, C., Drakakis, E., 2004. On the decadal increase in the tropical mean outgoing longwave radiation for the period 1984−2000. Atmospheric Chemistry and Physics 4, 1419−1425.

Houghton, J.T., et al. (Eds.), 2001. Climate Change 2001: The Scientific Basis. Contribution of Working Group I to the Third Assessment Report of the Inter-governmental Panel on Climate Change. Cambridge University Press, Cambridge, United Kingdom and New York, NY, USA, p. 881.

Jones, P.D., New, M., Parker, D.E., Martin, S., Rigor, I.G., 1999. Surface air temperature and its variations over the last 150 years. Reviews of Geophysics 37, 173−199.

Kiehl, J.T., Trenberth, K.E., 1997. The Earth's Radiation Budget. Bulletin of the American Meteorogical Society 78, 197.

Kiehl, J.T., Trenberth, K.E., 2009. Earth's global energy budget. Bulletin of the American Meteorogical Society 10.1175/2008BAMS2634.1.

Kimoto, K., 2009. On the confusion of Planck feedback parameters. Energy & Environment 20, 1057−1066.

Lindzen, R.S., 2007. Taking greenhouse warming seriously. Energy & Environment 18 (7−8), 937−950.

Lyman, J.M., Willis, J.K., Johnson, G.C., 2006. Recent cooling of the upper ocean. Geophysical Research Letters 33, L18604. 10.1029/2006GL027033.

Monckton of Brenchley, C.W., 2008. Climate sensitivity reconsidered. Physics & Society 37, 3.

Myhre, G., Highwood, E.J., Shine, K.P., Stordal, F., 1998. New estimates of radiative forcing due to well-mixed greenhouse gases. Geophysical Research Letters 25 (14), 2715−2718.

Pinker, R.T., Zhang, B., Dutton, E.G., 2005. Do satellites detect trends in surface solar radiation? Science 308, 850−854.

Rossow, W.B., Schiffer, R.A., 1991. ISCCP cloud data products. Bulletin of the American Meteorogical Society 72, 2.

Rossow, W.B., Schiffer, R.A., 1999. Advances in understanding clouds from ISCCP. Bulletin of the American Meteorogical Society 80, 2261.

Schwartz, S., 2007. Heat capacity, time constant, and sensitivity of Earth's climate system. Journal of Geophysical Research.

Soden, B.J., Held, I.M., 2006. An assessment of climate feedbacks in coupled ocean-atmosphere models. Journal of Climate 19, 3354−3360.

Solomon, S., et al. (Eds.), 2007. Climate Change 2007: The Physical Science Basis. Contribution of Working Group I to the Fourth Assessment Report of the Intergovernmental Panel on Climate Change. Cambridge University Press, Cambridge, United Kingdom and New York, NY, USA.

Spencer, R.W., Braswell, W.D., Christy, J.R., Hnilo, J., 2007. Cloud and radiation budget changes associated with tropical intraseasonal oscillations. Geophysical Research Letters 34, L15707. 10.1029/2007GL029698.

Wielicki, B.A., et al., 2002a. Evidence for large decadal variability in the tropical mean radiative energy budget. Science 295, 841−844.

Wielicki, B.A., et al., 2002b. Changes in tropical clouds and radiation: response. Science 296, 2095a.

Wild, M.A., et al., 2005. From dimming to brightening: Decadal changes in solar radiation at Earth's surface. Science 308, 847−850.

Willis, J.K., 2008. Is it me, or did the oceans cool? U.S. Climate Var. 6, 2.

Willis, J.K., Lyman, J.M., Johnson, G.C., Gilson, J., 2009. In-situ data biases and recent ocean heat content variability. Journal of Atmospheric & Oceanic Technology 26, 846–852.

Wong, T., Wielicki, B.A., Lee III, R.B., 2006. Reexamination of the observed decadal variability of the Earth radiation budget using altitude-corrected ERBE/ERBS nonscanner WFOV Data. Journal of Climate 19, 4028–4040.

Chapter 14

The Relationship of Sunspot Cycles to Gravitational Stresses on the Sun: Results of a Proof-of-Concept Simulation

Edward L. Fix

1395 Quail Ln., Beavercreek, OH 45434, USA

1. INTRODUCTION

The 11-year sunspot cycle is well known. With recent minima occurring in early 1965, 1976, 1986, and 1996, the sunspot cycle was expected to bottom out in 2006 or 2007, and begin an upswing into Solar Cycle 24. The sun, however, had other ideas. The numbers continued to decline through 2006, 2007, 2008, and 2009, only beginning a slow climb in late 2009 to early 2010. Since none of the many models had predicted this, sunspot prediction suddenly became a hot topic again. Understanding the sun's dynamics has become important in the space age, because sunspots bring solar storms, which affect satellites giving us communications, weather prediction, and navigation. And, of course, the sun's activity has a direct effect on the earth's climate, although the climate's exact sensitivity to solar activity is a subject of considerable controversy.

Evidence-Based Climate Science. DOI: 10.1016/B978-0-12-385956-3.10014-2

There are other characteristics of the sunspot cycles that are less well known. Sunspots have magnetic fields (Hale, 1908), and east—west oriented pairs of sunspots appear magnetically bound. The eastern sunspot of the pair might have a north-leading polarity (relative to the sun's rotation) while its western mate has a north-trailing polarity, and in the other hemisphere, that relationship will be reversed. During the next cycle, Hale discovered that the polarities of the sunspot pairs were reversed from the previous cycle (Hale et al., 1919). At first, he mistakenly attributed this to the difference in the sunspot's latitude, but it is now known that the vast majority of sunspots of each cycle have polarity reversed from those of the previous (and succeeding) cycle. This approximately 22-year cycle of pairs of 11-year sunspot cycle has become known as the Hale cycle.

The sun's total magnetic field is much more fluid than that of the earth. In the mid-twentieth century it was observed that the sun's global magnetic field reverses near the peak of each sunspot cycle (although the polarity of the sunspots themselves remain the same throughout the cycle) (Babcock, 1961).

Finally, sunspots have long been observed occurring in an 11-year cycle appear at the mid-latitudes (roughly midway between the equator and the pole in both hemispheres), and as the cycle progresses, they appear closer and closer to the equator. This is another way of distinguishing between sunspots of the previous cycle and those of the next, as sunspots often overlap during a sunspot minimum.

Motion of the planets has long been suspected as the driver of the 11-year sunspot cycle. Attempts to relate the two date back at least 150 years (Wolf, 1859).

Results of various analyses range from disappointing to tantalizing. Three models using essentially pattern matching techniques (Jose, 1965; Fairbridge and Shirley, 1987; Cole, 1973) did not predict the last half of the 20th century. Bracewell (1986) used precambrian varves to develop a sum-of-sines mathematical model which did a good job of predicting Solar Cycles 22 and 23. It matched the actual peak sunspot counts of the cycles, as well as the dates of the minima. Due to its nature, it was inherently incapable of predicting the long minimum between Cycles 23 and 24, and it failed to reproduce the Maunder Minimum. Bracewell also did no more than speculate on the genesis of the pattern. Many other models do not attempt to reproduce individual sunspot cycle amplitude or timing, but relate the motion of the sun about the barycenter of the solar system to large-scale patterns in the sunspot cycle such as the Maunder Minimum (e.g., Landscheidt, 1999).

It is important to distinguish between tidal forces on the sun, and the motion of the sun about the barycenter of the solar system. Tidal forces are due primarily to the inner planets, and involve differential gravitational force between the planet's subpoint on the sun, and the limb of the sun. Correlation between solar activity and tidal forces has been shown (Hung, 2007; Bigg, 1967), but not with the overall sunspot cycle. On the other hand, the sun's

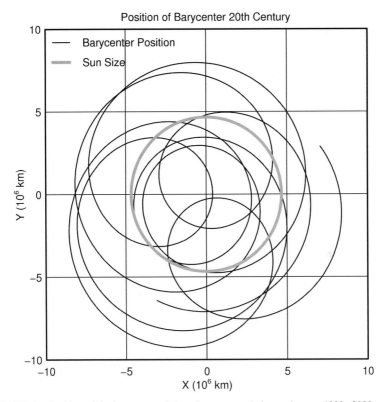

FIGURE 1 Position of the barycenter of the solar system relative to the sun, 1900–2000.

motion about the barycenter, shown in Fig. 1, is determined mainly by the four giant planets: Jupiter, Saturn, Uranus, and Neptune. This distinction is also emphasized by Fairbridge and Shirley (1987).

2. APPROACH

This study departs in a number of ways from earlier efforts. First was the selection of format for representing the historical sunspot data. Bracewell (1953) advocated displaying the sunspot cycle as a zero-mean sinusoid by reversing the sign of every second cycle, reflecting the 22.3-year Hale cycle. This conclusion was independently reached early in this current study. It was noted that the minima in the sunspot cycle tend to be sharper and better defined than the maxima, a characteristic of a graph of the absolute value of a sinusoid. This is not an entirely accurate analogy, owing to the fact that during the sunspot minima, spots from both the previous and succeeding cycle appear simultaneously. However, it is at least a good first approximation, and there seems no harm in plotting the sunspot data in this way, to see where it leads. Starting with the average monthly sunspot number data downloaded from the ftp server of the National Geophysical Data Center of the National Oceanic and Atmospheric Administration (NOAA/NGDC), Solar-Terrestrial Physics Division (ftp://ftp.ngdc.noaa.gov/STP/SOLAR_DATA/SUNSPOT_NUMBERS/MONTHLY), the sign of every second cycle is reversed, arbitrarily picking Cycle 23 as positive. I picked the crossover points by simple inspection. The result is called the "polarized sunspot number" in Fig. 2.

Another unique feature of this effort is the way the motion of the sun about the barycenter of the solar system is related to the sunspot cycle. Earlier efforts relate the solar cycle to parameters like spin-orbit coupling (Wilson et al., 2008; Javaraiah, 2005; Juckett, 2000) and changes in angular momentum, torque, or curvature of the sun's orbit (Paluš et al., 2007; Landscheidt, 1981, 1999; Wood and Wood, 1965; Javaraiah, 2005). These all primarily involve the sun's angular motion around its orbit, and the attendant changes in orbital speed.

In contrast, this study focuses primarily on the radial component of the sun's motion about the barycenter, and was initiated after making the following observation. An ephemeris of the barycenter of the solar system was obtained from the server at NASA JPL (http://ssd.jpl.nasa.gov/horizons.cgi) (Giorgini et al., 1996). For purposes of this study, the Z component of the barycenter's position was ignored—that is it was always assumed to lie on the ecliptic plane—and a month is taken as 1/12 of 365.25 days.

The barycenter's coordinates were plotted in polar format (r, θ), and various comparisons made to the polarized sunspot plot. The dark curve in Fig. 2 is the velocity $(\dot{r})$ of the radial distance of the barycenter relative to the sun; that is, the first derivative of the radial distance with respect to time, with dimension of km/month. The sunspot count and $\dot{r}$ data in Fig. 2 are from completely independent data sets. Intriguing similarities of the curves are evident between 1750

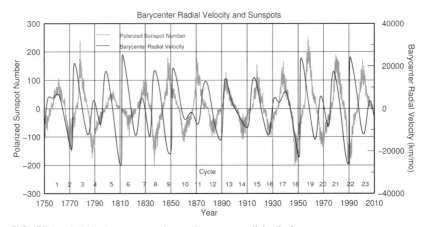

FIGURE 2   Polarized sunspot number vs. barycenter radial velocity.

and about 1780, and again between 1891 and about 1951. It seemed unlikely that the correlations could be coincidental.

The second derivative of the radial distance of the barycenter with respect to time ($\ddot{r}$) is acceleration, and acceleration is simply a force scaled by mass ($F = ma$). So, why not treat it as a force and see what happens?

A third departure of this effort, relative to previous work, was the use of a dynamic simulation to investigate the possible effect of changes in the sun's motion.

A variety of analytical techniques have been used to predict sunspot cycle activity. Some investigators have used more or less purely mathematical analyses—physically-based curve fitting efforts (Wolf, 1859; Bracewell, 1986). Others have performed frequency and/or phase analyses of the sunspot cycle (De Meyer, 1998; Paluš and Novotná, 1999; Mininni et al., 2000; Cole, 1973; Juckett, 2000). Many studies have sought to correlate features of the sun's orbit with features of the sunspot cycle (Landscheidt, 1981; Landscheidt, 1999; Wilson et al., 2008; Zaqarashvili, 1997; Charvátová, 1990a,b; Fairbridge and Shirley, 1987; Jose, 1965). These are mainly static analyses, and demonstrate correlation between the sun's orbit and the sunspot cycle.

However, correlation does not necessarily imply causation. A dynamic oscillatory model driven by the barycenter's motion, however, does show causation by its nature. A positive result would lend support to the notion that the sunspot cycle could be driven by the sun's motion.

A simple resonant oscillator was chosen as a test platform for this simulation. The equations describing a mass on a spring with damping are well known. The forces on an oscillating mass are the force of the spring pulling the mass back to neutral, and a damping force that increases with the mass' velocity. To this was added the radial acceleration (scaled linearly) as a forcing function.

There is no reason to expect that the output of this model could accurately predict the sunspot cycle. It is really just a placeholder to test the viability of the hypothesis that the radial acceleration could provide a driving force that could influence the sunspot cycle.

Consider the analogy of a young child bouncing a ball. If the child has not yet mastered the art of ball-bouncing, the ball's trajectory is likely to be erratic. The child might miss the ball on occasion, hit it on the way down instead of up, etc. If you wanted to build a model to recreate the ball's trajectory, there must be two components: the properties of a bouncing ball and the erratic forcing function of the child's hand. It would be relatively simple to build a ball simulation. If the mass, bounciness or squishiness of the ball, the air resistance, etc. were not quite right, we might still be able to at least model the gross features of the real ball's trajectory. However, if the timing and relative force of the child's hand were not essentially correct, the output of the model would not remotely match the timing and amplitude of the actual trajectory. While there have been several sunspot cycle models based on various types of oscillators (Polygiannakis et al., 1996;

Paluš and Novotná, 1999; Mininni et al., 2000), these were again static analyses, and little attempt has been made to identify the root cause of the oscillations.

3. METHOD

A simple simulation experiment was implemented on a spreadsheet. A conceptual "sunspot generator" within the sun was modeled as a driven one-dimensional damped spring, described by the equation:

$$F = -kx - bv \tag{1}$$

where for this purpose, x represents the polarized sunspot number, v is its current rate of change, k is the spring constant, and b is the damping constant. The spring constant k is determined by the resonant oscillation period:

$$k = 2m \left(\frac{2\pi}{T}\right)^2 \tag{2}$$

where m is normalized mass and T is the resonant period in months. It should be emphasized that this period T is not the Hale cycle; it is the intrinsic resonant frequency of the driven system. The driver forcing function has a fundamental frequency very near 20 years, determined by the combination of Jupiter and Saturn's orbits (the period between Saturn's superior oppositions to Jupiter is very close to 20 years), strongly modulated by the positions of Uranus and Neptune. It is the driven system's response to that forcing function that must produce the Hale cycle.

Now, this system must be driven by the outside force evidenced by the radial acceleration of the sun relative to the barycenter, $\ddot{r}$. This will be scaled by a linear scaling factor z called a "coupling constant" here. The final equation for the total force F_T on the hypothetical sunspot generator is:

$$F_T = -kx - bv + z\ddot{r} \tag{3}$$

and the acceleration produced by this force is:

$$a = \frac{F_T}{m} \tag{4}$$

Treating a as a constant (its time-dependent variation is independent of the response of the hypothetical "sunspot generator"), we can get an output by taking the second integral of Eq. (4) with respect to time:

$$x(t) = \frac{.5}{m}(F_r)t^2 + v_0 t + x_0 \tag{5}$$

To implement this as a computer model, a time unit must be selected. The historical sunspot record is recorded at 1-month intervals. Computing the output at 1-month intervals makes $t = t^2 = 1$ for each step, v_0 becomes

$x_{t-1} - x_{t-2}$ and x_0 becomes x_{t-1}. Substituting a discrete version of Eq. (3) for F_T yields this calculation at each 1-month interval:

$$x_t = \frac{.5}{m}\left(-kx_{t-1} - b(x_{t-1} - x_{t-2}) + z\ddot{r}\right) + 2x_{t-1} - x_{t-2} \qquad (6)$$

The next step was to select values for the constants m, T, b, and z. The mass number m was initially set to 1. Within this paradigm, we're assuming the system is oscillating at approximately 22 years, so an initial resonant period of 22 years was used. To get the model to oscillate, it needs to be assigned an initial x and v. In 1750, the sunspot cycle was at a maximum of about 90 sunspots, and that cycle (before cycle 1) has been assigned a negative sign. To help choose and refine the constants, the model was initially started with initial conditions of -90 in April, 1750. Because this model computes the rate of change of the previous iteration by subtracting x_{t-2} from x_{t-1}, March was also set to -90 to give a rate of change of 0.

NOTE: All empirically derived constants are treated as dimensionless numbers. The derivation of appropriate units is left as an exercise for the reader.

At first, b and z were set to 0, to verify that the model oscillates with the proper period. Then, b was increased to produce an initial damping effect. Finally, z was increased to give the resulting curve an amplitude comparable to the first couple of sunspot cycles after 1750. The amazing thing is that it did that.

3.1. Barycenter Ephemeris Preparation

Unfortunately, the ephemeris from the JPL server was not initially usable for this purpose. The ephemeris is provided from the server in a nearly unlimited variety of possible formats and computation intervals. This experiment used a 1-month interval from January, 1500 through January, 2200. The radial acceleration was computed by subtracting the previous radius from the current, yielding a velocity, and then subtracting the previous velocity from the current one to obtain acceleration. This straightforward method yields an acceleration in km/month^2, but it is also a small difference between large numbers. The results from the raw downloaded ephemeris show quasi-periodic spiky noise, shown in Fig. 3. This is most likely aliasing due to mismatched computation and sampling intervals, possibly with a side order of rounding error.

To alleviate this problem, a smoothed barycenter ephemeris was derived and used. The ephemeris of each of the four gas giant planets was downloaded. A smoothed approximation for the orbits of each was calculated by sinusoidally varying the radius between perihelion and aphelion, and similarly varying the orbital velocity (in degrees per month) between the minimum and maximum with a sine function. All the resultant planetary ephemerides were checked with the raw data, and found to remain within 0.5% or less of the planet's actual orbital distance, and within 0.5° or less of the planet's angular position, for the

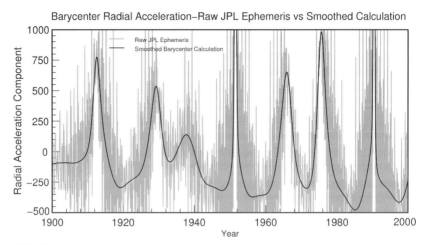

FIGURE 3   Comparison of radial accelerations from the JPL ephemeris and the smoothed outer planet data.

entire period from 1500 to 2200. This was judged to be sufficiently accurate, given that an earlier proof-of-concept experiment had produced promising results using a "homemade" ephemeris that assumed circular orbits.

These derived planetary ephemerides were then used to construct a five-body barycenter with the sun using masses provided by NASA's general educational website (http://solarsystem.nasa.gov/planets) for the entire period. From this, radial and angular velocities and accelerations could be computed. As a final check, the resulting barycenter position plot was compared to the one from the raw JPL ephemeris, shown in Fig. 1.

4. RESULT

The constants used in the model to produce the following plots were $m = 1$, $T = 25.5$ years, $b = 0.008$, and $z = 0.0002$. The first run with these constants, shown in Fig. 4, is ambiguous. The model seems to track reasonably well until about 1810. At that time, it loses synchronization with the actual data, and remains more or less completely uncorrelated until the early 20th century. At that time, it seems to regain synchronization until about 2006. The reason it lost and regained synchronization was enigmatic.

An examination of the sunspot data shows that in 1809 the sunspots dwindled to nothing, stayed at zero for 19 months, and restarted in mid-1811. The model, however, diverged sharply from the sunspot record by projecting continuing sunspot activity. Clearly, something happened that this simple little model cannot model. To account for this shortcoming of the model, its output was set to 0 during the period starting in March, 1807 when it crosses the 0 line until April, 1811 where the real data was 0 − essentially resetting

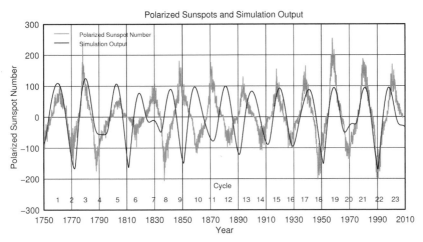

FIGURE 4   Model output vs. polarized sunspot number.

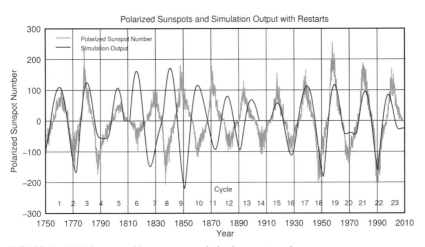

FIGURE 5   Model output with restarts vs. polarized sunspot number.

the initial condition. The same thing was done in the early 20th century, with the model forced to zero between November, 1904 until September, 1911 to cover that long minimum. This had the effect of putting the model's output in synchronization throughout the 19th century, but with the sign reversed, as shown in Fig. 5.

The next manipulation was to reverse the polarity sign of all the sunspot data between 1810 and 1912. The assignment of sign is arbitrary to begin with. If an oscillating body stops oscillating, it will restart in phase with whatever force starts it—without regard to the phase of its last swing. A pendulum has no

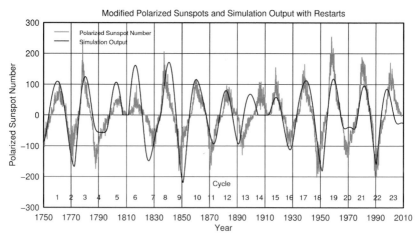

FIGURE 6   Model output with restarts vs. modified polarized sunspot data.

memory. The final result of this part of the exercise is shown in Fig. 6. Figure 7 shows the absolute values of both the sunspot data and the model's output. For an alternate scenario regarding this assumed phase reversal, see the Discussion.

It was noted in the course of the first run-through that, when the model gets out of sync with the historical data, it tends to regain synchronization. Until this point, the model had always been started with initial conditions that matched the sunspot data in 1750. This was necessary for initially picking the constants. However, once the constants are in place, is the initial condition still required?

To test this, the model was run starting in the year 1500 with initial conditions set to 0. As sunspots were known only by naked eye observation in 1500, it is impossible to know what the initial conditions actually were at that time. Figure 8 shows the model for two conditions: model started in 1750 with −90 value and 0 rate of change, and the model started in 1500 with value and rate of change both set to 0. As the figure shows, there is very little difference between the two model outputs; they are close in phase and amplitude from 1750, and nearly matched by Cycle 5. The models stay matched thereafter. Actually, there is not much difference when the model is started at any time earlier than about 1700 with initial conditions of zero.

More importantly, the model correlates with the sunspot cycle data. It maintains gross agreement with the timing, phase, frequency, and relative amplitude of the sunspot data, and in those instances where it gets out of phase or off amplitude, it always pulls back into agreement. However, many of the details of the sunspot record remain elusive, such as the exact timing of sunspot maxima and minima, and especially the times where the sun stopped producing spots altogether. This is not surprising given the simple nature of this model.

It is worth noting that this is a free-running model. Each monthly iteration depends only on the condition of the model in the previous iteration, and the value of the forcing function. At minimum, this result suggests that there are

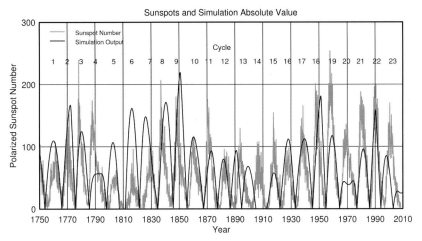

FIGURE 7   Absolute value of the model output vs. historical sunspot number.

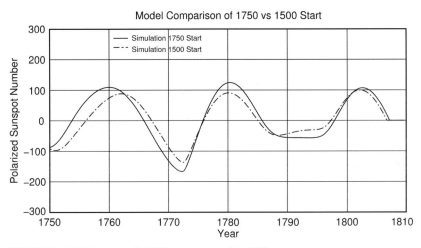

FIGURE 8   Model output with 1750 start compared to 1500 start.

periods when the acceleration of the radial component of the sun's orbit about the barycenter is the primary driving force of the sunspot cycle, and that the decades before the historical sunspot data begins around 1750, fortuitously, was one such period.

4.1. Details

This simple model misses many of the details of the historical sunspot activity. After another experiment, it became apparent that while the model does not depend on receiving exact initial conditions to mimic the gross features of the sunspot record, the fine features are another matter. They can depend greatly on the previous behavior of the model.

Figure 9 shows the model started from 0 in September, 1911. The constants remain the same. In addition, the barycenter's radial acceleration ($\ddot{r}$) is plotted. Note that in 1951, the model's peak is delayed from the sunspot data, and coincides with a sharp positive peak in the forcing function. In Fig. 10, that output peak is manually forced to -150 in May, 1948 to coincide with the historical data. The resulting graph shows how the model was affected for the next 50+ years. It even seems to show the recent delay in starting Cycle 24.

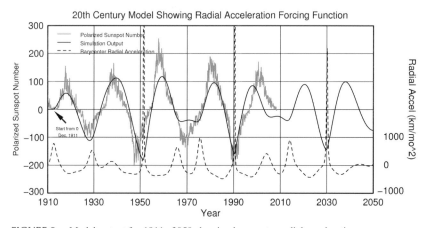

FIGURE 9   Model output for 1911−2050 showing barycenter radial acceleration.

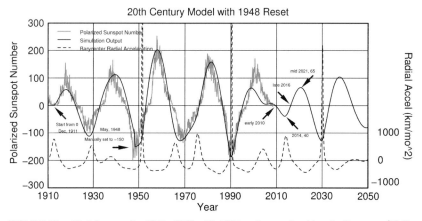

FIGURE 10   Model output for 1911−2050 with 1951 peak moved to historically correct 1948.

5. DISCUSSION

The purpose of this experiment was to test the hypothesis that the position and motion of the solar system's barycenter could be a driver of the sunspot cycle. The existence of a free-running model which, when driven by the barycenter's movements mimics the gross phase, timing, frequency, and relative amplitude of the historical sunspot cycle suggest that it is possible.

The model maintains gross agreement with the sunspot data, and when it loses synchronization tends to regain it even without resetting the initial conditions. It produces good results over a wide range of values for constants, meaning the result degrades gracefully if the oscillatory model is not exactly right. However, it would degrade catastrophically if the forcing function were wrong. It does not require knowledge of initial conditions or of the sunspot record to synchronize itself. In those times where it loses synchronization, such as the lulls between Cycles 5-6 and 14-15, restarting it with real-world initial conditions produces output that immediately matches the sunspot data.

It is interesting to note that the radial acceleration forcing function shown in Figs. 9 and 10 is not symmetrical. The positive peaks tend to be higher magnitude and shorter duration than the negative-going peaks. This matches the asymmetry first described by Gnevyshev and Ohl (1948). Their observation, known as the G-O Rule, states that the total sunspots of an odd-numbered cycle generally total more than the number of the previous even-numbered cycle.

5.1. Weaknesses

5.1.1. Rotation

This is an extremely simple model. It operates in one-dimensional space; the solar system occupies at least two dimensions. Thus far, the rotation of the system has not been considered, but there is reason to believe that it may have a profound effect.

In 1812, 1911, 1951, and 1990 for instance, the barycenter passed within a few thousand kilometers of the center of the sun. When it makes these looping passes through the sun's core, the azimuth of the force is suddenly changing much more rapidly than usual. It seems entirely possible that the force vector may slip ahead of the oscillation vector at such times. This would certainly reduce the influence it has on the oscillation, possibly reducing the amplitude of the next peak. It seems possible that if the slippage is great enough, the force vector may reverse phase altogether. It is interesting to note that the periods of greatest curvature or of greatest change in angular momentum that other researchers have used coincide exactly with the sharp positive-going peaks in the radial acceleration that this model uses.

Looked at another way, this model is a projection of a two-dimensional process onto one-dimensional space. The phase reversals that have been assumed between Cycles 5-6 and 14-15 may actually be phase reversals of the driving function, not of the bipolar sunspot cycle itself. This explanation is supported by the fact that magnetic phase of the sunspots of Cycle 15 was indeed reversed from those of Cycle 14; Cycle 15 was the first cycle after the magnetic field of individual sunspots was first observed, and the first time a magnetic phase reversal of the sunspots between cycles was observed (Hale et al., 1919).

5.1.2. The Oscillatory Model

This model is a simple driven, damped oscillator. It assumes an ideal spring analog with linear damping, and constant coupling between the forcing function and the oscillator. All these assumptions seem unlikely. Even real springs are not ideal springs. There is no reason to believe any of these processes are actually linear—so few natural processes are.

5.1.3. The Generator

There has been no attempt to define any relationship between the input and output of the hypothetical sunspot generator. Again, it is unlikely that there is a simple linear relationship between the oscillations of this "generator" and the number of sunspots it spits out as this model assumes. There are no threshold effects or other non-linearities included at all.

5.1.4. The Maunder Minimum

This seems to be everyone's first question; it certainly was mine. Unfortunately, this one-dimensional model does not replicate the Maunder Minimum. However, during the last half of the 17th century, the barycenter made three close passes through the sun's core in rapid succession; an occurrence which, under this paradigm can disrupt the orderly progression of the sunspot cycle. The next iteration of this model will take rotation into account.

5.2. Predictions

Referring again to Fig. 10, the model has apparently replicated the recent pause between Cycles 23 and 24. It then shows a weak resumption of sunspot activity in early 2010, reaching a peak of around 45 in early 2014, and quickly dying to a minimum in mid-2016. This is followed in late 2016 by Cycle 25, which peaks around 85 in mid-2020. However, this model is very simplistic and immature. While it is interesting to see what the model does with the next couple of cycles, it is too loosely correlated with the actual sunspot cycle to provide any useful predictions.

This model was run many times with different combinations of constants and even different properties of the model itself. As long as the output is held to zero through 2008, every run showed Cycle 24 with a delayed start, amplitude ranging from weak to entirely missing, and short duration, with Cycle 25 starting up around 2015–2016. The polarities continue to alternate as shown.

At this point, it is unknown how this might manifest. Cycle 24 did indeed begin to strengthen in early 2010 as this suggests, but it's unlikely the sunspots will actually go to 0 between a shortened Cycle 24 and Cycle 25. There may be a plateau or downswing before sunspots of opposite polarity begin appearing and numbers begin increasing. However, the barycenter will again pass through the sun in 2013, and its rapidly changing angle may cause the forcing function to flip polarity. This would mean the sunspots after 2016 would have the same polarity as those before. This might manifest as an abnormally long cycle with a pronounced upswing after 2013. In short, if this hypothesis is valid, the next cycle or two may well be very interesting.

6. CONCLUSION

The purpose of this experiment was to test the hypothesis that changes in the sun's orbit around the barycenter of the solar system, specifically its radial acceleration, might be an influence that governs the sunspot cycle. The existence of a model which, when driven by changes in the barycenter's position, mimics the sunspot cycle's timing, frequency, phase, and relative amplitude, supports the hypothesis, suggesting that there may be multi-cycle periods when that radial force is the primary driver.

REFERENCES

Babcock, H.W., 1961. The topology of the sun's magnetic field and the 22-year cycle: Astrophysical Journal 133, 572−587.

Bigg, E.K., 1967. Influence of the planet Mercury on sunspots. Astronomical Journal 72, 463−466.

Bracewell, R.N., 1953. The sunspot number series: Nature 171, 649−650.

Bracewell, R.N., 1986. Simulating the sunspot cycle. Nature 323, 516−519.

Charvátová, I., 1990a. The relations between solar motion and solar variability. Bulletin of the Astronomical Institute of Czechoslovakia 41, 56−59.

Charvátová, I., 1990b. On the relation between solar motion and solar activity in the years 1730−80 and 1910−60 A.D. Bulletin of the Astronomical Institute of Czechoslovakia 41, 200−204.

Cole, T.W., 1973. Periodicities in solar activity. Solar Physics 30, 103−110.

De Meyer, F., 1998. Modulation of the solar magnetic cycle. Solar Physics 181, 201−219.

Fairbridge, R.W., Shirley, J.H., 1987. Prolonged minima and the 179-yr cycle of the solar inertial motion. Solar Physics 110, 191−220.

Giorgini, J.D., Yeomans, D.K., Chamberlin, A.B., Chodas, P.W., Jacobson, R.A., Keesey, M.S., Lieske, J.H., Ostro, S.J., Standish, E.M., Wimberly, R.N., 1996. JPL's on-line solar system data service. Bulletin of the American Astronomical Society 28, 1158.

Gnevyshev, M.N., Ohl, A.I., 1948. About 22-year cycle of solar activity. Astronomical Journal 25, 18−20.

Hale, G.E., 1908. On the probable existence of a magnetic field in sun-spots. Astrophysical Journal 28, 315−343.

Hale, G.E., Ellerman, F., Nicholson, S.B., Joy, A.H., 1919. The magnetic polarity of sun-spots. Astrophysical Journal 49, 153−178.

Hung, C., 2007. Apparent Relations Between Solar Activity and Solar Tides Caused by the Planets. NASA/TM-2007-214817. NASA John H. Glenn Research Center, Cleveland, OH.

Javaraiah, J., 2005. Sun's retrograde motion and violation of even-odd cycle rule in sunspot activity. Monthly Notes of the Royal Astronomical Society 362, 1311−1318.

Jose, P.D., 1965. Sun's motion and sunspots: Astronomical Journal 70, 193−200.

Juckett, D.A., 2000. Solar activity cycles, north/south asymmetries, and differential rotation associated with solar spin−orbit variations. Solar Physics 191, 201−226.

Landscheidt, T., 1981. Swinging sun, 79-year cycle, and climate change. Journal of Interdisciplinary Cycle Research 12, 3−19.

Landscheidt, T., 1999. Extrema in sunspot cycle linked to sun's motion. Solar Physics 189, 413−424.

Mininni, P.D., Gómez, D.O., Mindlin, G.B., 2000. Stochastic relaxation oscillator model for the solar cycle. Physical Review Letters 85, 5476−5479.

Paluš, M., Kurths, J., Schwarz, U., Seehafer, N., Novotná, D., Charvátova, I., 2007. The solar activity cycle is weakly synchronized with the solar inertial motion: Physics Letters A 365, 421−428.

Paluš, M., Novotná, D., 1999. Sunspot cycle. a driven nonlinear oscillator? 83, 3406−3409.

Polygiannakis, J.M., Moussas, S., Sonett, C.P., 1996. A nonlinear RLC solar cycle model: Solar Physics 163, 193−203.

Wilson, I.R.G., Carter, B.D., Waite, I.A., 2008. Does spin−orbit coupling between the sun and the Jovian planets govern the solar cycle? Publications of the Astronomical Society of Australia, 25, pp. 85−93.

Wolf, R., 1859. Extract of a letter from Prof. R. Wolf of Zurich, to Mr. Carrington, dater Jan 12, 1859. Monthly Notes of the Royal Astronomical Society 19, 85−86.

Wood, R.M., Wood, K.D., 1965. Solar motion and sunspot comparison. Nature 208, 129−131.

Zaqarashvili, T.V., 1997. On a possible generation mechanism for the solar cycle. Astrophysical Journal 487, 930−935.

Part V

Modeling

A Simple KISS Model to Examine the Relationship Between Atmospheric CO_2 Concentration, and Ocean & Land Surface Temperatures, Taking into Consideration Solar and Volcanic Activity, As Well As Fossil Fuel Use

James P. Wallace, III [*], Anthony Finizza [†] and Joseph D'Aleo [**]

[*] President, Jim Wallace and Associates, LLC, [†] President, AJF Consulting, [**] Chief Meterologist, WeatherBell Analytics

Evidence-Based Climate Science. DOI: 10.1016/B978-0-12-385956-3.10015-4

353

1. INTRODUCTION

The notion that human-caused increases in atmospheric CO_2 levels are leading to serious adverse impacts on the Earth's Climate System is causing many governments to commit to CO_2 emission reduction plans. A key claim is that CO_2 generated by the complete combustion of any fossil fuel, naturally leads to increased atmospheric CO_2 levels, which then, via the so called Greenhouse Effect, must lead to a statistically significant increase in surface temperatures. These higher surface temperatures are then tied to adverse effects involving droughts, floods, hurricanes, and rising sea levels. Thus, the statistical significance of CO_2 as a predictor of ocean and land surface temperature is the key assumption underlying government policy decisions involving CO_2. Many papers in the literature have suggested a statistically significant, positive impact. To date, most of the arguments have been over *how positive*. These papers invariably used single equation regression or logically equivalent methodologies, which will be shown below to be inappropriate for the purpose intended.

This paper presents a simple set of equations[2], hereafter called the KISS (Keep It Simple Stupid) Model, which explains the CO_2 concentration—temperature interaction of the earth's climate system with surface area coverage well in excess of 70%. The analysis starts with first principles of model building and adheres to a set of generally accepted principles cataloged by forecasting professionals. The paper provides tentative conclusions and suggests areas for extension and research.

2. MODELING PRINCIPLES

This paper documents the results of a model development and validation process that proceeded according to a set of widely accepted principles.[3] The key principles include:

1. Following the (KISS) rule of parsimony, the model should be developed with as simple a structure as possible, using the smallest number of parameters that could predict the phenomenon.
2. The model should have a theoretical underpinning, with expected coefficient signs suggested *a priori* by the theory.
3. The data underlying the analysis should be free as practical of observational bias.
4. The model should be estimated by sound statistical procedures.
5. The model should perform well within sample, that is, provide forecasting value using a hold-out sample.

---

2. A complete data set is available upon request.

3. These can be found at www.forecastingprinciples.com. These represent the work of 120 experts.

6. In order to be useful for policy analysis, the model should have predictive power, particularly regarding turning points, as evidenced by its ability to outperform alternative models.
7. The analysis results should be transparent with all data, references, and relevant material being available to the reader.

Not all global climate forecasting efforts adhere to these principles.

3. SCIENTIFIC BASIS FOR THE MODEL

The simplest KISS Model to examine the interaction of CO_2 atmospheric concentration levels and surface temperatures contains three temperature equations and a carbon concentration equation, each representing a state variable in the system.

The representative temperature equation is:

$$T = f_1 [g_1(S), g_2(C), g_3(S * V)], \tag{1}$$

where T = temperature; S = solar activity; C = atmospheric CO_2 concentrations; V = volcanic activity; $S*V$ = interaction between solar and volcanic activity; $g_i(.)$ are distributed lag functions of the indicated variables.

The expected signs are: positive for S and C, and negative for $S*V$, if the volcanic activity is expressed in positive terms. Versions of this equation will be used for the Atlantic/Indian Ocean, Pacific Ocean, and the U.S.

The CO_2 equation is:

$$(\Delta C - c_{fossil}) = f_2[g_5(T), g_6(L)], \tag{2}$$

where $C = \Delta C + C_{-1}$ and the dependent variable, $\Delta C - c_{fossil}$, is the efflux of Net non-fossil fuel CO_2 emissions from the oceans and land into the atmosphere. T = temperature; L = land use; $c_{fossil} = CO_2$ emissions from fossil fuel use.

The expected signs are: positive for T and uncertain for L, depending on the particular measure used.

3.1. Data Analysis

The first step in an applied statistics/econometric analysis is to examine the endogenous and exogenous variables of the system. The model formulated above has data for the common window 1960−2008. The model is estimated over the period from 1960 to 2000.

3.1.1. Endogenous/Dependent Variables of the KISS System

The endogenous variables are three temperature variables representing Atlantic/Indian and Pacific Ocean surface temperatures, U.S. land surface temperature, and atmospheric CO_2 concentrations. Because of the well known Atlantic/Indian ocean conveyor belt, as shown in Fig. 1, the AMO is best

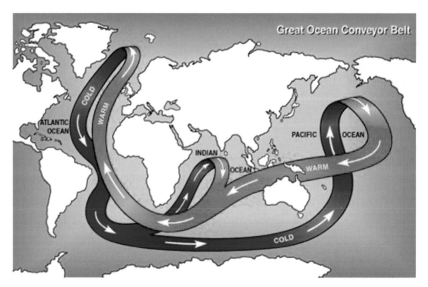

FIGURE 1 Thermohaline circulation flow. Source: thermohaline circulation from NASA.

thought of as a proxy for these two oceans behaving as one in terms of their-impact on atmospheric CO_2 concentrations. Figure 2 displays the temperature time series in level form. NINO3.4[4] represents the sea surface temperatures in the east central tropical Pacific, PDO[5] reflects the temperature for the North Pacific Basin, AMO[6] is the temperature for the North Atlantic, and U.S.[7] is the surface temperature of the United States.

For example, according to Feng and Hu (2008), "Observations have indicated that the North Atlantic SST variations have persistent effects on the Indian summer rainfall at multi-decadal and longer timescales."[8]

The various temperature time series exhibit significant differences in variability. Table 1 shows the coefficient of variation (the ratio of standard deviation to the mean of each time series). The North Pacific temperature time series has the largest variation and the Atlantic by far, the smallest.

The KISS Model choice of temperature series allows good coverage of the earth's surface. The Pacific, Atlantic, and U.S. coverage are about 57% of the

---

4. NINO34 is the NINO 3.4 time series of East Central Tropical Pacific SST (5N−5S)(170−120W) − http://www.cdc.noaa.gov/data/correlation/nina34.data.

5. PDO is the PDO Index which is standardized values for the leading PC of monthly SST anomalies in the North Pacific, poleward of 20°N − http://jisao.washington.edu/pdo/PDO.latest.

6. AMO_Raw is the raw Atlantic Multi-decadal Oscillation; that is, not detrended − http://www.cdc.noaa.gov/data/correlation/amon.us.long.mean.data.

7. GISS-U.S. are from http://data.giss.nasa.gov/gistemp/graphs/Fig.D.txt and are anomalies for the U.S. only. Converted to level form.

8. Feng and Hu (2008), page 1.

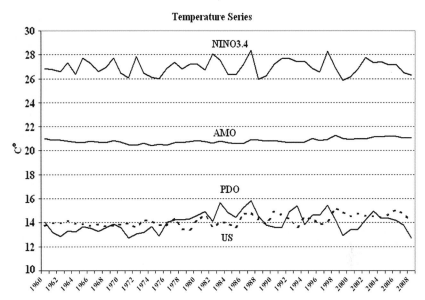

FIGURE 2 Temperature time series.

TABLE 1 Coefficient of Variation in the Temperature Time Series
(1960−2000)

	AMO	NINO3.4	PDO	U.S.
Mean (°C)	20.8	27.0	14.0	14.2
Coefficient of variation (ratio standard deviation/mean, %)	1.0	2.3	5.6	3.2

world's surface area. If we add in the surface area of the Indian Ocean, KISS model coverage, very loosely speaking, extends to 72%.[9] It should be noted that one cannot simply average the separate temperature forecasts for different geographic areas to obtain a global average temperature forecast.

As shown in Table 2, the Nino3.4 temperature series does not have a statistically significant trend, which remains true even over its entire 1950−2008 history. The AMO, PDO, and U.S. all have statistically significant, positive trends over the 1960−2000 period.

From the correlations shown in Table 3, and looking ahead to the modeling of temperature, it would appear that the equation structure that works for the

---

9. Wikipedia.

TABLE 2 Estimates of Linear Trend in the Temperature Data, 1960–2000

	AMO	NINO3.4	PDO	U.S.
Slope coefficient	0.0047	0.0031	0.0333	0.0184
t-statistic	2.10	0.35	3.51	3.63
p-value	0.042	0.725	0.001	0.001

TABLE 3 Correlation Among Temperature Times Series

	AMO	NINO3.4	PDO	U.S.
AMO	1.0	0.059	0.167	0.479
NINO3.4		1.0	0.454	0.010
PDO			1.0	0.119
U.S.				1.0

AMO might be a starting point for the U.S., but that it is highly unlikely to work at all for the PDO. This will turn out to be correct.

Atmospheric concentrations of CO_2 have increased at an average annual rate of 0.4% from 1960 through 2000.

3.1.2. Exogenous/Independent Variables of the System

The exogenous variables in the entire KISS Model system of equations are: Total Solar Irradiance[10] (TSI), volcanic activity[11] (DVI), the Southern Oscillation Index (SOI) "Central Tendency" (defined in Section 4.3 below), and

---

10. Total Solar Irradiance (TSI) data are from Willie Soon (Hoyt and Schatten (1993), scaled to fit ACRIM).

11. There are two competing variables that describe volcanic activity, the Dust Veil Index (reported as a positive number) due to Mann and Volcanic due to NOAA (reported as a negative number). Since, volcanic activity is expected to reduce temperature, other things being equal, the expected coefficient in the equation is negative for the first of the pair and positive for the second of the pair. DVI is the 'Weighted' Dust Veil Index from Mann et al. (1998). http://bobtisdale.blogspot.com/2008/04/mann-et-al-weighted-dust-veil-index.html. The alternative series, volcanic, is the forcing (W m^{-2}) due to volcanic activity — ftp://ftp.ncdc.noaa.gov/pub/data/paleo/gcmoutput/crowley2000/forc-total-4_12_01.txt.

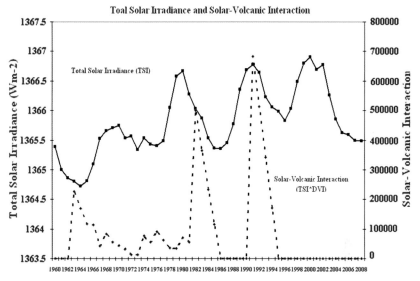

FIGURE 3   Total Solar Irradiance and the TSI—Volcanic interaction variables.

fossil fuel emissions into the atmosphere.[12] The annual TSI composite record was constructed by Hoyt and Schatten (1993) (and updated in 2005) utilizing all five historical proxies of solar irradiance including sunspot cycle amplitude, sunspot cycle length, solar equatorial rotation rate, fraction of penumbral spots, and decay rate of the 11-year sunspot cycle. Note the regular *solar seasonal* cycle of about 11 years, which is superimposed on the much longer term Trend Cycle in the data. (See Abdussamov (2008).)

Major volcanism is a factor that alters global climate for up to several years. Lamb (1970, 1977, 1983) formulated the Dust Veil Index (DVI) as a numerical index that quantified the impact of a particular volcanic eruption's release of dust and aerosols over the years following the event on the Earth's energy balance of changes in atmospheric composition due to explosive volcanic eruptions. The series plotted in Fig. 3 shows two of the exogenous variables, Total Solar Irradiance, and a variable that is the product of TSI and DVI, used to represent the interaction of the two exogenous variables.

Figure 4 compares fossil fuel CO_2 emissions with changes in atmospheric CO_2 concentration. Note that annual fossil fuel emissions are larger than the changes in CO_2 emissions suggesting that oceans and land surface flora are absorbing some fossil fuel-related CO_2 emissions. Over this period, about 44% of the fossil emissions did not show up in the atmosphere.

---

12. Global CO_2 Emissions from Fossil Fuel Burning, Cement Manufacture, and Gas Flaring: 1751—2005. http://cdiac.ornl.gov/ftp/ndp030/global.1751_2005.ems. Conversion to CO_2 was calculated on the basis of the carbon content of each fossil fuel.

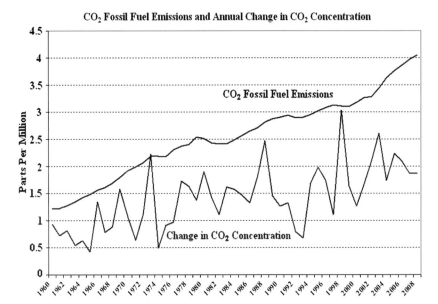

FIGURE 4 Fossil fuel CO_2 emissions and changes in atmospheric CO_2 concentration.

It is straightforward to calculate the Net non-fossil CO_2 additions placed into the atmosphere by subtracting fossil fuel emissions from the change in atmospheric CO_2 concentration. This arithmetic difference is plotted in Fig. 5 along with AMO temperature. While the Net non-fossil additions exhibit step-changes as indicated by the horizontal lines, the high correlation in *the annual data* is obvious and seems remarkable on its face.

Later in the paper, the Southern Oscillation Index $(SOI)^{13}$, the Tahiti minus Darwin Pressure anomalies, will be introduced as an instrumental variable to explain the impact of changes in solar/magnetic activity on Pacific tempera-tures. Note the high correlation (0.93) in Fig. 6 between SOI and NINO3.4.

4. MODEL ESTIMATION AND RESULTS

4.1. AMO Temperature Equation Estimation

The implicit KISS Model presented in Section 3 is estimated explicitly in this section. The explicit temperature equation for AMO is given in eq. (3).

$$\text{AMO} = c + b * \text{pdl(TSI}, 15, 3, 1) + d * \text{pdl(TSI} * \text{DVI}, 3, 3, 2) + e * C, \quad (3)$$

---

13. SOI is the Southern Oscillation Index, the Tahiti minus Darwin Pressure anomalies — ftp:// ftp.cpc.ncep.noaa.gov/wd52dg/data/indices/soi.his and http://www.cpc.ncep.noaa.gov/data/ indices/soi.

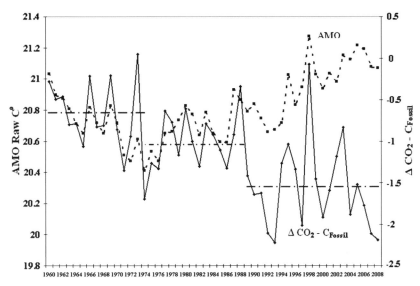

FIGURE 5 Net non-fossil CO_2 additions to atmosphere vs. temperature.

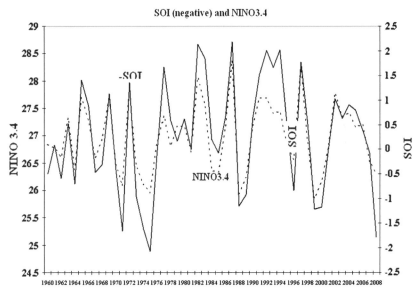

FIGURE 6 Southern Oscillation Index (SOI) vs. Nino3.4 anomalies.

where in pdl(independent variable x,y,z), x is the number of lags, y is the polynomial order, and z is the constraint (in the KISS model, TSI*DVI is constrained at the far end, but no constraints on TSI are assumed)

Since the KISS Model is a system of simultaneous equations, it is possible to start with any equation. This analysis started with the temperature eq. (3), using the Atlantic sea surface temperature series AMO as the dependent variable. Since the system is simultaneous, using ordinary least squares, or direct LS, on this equation would result in biased and inconsistent (i.e., worthless for policy decision making) estimates of its coefficients. To obtain unbiased and consistent parameter estimates, it is necessary to estimate the coefficients by a simultaneous equation estimation technique. There are a number of such techniques, but one of the easiest to understand and apply is called Two Stage Least Squares (TSLS).[14] TSLS is carried out by first regressing CO_2 that appears on the right-hand side of eq. (3) on all the exogenous variables of the entire system: TSI, TSI*DVI, SOI-CT, and c_{fossil}. The fitted values of this variable, CO_2Fit, are then substituted for the actual CO_2 values in the estimation of this equation by direct least squares. The results of this estimation process are shown in Fig. 7. Note that the coefficient of the CO_2Fit variable is statistically insignificant; while the other exogenous variables are highly significant (the t-statistic of the sum of each lag distribution is greater than 2 in absolute value).

This result is very important and allows us to conclude that the set of equations given in Section 3 is recursive and therefore each equation can be estimated by ordinary least squares.[15] Strictly speaking, to show that the model is recursive, it is necessary to carryout the same TSLS process for the other two regions (i.e., Pacific and U.S.) regarding their temperature history, which was done with the same result. CO_2 was not a statistically significant variable using TSLS for any of the three temperature equations.

The choice of the order of the polynomial and the number of lags in the Polynomial Distributed Lag Structure (PDL) formulation was determined by considering alternative formulations and choosing the one with the largest adjusted R-squared.[16]

---

14. See Theil, Henri. *Introduction to Econometrics*, Prentice-Hall, 1978, pages 328–342 and Goldberger, A.S., *Econometric Theory*, 1964, pages 329–348.

15. See Theil, Henri. *Introduction to Econometrics*, Prentice-Hall, 1978, pages 346–349 and Goldberger, A.S., *Econometric Theory*, 1964, pages 354–355. The model was estimated in EVIEWS. For a description of Polynomial Distributed Lags, see Batten, Dallas and Daniel Thornton, "Polynomial Distributed Lags and the Estimation of the St. Louis Equation," St. Louis Review, April 1983, pages 13–25.

16. This is a robust test of goodness-of-fit whenever the equations compared have the same dependent variable. For example,

Example of Determination of Order and Lag in AMO Temperature Equation

Lag of TSI variable	Order of polynomial	Adjusted R-squared
13	3	0.598
14	3	0.593
15	3	0.593
16	3	0.590
17	3	0.581
12	2	0.587

Dependent Variable: AMO_RAW__C_
Method: Least Squares
Date: 08/15/09   Time: 11:15
Sample: 1960 2000
Included observations: 41

Variable	Coefficient	Std. Error	t-Statistic	Prob.
C	-470.4800	142.8171	-3.294283	0.0023
CO2FIT1	0.002158	0.002703	0.798320	0.4302
PDL01	-0.003148	0.007867	-0.400169	0.6915
PDL02	0.001525	0.001343	1.136264	0.2638
PDL03	-8.08E-05	5.78E-05	-1.396980	0.1715
PDL04	-1.54E-07	5.47E-08	-2.821001	0.0079
PDL05	7.26E-08	6.41E-08	1.132226	0.2655

R-squared	0.650580	Mean dependent var		20.75417
Adjusted R-squared	0.588917	S.D. dependent var		0.175022
S.E. of regression	0.112217	Akaike info criterion		-1.382516
Sum squared resid	0.428149	Schwarz criterion		-1.089955
Log likelihood	35.34157	Hannan-Quinn criter.		-1.275981
F-statistic	10.55067	Durbin-Watson stat		1.732462
Prob(F-statistic)	0.000001			

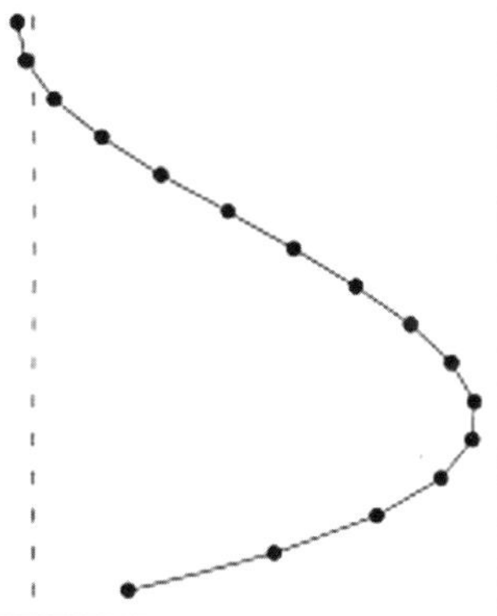

Lag Distribution of TSI__WM_2	i	Coefficient	Std. Error	t-Statistic
	0	-0.00170	0.00665	-0.25625
	1	-0.00084	0.01110	-0.07577
	2	0.00210	0.01373	0.15313
	3	0.00664	0.01491	0.44553
	4	0.01229	0.01502	0.81825
	5	0.01857	0.01446	1.28409
	6	0.02499	0.01359	1.83856
	7	0.03107	0.01271	2.44371
	8	0.03631	0.01197	3.03273
	9	0.04025	0.01132	3.55607
	10	0.04239	0.01052	4.02963
	11	0.04224	0.00935	4.51922
	12	0.03933	0.00796	4.94082
	13	0.03316	0.00787	4.21226
	14	0.02326	0.01170	1.98821
	15	0.00913	0.01965	0.46473
Sum of Lags		0.35920	0.10512	3.41715

Lag Distribution of TSI_DVI	i	Coefficient	Std. Error	t-Statistic
	0	-2.3E-07	9.4E-08	-2.48657
	1	-1.5E-07	5.5E-08	-2.82100
	2	-8.9E-08	6.8E-08	-1.30886
	3	-3.7E-08	5.5E-08	-0.68099
Sum of Lags		-5.1E-07	1.8E-07	-2.82100

FIGURE 7   Estimation of the AMO temperature equation by Simultaneous Estimation Techniques (Two Stage Least Squares).

4.2. CO$_2$ Equation Estimation

With CO$_2$ determined to be not statistically significant in the temperature equations, the recursive nature of the equation system allows estimation of the CO$_2$ equation in the system by ordinary or direct least squares.

The explicit form of the CO$_2$ equation is:

$$(\Delta C - c_{\text{fossil}})_t = a + b * \text{AMO}_t + c * \text{CO}_{2,t-1}, \qquad (4)$$

where $(\Delta C - c_{\text{fossil}})_t$ is the efflux of Net non-fossil fuel CO$_2$ emissions from the oceans and land into the atmosphere; AMO$_t$ is Atlantic sea surface temperature. The expected sign is positive. CO$_{2,t-1}$ on the right-hand side is a proxy for Land use. The expected sign is negative, because as CO$_2$ levels rise, other things equal, the CO$_2$ absorption of the flora increase.[17,18]

---

17. The authors wish to acknowledge a major contribution to the formulation of this equation and the following footnote resulting from discussions with A.J. Meyer (ajmeyer@optonline.net).

18. The number of carbon atoms on/in the Earth is fixed, except for a small percentage that arrive from extraterrestrial sources via bombardment of comets, meteors and space dust. This means that essentially a fixed amount of carbon is continually traveling back and forth at various rates between the biosphere (flora and fauna, including mankind), the oceans, lakes, land (on or under the Earth's crust), and the atmosphere via the processes of photosynthesis, respiration, combustion, decay, weathering, subduction, and volcanism.

The published data on the amount of CO$_2$ flowing between the sources and sinks of the carbon cycle is not in balance. There should be more CO$_2$ in the atmosphere. Where is it all going? The deep oceans are the primary candidates as extra carbon sinks. But in addition to the oceans, it appears that the forests of the northern hemisphere are fixing carbon at a much faster rate than has previously been estimated. To gain some perspective, just the annual seasonal CO$_2$ flux, driven by the deciduous trees in the northern hemisphere, **due only to leaves** sprouting in the spring and falling in the fall, is equal to about 220 billion tonnes of CO$_2$ per year.

If the atmospheric concentration of CO$_2$ increases, then the rate at which CO$_2$ is extracted from the atmosphere by the Earth's **flora** will also increase. In many if not most climate models, variable carbon fixing rates, especially the rates of the Northern Hemisphere's forests, have either been ignored altogether or greatly underestimated. co2science.org has a list of references to peer reviewed publications on Carbon sequestration in forests. See, http://www.co2science.org//subject/c/carbonforests.php. "Thus, it would appear that elevated CO$_2$ typically reduces or has no effect upon plant litter decomposition rates. In addition, it is important to note that none of these decomposition studies looked at wood, which can sequester carbon for long periods of time, even for millennia (Chambers et al., 1998), provided it is not burned. Based upon several different types of empirical data, a number of researchers have concluded that current rates of carbon sequestration are robust and that future rates will increase with increasing atmospheric CO$_2$ concentrations. In Fan et al. (1998) based on atmospheric measurements, for example, the broad-leaved forested region of North America between 15 and 51°N latitude was calculated to possess a current carbon sink that can annually remove all the CO$_2$ emitted into the air from fossil fuel combustion in both Canada and the United States. On another large scale, Phillips et al. (1998) used data derived from tree basal area to show that average forest biomass in the tropics has increased substantially over the last 40 years and that growth in the Neotropics alone can account for 40% of the missing carbon of the entire globe. And in looking to the future, White et al. (2000) have calculated that coniferous and mixed forests north of 50°N

Applying ordinary least squares to this equation, yields a high adjusted *R*-square (0.53), considering that the equation is estimated on an annual basis. The coefficients have the correct signs and are statistically significant at the 95% confidence level.

There is a useful validation test for all of the estimated parameters of this equation. In equilibrium, if there were no fossil fuel emissions and no change in sea surface temperature (T_0), then there would be no change in the concentration of atmospheric CO_2, so that:

$$C_t = C_{t-1} = C_{\text{equilibrium}}. \tag{5}$$

Using eq. (4), we find that:

$$0 = a + b * T_0 + c * C_{\text{equilibrium}}. \tag{6}$$

Or, rearranging,

$$C_{\text{equilibrium}} = (a + b * T_0)/(-c). \tag{7}$$

---

latitude will likely expand their northern and southern boundaries by about 50% due to the combined effects of increasing atmospheric CO_2, rising temperature, and nitrogen deposition."

It is important to note that young forests on recently cleared land fix atmospheric carbon (grow) at a rate at least an order of magnitude faster than do forests in a later (a decade or so) stage of re-growth, whilst tropical old growth forests fix carbon at essentially the same rate as they emit it back into the atmosphere, primarily through insect (termite) flatulence and rapid soil bacterial metabolism. (A study, published in *Science* in the early 1980s, indicated that the annual amount of CO_2, emitted by termite flatulence is over twice the amount produced by all anthropogenic carbon emissions. **[Science, Vol. 218, Nov. 5, 1982, Zimmerman et al.]** As is well known, termites emit prodigious amounts of methane. However, what may be less well known is that termites also annually emit about 52.4 billion tonnes of CO_2, equivalent to around 14.3 billion tonnes of combusted carbon. Whereas humanity's annual atmospheric contribution is about 6 billion tonnes of carbon or 22 billion tonnes of CO_2.) That is, tropical old growth rain forests are in a state of quasi-equilibrium. In northern forests, due to colder temperatures, more of the annual carbon detritus (dead wood, leaves, etc.) is fixed in the soil, or drops into in anoxic bogs, which can slowly reduce the detritus to peat (another long-term carbon sink). Much of the carbon that is harvested does not return to the atmosphere, since it becomes stored in building materials, furniture, etc., or ends up in the form of paper and packaging material buried in landfills which are essentially anoxic carbon sinks. Also, all plastic products, packaging material and plastic bags, unless incinerated are practically 'eternal' carbon sinks.

In addition, greater atmospheric concentrations of CO_2 along with any minimal warming, due to whatever cause, should trigger greater forest growth rates and should also extend the tree line of the Boreal forest further north into the tundra. ("The boreal forest is a major carbon sink, absorbing an average of between 700 million and 1.3 billion metric tonnes of carbon each year between 1970 and 1990. This is the equivalent of the net addition of 750 million cubic meters of wood each year. This sink is 13−24% of the 5.5 billion tonnes of carbon released annually from the burning of fossil fuels during 1980−1989." http://archive.greenpeace.org/climate/arctic99/reports/forests.html).

A rough back-of-the-envelope-calculation quickly shows that an extension of the northern boundary between forests and tundra by 1° latitude will result in an annual reduction of about 0.5 ppmv in atmospheric CO_2 concentration. This, of course, does not include the increased carbon fixing response rates due to a greater atmospheric concentration of CO_2 by the rest of the world's flora.

Dependent Variable: DELTA_C___CMAN Method: Least Squares Date: 06/22/09 Time: 09:09 Sample: 1960 2000 Included observations: 41				
Variable	Coefficient	Std. Error	t-Statistic	Prob.
C	-27.88501	7.223857	-3.860127	0.0004
AMO_RAW__C_	1.720964	0.367663	4.680819	0.0000
CO2_LAG	-0.026181	0.004027	-6.500690	0.0000
R-squared	0.558018	Mean dependent var		-1.033377
Adjusted R-squared	0.534756	S.D. dependent var		0.551963
S.E. of regression	0.376487	Akaike info criterion		0.954491
Sum squared resid	5.386222	Schwarz criterion		1.079874
Log likelihood	-16.56706	Hannan-Quinn criter.		1.000148
F-statistic	23.98820	Durbin-Watson stat		1.923067
Prob(F-statistic)	0.000000			

FIGURE 8 Estimation of the recursive CO_2 equation by direct least squares.

Substituting the estimated coefficients from Fig. 8 and substituting the average temperature observed over 1960−2000 for AMO (20.75) into eq. (7), yields:

$$C_{\text{equilibrium}} = (-27.88501 + 1.720964 * 20.75)/(.026181) = 298.88.$$

Thus, in equilibrium, without any Fossil Fuel consumption, and assuming a 20.75 °C AMO, atmospheric CO_2 concentrations would average around 300 ppm.

As an additional validation of the CO_2 equation, it can be shown that the equation suggests that the fraction of CO_2 not absorbed by the land and ocean, that is, the fraction of CO_2 from fossil fuel emissions that remains in the atmosphere, is about 53%, which roughly speaking, agrees with historical observation of about 56%. It must be noted, however, that this ratio over a forecast horizon is a function of the assumed growth rate of fossil fuel consumption and is not a constant.

4.3. The AMO Equation

Figure 9 shows the AMO equation estimated by ordinary least squares. The coefficients have the expected signs, the adjusted R-square is high, and the variables are statistically significant at the 95% confidence level. The lagged terms for the TSI variable indicate a maximum impact at about 10 years, and the TSI−DVI interaction variable, as expected, has a statistically significant negative impact, fading out over 3 or 4 years.

Dependent Variable: AMO_RAW__C_
Method: Least Squares
Date: 08/15/09 Time: 10:46
Sample: 1960 2000
Included observations: 41

Variable	Coefficient	Std. Error	t-Statistic	Prob.
C	-562.2999	84.22411	-6.676234	0.0000
PDL01	0.000981	0.005897	0.166385	0.8688
PDL02	0.001057	0.001201	0.879767	0.3850
PDL03	-6.96E-05	5.58E-05	-1.246580	0.2208
PDL04	-1.45E-07	5.33E-08	-2.729881	0.0098
PDL05	7.42E-08	6.37E-08	1.163774	0.2524

R-squared	0.644030	Mean dependent var	20.75417
Adjusted R-squared	0.593177	S.D. dependent var	0.175022
S.E. of regression	0.111634	Akaike info criterion	-1.412725
Sum squared resid	0.436174	Schwarz criterion	-1.161958
Log likelihood	34.96086	Hannan-Quinn criter.	-1.321410
F-statistic	12.66458	Durbin-Watson stat	1.688949
Prob(F-statistic)	0.000000		

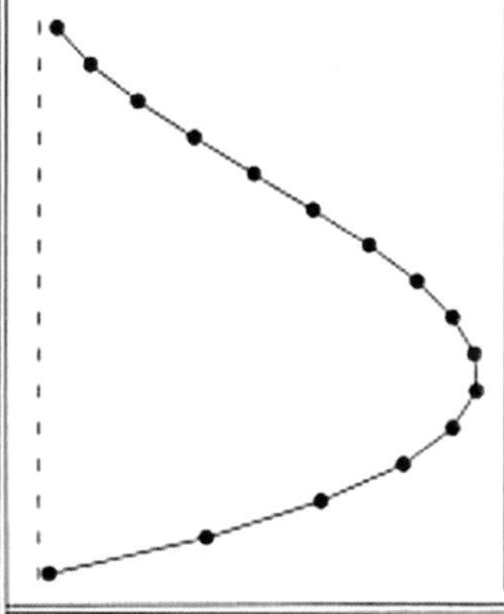

Lag Distribution of TSI__WM_2_i		Coefficient	Std. Error	t-Statistic
	0	0.00197	0.00478	0.41212
	1	0.00563	0.00754	0.74658
	2	0.01057	0.00867	1.22002
	3	0.01638	0.00854	1.91866
	4	0.02262	0.00760	2.97645
	5	0.02889	0.00645	4.47961
	6	0.03477	0.00586	5.93588
	7	0.03984	0.00636	6.26619
	8	0.04368	0.00759	5.75512
	9	0.04588	0.00881	5.20957
	10	0.04602	0.00944	4.87621
	11	0.04367	0.00913	4.78539
	12	0.03843	0.00784	4.90242
	13	0.02987	0.00668	4.47551
	14	0.01759	0.00925	1.90221
	15	0.00115	0.01683	0.06830

Sum of Lags	0.42697	0.06167	6.92297

Lag Distribution of TSI_DVI	i	Coefficient	Std. Error	t-Statistic
	0	-2.3E-07	9.3E-08	-2.44560
	1	-1.5E-07	5.3E-08	-2.72988
	2	-8.0E-08	6.7E-08	-1.19999
	3	-3.1E-08	5.4E-08	-0.58054

Sum of Lags	-4.8E-07	1.8E-07	-2.72988

FIGURE 9 Estimation of the AMO temperature equation by direct least squares.

Dependent Variable: PDODT
Method: Least Squares
Date: 07/26/09   Time: 08:43
Sample: 1960 2000
Included observations: 41

Variable	Coefficient	Std. Error	t-Statistic	Prob.
C	-1262.764	554.9976	-2.275260	0.0285
PDL01	0.019477	0.008466	2.300543	0.0269

R-squared	0.119490	Mean dependent var	14.03199
Adjusted R-squared	0.096913	S.D. dependent var	0.815785
S.E. of regression	0.775248	Akaike info criterion	2.376283
Sum squared resid	23.43937	Schwarz criterion	2.459872
Log likelihood	-46.71381	Hannan-Quinn criter.	2.406722
F-statistic	5.292500	Durbin-Watson stat	0.898113
Prob(F-statistic)	0.026851		

Lag Distribution of TSI__WM_2_i		Coefficient	Std. Error	t-Statistic
	0	0.01833	0.00797	2.30054
	1	0.03437	0.01494	2.30054
	2	0.04812	0.02092	2.30054
	3	0.05958	0.02590	2.30054
	4	0.06874	0.02988	2.30054
	5	0.07562	0.03287	2.30054
	6	0.08020	0.03486	2.30054
	7	0.08249	0.03586	2.30054
	8	0.08249	0.03586	2.30054
	9	0.08020	0.03486	2.30054
	10	0.07562	0.03287	2.30054
	11	0.06874	0.02988	2.30054
	12	0.05958	0.02590	2.30054
	13	0.04812	0.02092	2.30054
	14	0.03437	0.01494	2.30054
	15	0.01833	0.00797	2.30054
Sum of Lags		0.93490	0.40638	2.30054

FIGURE 10   PDO temperature equation estimation with AMO functional form.

Figure 10 shows the estimation of the PDO temperature equation using the same form as the AMO equation.[19] Note that while the TSI variable is statistically significant, the interaction term is not and the equation has extremely

19. All temperature equations were initially estimated by TSLS, using the appropriate set of exogenous variables. None of the variables for the fitted values of CO_2 were statistically significant.

TABLE 4 Correlation Between SOI and Various Temperature Series

	AMO	PDO	NINO3.4	U.S.
SOI	−0.077	−0.543	−0.925	−0.084

poor explanatory power (an adjusted R-square of 0.1). This result is not surprising, given the low correlation between PDO and the AMO and U.S. time series. (See Table 3.)

Some assume that solar activity and a disturbance of the Earth's magnetic field may be considered as an external force for excitement of ENSO variability, for example as measured by the Southern Oscillation Index (SOI). The SOI is a function of the standardized data of Sea Level Pressure (SLP) difference between Tahiti and Darwin (Australia). As noted in Nuzhdina (2002) [20], "cyclic dynamics of ENSO phenomena are due to solar activity and geomagnetic variations. It is background long-period variations on which high frequency oscillations are imposed." Since these individual exogenous impacts are not well understood, the Southern Oscillation Index (SOI)[21] is used here as a proxy, or instrumental variable, to capture this complex solar/magnetic field influence on temperatures. The correlation between SOI and the temperature series are given in Table 4. Note that this exogenous variable's impact is insignificant for the AMO and the U.S., but critical regarding both the PDO and NINO3.4.

The impact of SOI is modeled as a "Central Tendency," that is, a dummy variable depending on whether the SOI series is indicating a La Nina central tendency (0) or an El Nino tendency (+1). If the central tendency is La Nina, then La Nina conditions are more frequent/important and vice versa. The dummy variable is presented with the SOI series in Fig. 11.

Note: the SOI Dummy = 1 from 1976 to 1998, 0 elsewhere. The alternative using values of 1 for 1999 and 2000 has poorer statistical properties, e.g., adjusted R-squared.[22]

---

20. Nuzhdina (2002), page 88.

21. SOI is the Southern Oscillation Index, the Tahiti minus Darwin Pressure anomalies − ftp://ftp.cpc.ncep.noaa.gov/wd52dg/data/indices/soi.his and http://www.cpc.ncep.noaa.gov/data/indices/soi.

22. This method of determining whether a switch in Central Tendency may have happened seems worth further analysis since looking at the 2001−2008 seems to confirm the switch. Since this PDO behavior is clearly chaotic and independent of CO_2, some means selecting at least the near term outlook the SOI Dummy is always needed.

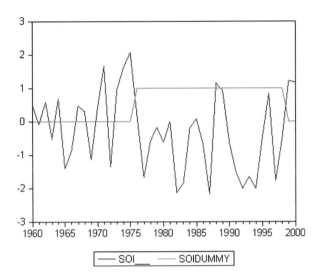

FIGURE 11 SOI and "Central Tendency"

Using the SOI Dummy as a proxy for such complex solar/magnetic field influences, the PDO temperature equation estimation is shown in Fig. 12.[23]

The adjusted R-squared is at roughly the same level as for the AMO equation and the coefficient sign is correct with a very high t-statistic. It should be noted that CO_2Fit was not statistically significant if added to this model formulation or any other considered.

4.4. CO_2 Equation Estimation Adding the Pacific Ocean (PDO) Impact

In the CO_2 model estimation shown in Fig. 13 below the SOI Dummy has been added as the only change from the estimate in Fig. 8. The adjusted R-squared is slightly higher (0.56 vs. 0.53) suggesting this is the preferred model. All the signs are as expected and statistically significant, particularly with a DW statistic of 1.92. When the equilibrium atmospheric CO_2 concentration calculation is repeated, the result is a nearly identical 301 vs. 299 ppm! Another interesting result of this formulation is that it allows estimation of the impact of a shift in Central Tendency; the impact is 8 ppm.

4.4.1. U.S. Temperature Equation

As mentioned above, it was expected that the U.S. surface temperature could be successfully modeled using the same basic structure as for AMO, but that

---

23. It should be noted that only a detrended version of the PDO time series was available to the authors. Adding any reasonable trend, however, had no impact on the lack of statistical significance of CO_2.

Dependent Variable: PDODT				
Method: Least Squares				
Date: 08/03/09 Time: 12:46				
Sample: 1960 2000				
Included observations: 41				

Variable	Coefficient	Std. Error	t-Statistic	Prob.
C	13.33949	0.126495	105.4548	0.0000
SOIDUMMY	1.234459	0.168889	7.309297	0.0000

R-squared	0.578040	Mean dependent var	14.03199
Adjusted R-squared	0.567221	S.D. dependent var	0.815785
S.E. of regression	0.536672	Akaike info criterion	1.640692
Sum squared resid	11.23267	Schwarz criterion	1.724281
Log likelihood	-31.63419	Hannan-Quinn criter.	1.671131
F-statistic	53.42582	Durbin-Watson stat	1.620538
Prob(F-statistic)	0.000000		

FIGURE 12   PDO temperature equation estimation using SOI Dummy.

Dependent Variable: DELTA_C_CMAN				
Method: Least Squares				
Date: 11/12/09 Time: 12:39				
Sample: 1960 2000				
Included observations: 41				

Variable	Coefficient	Std. Error	t-Statistic	Prob.
C	-28.60176	7.073216	-4.043671	0.0003
AMO_RAW	1.838641	0.366190	5.020997	0.0000
CO2_LAG	-0.031697	0.005141	-6.165855	0.0000
SOIDUMMY	0.253568	0.152008	1.668126	0.1037

R-squared	0.588933	Mean dependent var	-1.033377
Adjusted R-squared	0.555603	S.D. dependent var	0.551963
S.E. of regression	0.367955	Akaike info criterion	0.930758
Sum squared resid	5.009476	Schwarz criterion	1.097936
Log likelihood	-15.08054	Hannan-Quinn criter	0.991635
F-statistic	17.66990	Durbin-Watson stat	1.929856
Prob(F-statistic)	0.000000		

FIGURE 13 CO_2 equation estimation with Pacific impact.

different lag structure would result. For example, it would be expected that the solar influences would occur more rapidly with air than water temperature — which is exactly the result obtained. The impact of volcanic action is similar in lag structure but more muted than was the case for the AMO. Given the on going

PART | V   Modeling

Dependent Variable: GISS_US_TEMP
Method: Least Squares
Date: 09/11/09   Time: 09:40
Sample: 1960 2000
Included observations: 41

Variable	Coefficient	Std. Error	t-Statistic	Prob.
C	-1701.104	476.2577	-3.571814	0.0010
PDL01	0.040946	0.015432	2.653330	0.0117
PDL02	-0.002977	0.001578	-1.886781	0.0671
PDL03	-4.27E-07	2.80E-07	-1.523951	0.1360

R-squared	0.300036	Mean dependent var	57.38659
Adjusted R-squared	0.243282	S.D. dependent var	0.789190
S.E. of regression	0.686513	Akaike info criterion	2.178085
Sum squared resid	17.43811	Schwarz criterion	2.345263
Log likelihood	-40.65074	Hannan-Quinn criter.	2.238962
F-statistic	5.286611	Durbin-Watson stat	1.879781
Prob(F-statistic)	0.003901		

Lag Distribution of TSI	i	Coefficient	Std. Error	t-Statistic
	0	0.03797	0.01391	2.73025
	1	0.06998	0.02479	2.82311
	2	0.09604	0.03270	2.93672
	3	0.11615	0.03774	3.07738
	4	0.13030	0.04006	3.25233
	5	0.13849	0.03996	3.46591
	6	0.14074	0.03801	3.70263
	7	0.13702	0.03541	3.86924
	8	0.12736	0.03449	3.69251
	9	0.11174	0.03841	2.90866
	10	0.09016	0.04896	1.84165
	11	0.06263	0.06555	0.95546
	12	0.02914	0.08706	0.33477
Sum of Lags		1.28772	0.34874	3.69251

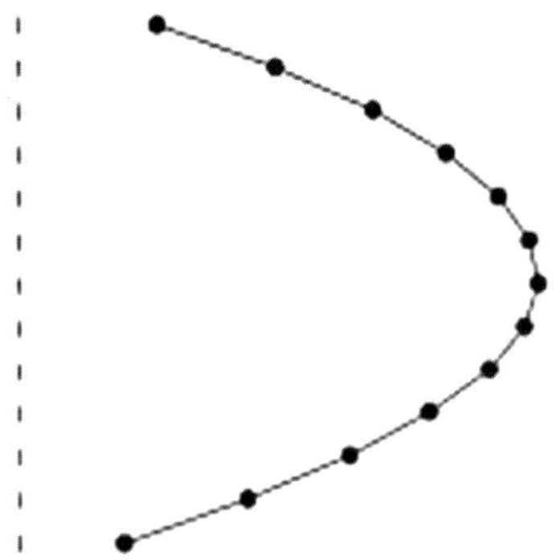

Lag Distribution of TSI_DVI	i	Coefficient	Std. Error	t-Statistic
	0	-5.7E-07	3.7E-07	-1.52395
	1	-4.3E-07	2.8E-07	-1.52395
	2	-2.8E-07	1.9E-07	-1.52395
	3	-1.4E-07	9.3E-08	-1.52395
Sum of Lags		-1.4E-06	9.3E-07	-1.52395

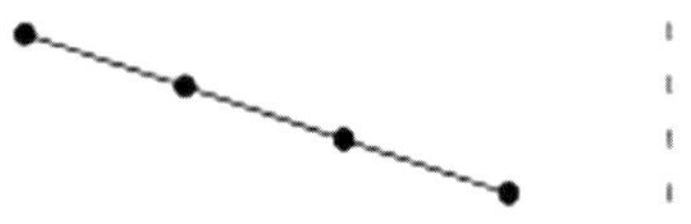

FIGURE 14   Estimation of U.S. temperature equation GISS [NASA].

controversy regarding errors of observation in this data set, it is not surprising that this result has an adjusted R-squared less than half that of AMO and PDO (See Fig. 14.). And, it should again be noted that when CO_2Fit was added to this or any other formulation tested, it was not statistically significant.

4.5. KISS Model Forecasts and Validation

While all of the rules for model development and validation spelled out above have thus far been followed and with success, a critical step remains; testing whether or not the KISS Model has any hope of providing forecasts that are truly useful as input for policy makers. In this test, the model, as estimated over the period 1960−2000, is examined regarding its ability to forecast the period 2001−2008. This is a far more stringent test than simply undertaking a within-sample test.

In this analysis, actual values of the exogenous variables for the out-of-sample period, 2001−2008 are used as model input. In Section 4.6, KISS Model forecasts will be made using an array of likely exogenous variable that might have reasonably been constructed in the year 2000.

Substituting known values for TSI and DVI, the forecasts for AMO are shown in Fig. 15. The expected value of the forecast is for a decline in temperature, much like what actually happened. How the KISS Model does relative to a simple naïve models is also a crucial test. The naïve models used here are: a linear time trend fitted to the period 1960−2000 and a linear trend fitted to the period 1976−2000. Figure 15 shows that the KISS Model

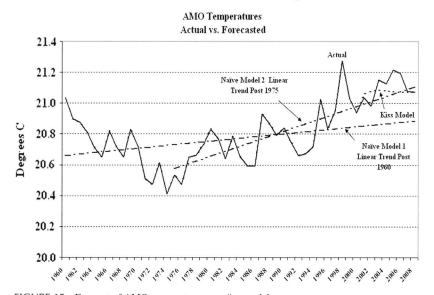

FIGURE 15 Forecast of AMO temperature vs. naïve models.

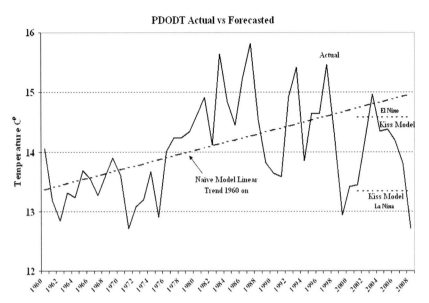

FIGURE 16   Forecast of PDO temperature vs. naïve models.

outperformed both of these naïve models.[24] Figures 16−18 show the same performance for the other dependent variables in the KISS Model.

Table 5 compares the KISS Model forecasts with the various naïve or benchmark model forecasts on the basis of three widely used evaluation metrics: the Bias (how close were the forecasts on average), the Mean Absolute percent Error (MAPE)[25], and turning point errors (did the forecast capture the turning point, if there was one). For all endogenous variables, the KISS Model outperformed the naive models on the basis of Bias and MAPE. The model also correctly captured the turning point in all three temperature series, whereas the naïve models did not. As expected, the MAPE for the chaotic temperature, PDO, is significantly higher than for the other temperatures. The ranking of the MAPE measures corresponds to the coefficient of variation hierarchy identified earlier.

4.6. KISS Model 2001−2030 Simulations

This section illustrates the use of the KISS Model for policy analysis purposes. Note that, in doing this, the KISS Model is subjected to a strong validation test, by starting the forecasts in the year 2000, not in the year 2008. The model was simulated for the 30-year period 2001−2030, using a plausible set of exogenous variables specified using only information available when year 2000 was published.

---

24. The AMO time series was stationary of order one, so a linear trend is a reasonable naïve model.

25. The MAPE is used to compare among different variables.

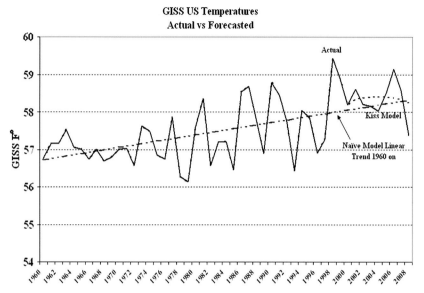

FIGURE 17   Forecast of GISS U.S. temperature vs. naïve models.

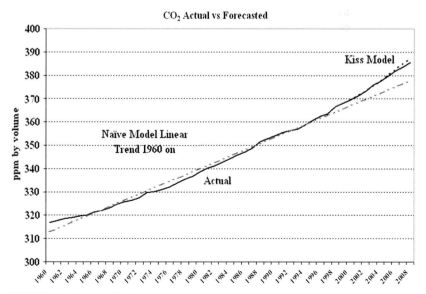

FIGURE 18   Forecast of CO_2 concentration vs. naïve models.

TABLE 5 Comparison of KISS Model Forecasts with Naïve Models

	BIAS		MAPE		Captured turning point	
	KISS Model	Naïve Model (Trend '60–'00)	KISS Model	Naïve Model (Trend '60–'00)	KISS Model	Naïve Model (Trend '60–'00)
AMO	0.033	0.239 0.056[a]	0.294	1.13 0.355[a]	Yes	No
PDO La Nina	0.667	−0.842	5.796	6.542	Yes	No
PDO El Nino	−0.568	−0.842	4.491	6.542	Yes	No
GISS U.S.	−0.075	0.124	0.636	0.680	Yes	No
CO_2	−0.459	5.534	0.156	1.457	NR	NR

[a] = Trend from 1975 to 2000.

The first exogenous variable, TSI, has an observed *solar seasonal* pattern of about 11 years superimposed on a longer term trend cycle. To represent a High Side TSI case, using data from 1960 to 2000 only, a trend-cycle that would continue upward for the 30 years after 2000 would seem plausible. Alternatively, as a Base Case, a trend-cycle using a longer set of historical data, from 1910 to 2000 say, would support *a no linear trend* outlook with only the solar seasonal pattern superimposed. Finally, a simulated return of a Maunder Minimum-type pattern is not inconceivable.[26] These three scenarios are shown in Fig. 19.

The second exogenous variable, volcanic activity, if any, would tend to decrease temperatures. For these scenarios, this interaction term was set equal to zero in all three scenarios. Significant volcanic action reduces the average temperatures somewhat. If the volcanic activity of the early 1990s (Pinatubo) coupled with the behavior of TSI at that time (coming off an 11-year cycle peak) were to occur again, the average AMO temperature, over the 4-year period, would fall about 0.16 °C.

The third exogenous variable, c_{fossil}, emissions from fossil fuel burning, is assumed to grow at its historical rate of 2.5% per annum in all three scenarios. If technological change or policy actions reduce this rate of growth of fossil fuel emissions, the forecasted CO_2 concentrations would be overstated. **According**

26. Abdussamatov (2008), page 4.

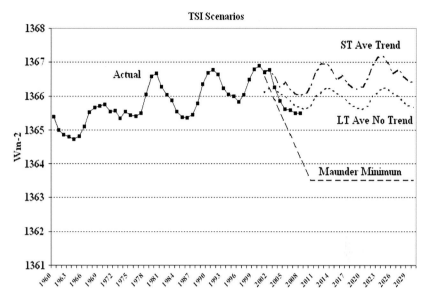

FIGURE 19   Solar irradiance scenarios.

TABLE 6   Exogenous Assumptions, 2001–2030

	High side case	Base case	Maunder Minimum
Exogenous variable			
Total Solar Irradiance	Trend-cycle calculated over 1960–2000. (Has a trend)	Cycle calculated over 1910–2000 time period (does not have a trend)	TSI corresponding to 1363.37, the lowest observable point in the time series (in 1890)
Volcanic activity	No significant volcanic activity		
Fossil fuel emissions	Historical growth rate = 2.5% per year		
Pacific Central Tendency	La Nina Central Tendency		

to the KISS Model, the effects of such policy action will have no impact on the AMO, PDO and U.S. temperature outlooks.

The final exogenous variable, the Central Tendency in the Pacific, is assumed to be La Nina for all three scenarios. If the tendency were to revert to

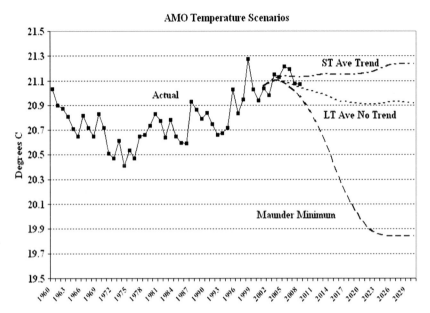

FIGURE 20   AMO temperature simulations.

the El Nino Central Tendency, the temperatures in the Pacific would be about 1.2 °C higher. These cases are summarized in Table 6.

As shown in Fig. 20, under the Base Case assumptions, AMO temperatures would fall over the forecast period and flatten out at 20.9 °C, or about 0.15 °C above the average for 1960–2000. In the High Case, AMO temperature rises to 21.2 °C, or 0.4 °C above the average over the estimation period. In the Maunder Minimum Case, AMO temperature falls precipitously to 19.9 °C, which is below any point in the 1960–2000 period.

In the KISS Model, the forecast for PDO depends solely on the assumption as to the SOI Central Tendency in the Pacific. As shown in Fig. 21, using the assumption of a continuation of the "La Nina" Central Tendency, Pacific temperatures are forecast to average 12.2 °C under all scenarios. If that were to prove wrong and the Central Tendency were to be El Nino, then the PDO would be 1.2 °C higher.

Figure 22 shows the U.S. temperature outlook under each of the three scenarios.

If these scenarios, adequately bracket the outlook, there would certainly not seem to suggest serious upside temperature risk.

Under all three scenarios, given that man-made fossil fuel emissions continue unabated by assumption, as shown in Fig. 23, CO_2 concentrations continue to rise. But they stay under 460 ppm through 2030.

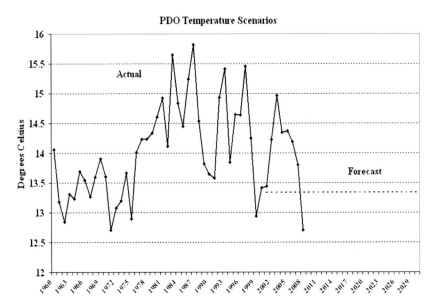

FIGURE 21   PDO temperature simulations.

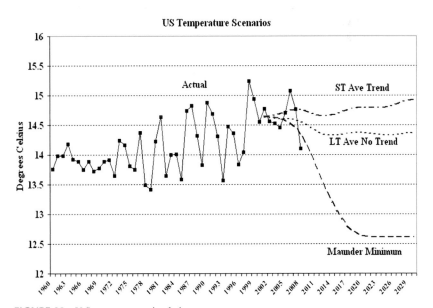

FIGURE 22   U.S. temperature simulations.

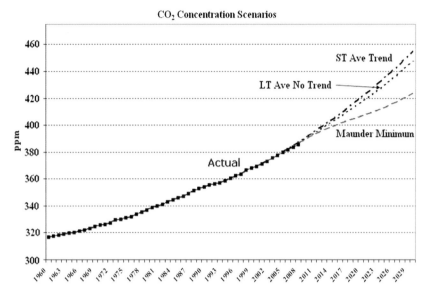

FIGURE 23 CO^2 simulations.

5. CONCLUSIONS AND RAMIFICATIONS

Our analysis suggests the following conclusions and ramifications:

The simplest model that can characterize the relationship between atmospheric CO_2 concentration levels and temperature levels must contain at least two simultaneous equations, one for each of these two state variables. Therefore, the climate system must be analyzed using simultaneous equation estimation techniques. Otherwise the parameter estimates of any structural equations will be both biased and inconsistent, which implies they are useless for policy analysis purposes. The existence of a robust atmospheric CO_2 equation has been amply demonstrated, thus guaranteeing that **ANY** modeling system designed to forecast temperature must include at least two equations.

Using the KISS Model and seeking to link temperature to CO_2 concentrations, the null hypothesis that atmospheric CO_2 levels have a statistically significant impact on temperature levels is rejected. In fact, it is rejected three times, i.e., for the AMO, PDO, and U.S. temperatures.

The KISS climate modeling system is recursive. It has been shown that Ocean and U.S. Surface temperatures may be modeled and forecast reasonably well using only exogenous variables that relate to solar/magnetic activity, volcanic activity, and ocean oscillations. Then, atmospheric CO_2 concentrations can be modeled and forecast reasonably well using these forecast ocean temperatures as well as forecast fossil fuel CO_2 emissions.

The KISS Model forecasts of AMO, PDO, and U.S. surface temperatures and CO_2 concentrations for the period 2001–2008, using the model parameters estimated over the period from 1960 to 2000, and with actual exogenous variable values through 2008, capture the observed decline in ALL THREE temperatures and capture the continued upward movement in CO_2. The KISS Model forecasts outperform all naïve model forecasts over the same period for all four dependent variables.

Since the KISS Model was developed and estimated following the fundamental professional guidelines spelled out in Section 2, and passed all of the standard statistical tests, it can be used for policy analysis to simulate alternative AMO, PDO, and U.S. temperature and CO_2 concentration futures. Since the geographic coverage is at least 70%, it is highly unlikely that *actual* global average temperatures rise beyond that which would be consistent with reasonable KISS Model simulations. Under the assumptions that solar irradiance follows its linear trendless 1910–2000 historical trend cycle or continues falling to its 200-year minimum and the PDO exhibits a La Nina Central Tendency, all three temperatures are expected to decline through 2030, even as CO_2 concentrations continue to rise. Under the assumption that solar irradiance rises at its 1975–2000 trend-cycle rate and volcanic activity is again benign, U.S. temperatures rise modestly, on the order of 1 °C through 2030. However, given the recent behavior of the sunspot activity this scenario seems highly unlikely.

Finally, it seems extremely critical to note that, over the period 1975 through roughly 1998, there were only two significant volcanic eruptions, the TSI trend cycle was upward sloping and the SOI was in its Hot El Nino Central Tendency Mode. This situation led to rising AMO, PDO, and U.S. temperatures. Because CO_2 was also rising, some analysts claimed causation. The KISS Model results categorically refute this claim. Despite the continued rise in CO_2, the KISS Model forecasts the 2000–2008 decline in temperature for all three regions of the world, which together cover over 70% of its surface area.

Another ramification of all this is that any rational modeling activity, designed to provide a global average surface temperature outlook, would need to develop and validate separate temperature equations covering the bulk of the surface area of the Earth, including both polar regions. In this paper, a means of selecting and estimating the parameters of the appropriate functional form of the temperature equations for additional geographic areas was indicated by the PDO analysis. Follow on work might include determining a reasonable mutually exclusive, collectively exhaustive set of regions that meet the KISS model development and validation rules set as outlined in Section 2.

ACKNOWLEDGMENTS

The authors wish to thank Dr. William Niskanen for his review of the paper.

REFERENCES

Abdussamatov, H., 2008. The Sun defines the climate. (Translated from Russian by Lucy Hancock).

Chambers, J.Q., et al., 1998. Ancient trees in Amazonia. Nature 391, 135−136.

Fan, S., et al., 1998. A Large Terrestrial Carbon Sink in North America Implied by Atmospheric and Oceanic Carbon Dioxide Data and Models. Science 282, 442−446.

Feng, S., Hu, Qi, 2008. How the North Atlantic Multidecadal Oscillation may have influenced the Indian summer monsoon during the past two millennia. Geophysical Research Letters 35.

Hoyt, D.V., Schatten, K.H., 1993. A discussion of plausible solar irradiance variations. Journal of Geophysical Research 98A, 18895−18906.

Lamb, H.H., 1970. Volcanic dust in the atmosphere; with a chronology and assessment of its meteorological significance: Philosophical Transactions of the Royal Society of London. Series A 266, 425−533.

Lamb, H.H., 1977. Supplementary volcanic dust veil assessments. Climate Monitor 6, 57−67.

Lamb, H.H., 1983. Update of the chronology of assessment of the volcanic dust veil index. Climate Monitor 12, 79−90.

Mann, M.E., et al., 1998. Global-Scale Temperature Patterns and Climate Forcing over the Past Six Centuries. Nature 392, 779−787.

Nuzhdina, M.A., 2002. Connection between ENSO phenomena and solar and geomagnetic activity. Natural Hazards and Earth Sciences 2, 83−89.

Phillips, O.L., et al., 1998. Changes in the carbon balance of tropical forests: evidence from long-term plots. Science 282, 439−442.

White, A.M., et al., 2000. The high-latitude terrestrial carbon sink: a model analysis. Global Change Biology 6, 227−245.

Index

Note: Page numbers followed by f indicate figures and t indicate tables.

A